Mechanical Engineering
Fundamentals

Mechanical Engineering: Fundamentals

Devendra Vashist
Associate Professor
Manav Rachna International University
Faridabad, Haryana

©Copyright 2019 I.K. International Pvt. Ltd., New Delhi-110002.

This book may not be duplicated in any way without the express written consent of the publisher, except in the form of brief excerpts or quotations for the purposes of review. The information contained herein is for the personal use of the reader and may not be incorporated in any commercial programs, other books, databases, or any kind of software without written consent of the publisher. Making copies of this book or any portion for any purpose other than your own is a violation of copyright laws.

Limits of Liability/disclaimer of Warranty: The author and publisher have used their best efforts in preparing this book. The author make no representation or warranties with respect to the accuracy or completeness of the contents of this book, and specifically disclaim any implied warranties of merchantability or fitness of any particular purpose. There are no warranties which extend beyond the descriptions contained in this paragraph. No warranty may be created or extended by sales representatives or written sales materials. The accuracy and completeness of the information provided herein and the opinions stated herein are not guaranteed or warranted to produce any particulars results, and the advice and strategies contained herein may not be suitable for every individual. Neither Dreamtech Press nor author shall be liable for any loss of profit or any other commercial damages, including but not limited to special, incidental, consequential, or other damages.

Trademarks: All brand names and product names used in this book are trademarks, registered trademarks, or trade names of their respective holders. Dreamtech Press is not associated with any product or vendor mentioned in this book.

ISBN: 978-93-89583-67-0

Edition: 2019

Printed at: Shree Maitrey Printech Pvt. Ltd., Noida

Preface

The purpose of this book is to provide a fundamental knowledge of basic Mechanical Engineering for undergraduate students of all branches of engineering.

The various topics of Mechanical Engineering that are being taken care of in the book are as follows:
- Machine tool, and fabrication processes
- Thermodynamics, I.C. engines and steam turbines
- Hydraulic turbines and pumps
- Refrigeration and air-conditioning
- Power transmission methods and devices
- Stresses, strain, shear force and bending moment diagrams.
- Numerical control machines. (NC and CNCs).
- Applied mechanics

Objective type questions are provided at the end of each unit so that students can check their knowledge before appearing in a viva voce.

I have taken extra care to avoid mistakes but as no body is perfect some mistakes might have crept in. I will be obliged if these are brought to my notice. Constructive suggestions will be warmly welcomed for further improvement of the book.

I am thankful to students and colleagues who have given valuable suggestions while writing this book.

My sincere thanks to I.K. International Publishers for excellent work done in publishing this book. Finally I sincerely hope that the book will meet the need of the readers.

Devendra Vashist

Contents

Preface v

Unit 1: Introduction to Commonly Used Machine Tools in a Workshop 1
    1.1   Lathe 1
    1.2   Shaper 4
    1.3   Planer 6
    1.4   Milling Machine 9
    1.5   Drilling Machine 10
    1.6   Slotter 13
    1.7   Introduction to Metal Cutting 14
    1.8   Cutting Tool Nomenclature 17
    1.9   Cutting fluid 18
    Exercise 19

Unit 2: Basic Concepts of Thermodynamics 23
    2.1   Basic Concepts 23
    2.2   Thermodynamic Properties 24
    2.3   Zeroth Law of Thermodynamics 29
    2.4   Introduction to First Law 30
    2.5   Limitations of the First Law and Introduction to the Second Law 32
    2.6   Carnot Cycle 34
    2.7   Entropy 37
    2.8   Third Law of Thermodynamics 39
    Exercise 43

Unit 3: Properties of Steam and Steam Generators (Boilers) 48
    3.1   Properties of Steam And Boilers 48
    3.2   Steam Tables 55
    3.3   Methods of Determining the Dryness Fraction of Steam 59
    3.4   Boiler 64
    Exercise 92

Unit 4: Refrigeration and Air-Conditioning — 97
- 4.1 Introduction — 97
- 4.2 Rating of Refrigeration Machine — 100
- 4.3 Simple Refrigeration Vapour Compression Cycle — 100
- 4.4 Air Conditioning — 102
- 4.5 Psychrometric Chart — 104
- 4.6 Plasma Air Purifier — 106
- 4.7 Comfort Charts — 106
- Exercise — 110

Unit 5: Hydraulic Turbines and Pumps — 115
- 5.1 Introduction — 115
- 5.2 Classification — 116
- 5.3 Terminology Used for Turbines — 117
- 5.4 Pelton Wheel Constructional Details — 119
- 5.5 Francis Turbine Constructional Details — 125
- 5.6 Kaplan Turbine — 130
- 5.7 Specific Speed — 133
- 5.8 Unit Quantities — 135
- 5.9 Turbine Selection — 138
- 5.10 Pumps — 139
- 5.11 Centrifugal Pumps — 139
- 5.12 Reciprocating Pump — 145
- 5.13 Rotary Pumps — 145
- 5.14 Hydraulic Lift — 149
- 5.15 Hydraulic Jack — 151
- Exercise — 152

Unit 6: Power Transmission Methods and Devices — 157
- 6.1 Introduction — 157
- 6.2 Belt Drives — 158
- 6.3 Chain Drives — 170
- 6.4 Gears — 172
- 6.5 Gear Trains — 179
- 6.6 Clutches — 182
- 6.7 Brakes — 187
- 6.8 Dynamometer — 191
- Exercise — 197

Unit 7: Stresses and Strain — 202
- 7.1 Introduction — 202
- 7.2 Strain — 204
- 7.3 Poisson Ratio — 205

	7.4	Stress and Strain in Simple and Compound bar under Axial Loading	206
	77.5	The Stresses in Bars of Uniformly Tapering Circular Cross-section	206
	7.6	Extension of Bar due to self Weight	210
	7.7	Stress-Strain Diagram	211
	7.8	Hooke's Law and Elastic Constants	213
	Exercise		220
Unit 8:	**Manufacturing System**		**225**
	8.1	Introduction	225
	8.2	Classification of Manufacturing System	227
	8.3	Flexibility in Manufacturing System (FMS)	228
	8.4	Fundamentals of Numerical Control (NC)	228
	Exercise		237
Unit 9:	**Shear Force and Bending Moment Diagram**		**240**
	9.1	Beam	240
	9.2	Shear force	240
	9.3	Types of Support	242
	9.4	Types of Beams	242
	9.5	Types of Loading	244
	9.6	Method to Draw the SFD and BMD for Simply Supported Beam	246
	9.7	Point of Contraflexure (or Point of Inflection)	249
	9.8	Method to Draw SFD and BMD for Cantilever Beam	250
	Exercise		259
Unit 10:	**Fabrication Process**		**263**
	10.1	Introduction	263
	10.2	Effect of Chemical Elements on Iron	264
	10.3	Steels	265
	10.4	Introduction to Foundry	267
	10.5	Pattern	267
	10.6	Types of Moulding Sand	268
	10.7	Types of Pattern	269
	10.8	Moulding Processes based on Sand use	272
	10.9	Casting Method	275
	10.10	Forging	279
	10.11	Rolling	281
	10.12	Extrusion	281
	10.13	Drawing	282
	10.14	Bending	282

10.15	Welding	282
10.16	Soldering	286
10.17	Brazing	286
Exercise		286

Unit 11: Applied Mechanics — 291

11.1	Introduction	291
11.2	Laws of Mechanics	292
11.3	Concept of Moment	300
11.4	Varignon's Theorem	302
11.5	Particle Dynamics	302
11.6	Kinematic of Rigid Bodies	303
11.7	General Plane Motion	310
11.8	Free Vibration	316
Exercise		322

Unit 12: Steam Turbines, Condensers, Cooling Towers, I.C. Engines and Gas Turbine — 328

12.1	Introduction of Steam Turbines	328
12.2	Classification of Turbines	329
12.3	Impulse Turbine	329
12.4	Impulse-Reaction Turbine	331
12.5	Compounding of Impulse Turbine	332
12.6	Steam Condenser	337
12.7	Cooling Water Supply	344
12.8	Cooling Towers	345
12.9	Internal Combustion Engine	349
12.10	Constructional Detail of I.C. Engines	356
12.11	Constructional details of 2-stroke SI Engines	361
12.12	Constructional details of C.I. Engines	362
12.13	Working of Internal Combustion Engines	365
12.14	Air Standard Cycles	370
12.15	Working Principle of a Gas Turbine	382
Exercise		388

Steam Tables — 392

Index — 425

UNIT I

Introduction to Commonly Used Machine Tools in a Workshop

1.1 LATHE

Lathe is a machine which is used to remove metal from a piece of work to give it the required shape and size. This is done by holding the work rigidly on the machine and then turning it against cutting tool which will remove metal from the work in the form of chips. The cutting tool should be harder than the material of the workpiece. The tool should be rigidly held on the machine and should be fed or progressed in a definite way relative to the work.

The different types of lathe used are:
(1) Speed lathe
(2) Engine lathe
(3) Tool room lathe
(4) Capstan and turret lathe
(5) Special purpose lathe
(6) Automatic lathe
(7) Bench lathe

The different parts of lathes are:
(1) Bed
(2) Headstock
(3) Tailstock
(4) Carriage
(5) Feed mechanism
(6) Screw cutting mechanism

Bed

The lathe bed forms the base of the machine. The headstock and tailstock are located at either end of the bed and the carriage rests over the lathe bed and slides

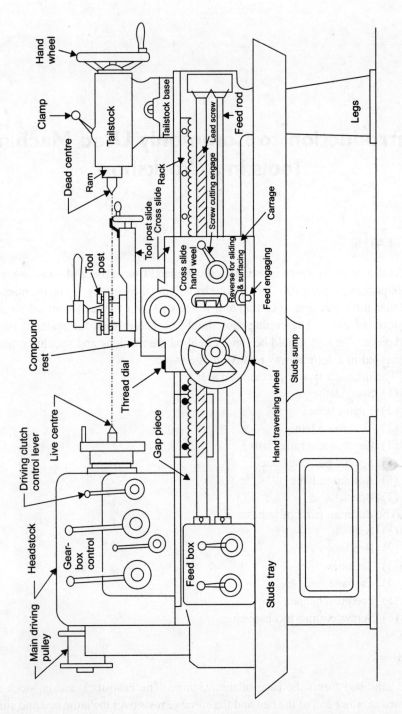

Fig. 1.1: Showing details of lathe.

on it. The bed should be sufficiently rigid to prevent deflection under tremendous cutting pressure transmitted through the tool post and carriage to the lathe bed.

Headstock

The headstock is secured permanently on the innerways at the left-hand end of the lathe bed, and it provides mechanical means of rotating the work at multiple speeds. It comprises a hollow spindle and mechanism for driving and altering the spindle speed.

Tailstock

The tailstock supports the other end of the work when it is being machined between centres. It can also hold tool for performing operations such as drilling, reaming, tapping, etc. To accommodate different lengths of work, the body of the tailstock can be adjusted along the ways chiefly by sliding it to the desired position where it can be clamped by bolts and plates.

The tailstock handwheel is used to move the tailstock spindle in or out of the tailstock casting and a spindle binding (clamping) lever or lock handle is used to hold the tailstock spindle in a fixed position.

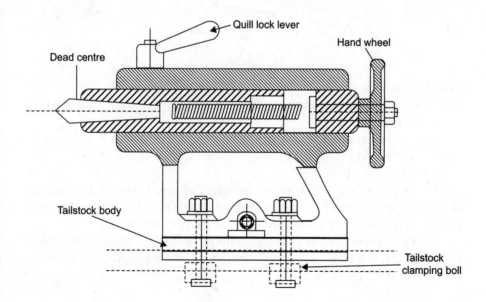

Fig. 1.2: Tailstock.

Carriage

The carriage controls and supports the cutting tool. By the help of this, the tool moves away or towards the headstock. It has five major parts:
  (i) Saddle
  (ii) Cross slide
  (iii) Compound rest
  (iv) Tool post
  (v) Apron.

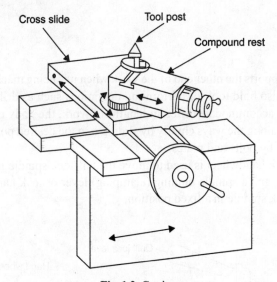

Fig. 1.3: Carriage.

Feed Mechanism

The movement of the tool relative to the work is known as feed. A lathe tool may have three types of feed.
  (a) Longitudinal feed
  (b) Cross feed
  (c) Angular feed

The size of a lathe is determined by height of centres over bed, maximum swing over bed, maximum swing over carriage, length of bed, maximum length of the work that can be accommodated between the lathe centres, etc.

The various operations that lathe can perform are centring, facing, plain turning, step turning, taper turning, drilling, reaming, boring, grooving, threading, knurling, parting off, eccentric turning.

1.2 SHAPER

Shaper is a reciprocating machine tool in which the ram moves the cutting tool backward and forward in a straight line to generate the flat surface. The flat surface may be horizontally inclined or vertical.

In a shaper, a single point cutting tool reciprocates over the stationary workpiece. The workpiece is rigidly held in a vice or clamped directly on the table. The tool is held in the tool head mounted on the ram of the machine. When the ram moves forward, cutting of material takes place. So, it is called cutting stroke. When the ram moves backward, no cutting of material takes place hence called idle stroke. The time taken during the return stroke is less as compared to forward stroke and this is obtained by quick return mechanism. The depth of cut is adjusted by moving the tool downward towards the workpiece.

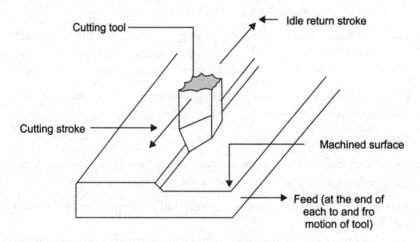

Fig. 1.4: Cutting action of a shaper.

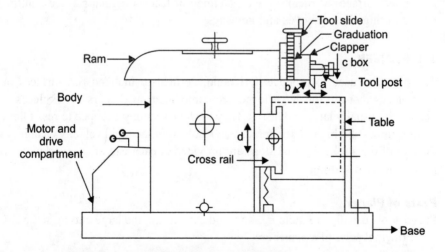

Fig. 1.5: Principal parts of a shaper.
a: Cutting and return motion; b: feed motion; c: Depth of cut (for horizontal machining); d: Motion for accommodating job

Principal parts of a shaper are:
(1) Base
(2) Column (Body)
(3) Cross rail
(4) Table
(5) Ram
(6) Tool head

(1) Base: It is a heavy and robust cast iron body which acts as a support for all other parts of the machine which are mounted over it.

(2) Column (Body): It is a box type iron body mounted upon the base. It acts as a housing for the operating mechanism of the machine electrical, cross rail and ram. On the top of it is has two guideways on which the ram reciprocates.

(3) Cross rail: It is a heavy cast Iron construction, attached to the column at its front of the vertical guideways.

(4) Table: It is used for holding workpiece.

(5) Ram: It reciprocates on the guideways provided above the column. It carries the tool head and mechanism for adjusting the stroke length.

(6) Tool head: It is attached to the front portion of the ram and is used to hold the tool rigidly.

The shapers are specified by maximum length of stroke, maximum horizontal travel of table, maximum vertical travel of table size of table, etc.

The different operations that can be performed on a shaper are machining horizontal surfaces, vertical surfaces, inclined surfaces, irregular surfaces, cutting gears, cutting slots, grooves and keyways.

1.3 PLANER

The planer is used to produce horizontal, vertical or inclined flat surfaces on workpieces that are too large to be accommodated on shapers. Workpieces in planers are usually large and heavy. They must be securely clamped to resist large cutting forces and high inertia forces that result from the rapid velocity changes at the end of the stroke. Special stops are provided at each end of the workpiece to prevent it from shifting.

Parts of Planer
Planer tools usually are quite massive and can sustain the large cutting forces.
Principal parts of a planer are:

(1) Bed: The bed of the planer is the heavy cast iron structure which provides the foundation for the machine and supports the housing and all other moving parts. At its top, V type guideways are provided on which the table slides.

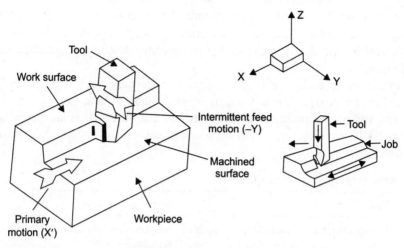

Fig. 1.6: Principle of planer.

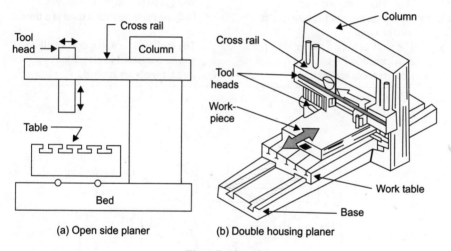

(a) Open side planer   (b) Double housing planer

Fig. 1.7: Planers.

(2) Table: Table reciprocates along the ways of the bed and supports the work. At its top, it carries longitudinal T slots and holes to accommodate the clamping bolts and other devices.

(3) Housing or columns: These are rigid castings placed on each side of the bed in case of double housing planer and on one side only in case of open side planer. It carries cross rail elevating screws, vertical feed shaft and cross feed bar to transmit the power to the upper parts of the machine. The front face carries the vertical ways along which the cross rail slides up and down.

(4) Cross rail: The cross rail mounted on the vertical guideways of the two housings. It can be raised or lowered.

(5) Tool heads: Two tool heads are mounted on the cross rail and the other two on the vertical columns. Each column carries one side tool head. All the four tool heads work independently, such that they can operate separately or simultaneously as desired.

Planes are generally specified by the width of the table, maximum distance of the table to the cross rail, maximum stroke of the table in mm, etc.

The various operations that can be performed on a planer are machining horizontal flat surfaces, vertical flat surfaces, angular surfaces, including downtails, slots and grooves.

Difference between planer and shaper.

	Planer	Shaper
1.	It is a heavier, more rigid and costlier machine.	It is comparatively lighter and cheaper machine.
2.	It requires more floor area.	It requires less floor area.
3.	It is used for large works.	It is used for small works.
4.	Tool is fixed and work moves.	Work is fixed and the tool moves.
5.	More than one cutting tool can be used at a time.	Only one cutting tool is used at a time.
6.	The tools used on a planer are larger, heavier and stronger.	The tool used on a shaper is small in size as compared to planer tool.
7.	Heavier feeds are applied.	Lighter feeds are applied.

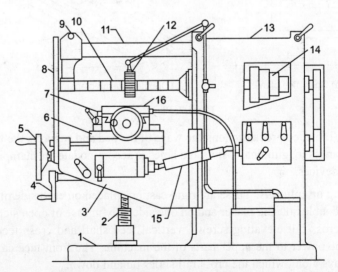

Fig. 1.8: Column and knee type milling machine.
1. Base, 2. Elevating screw, 3. Knee, 4. Knee elevating handle, 5. Crossfeed handle, 6. Saddle, Table 8. Front brace, 9. Arbor support, 10. Conepulley, 15 Telescopic feed shaft.

1.4 MILLING MACHINE

A milling machine is a machine tool that removes metal as the work is fed against a rotating multipoint cutter. The cutter rotates at a high speed and because of the multiple cutting edges, it removes metal at a very fast rate. The machine can also hold one or more number of cutters at a time. This is why a milling machine finds wide applications in production work. This is superior to other machines as regards to accuracy and better surface finish.

The usual classification according to the general design of the milling machine are:
  (1) Column and knee type
  (2) Fixed bed type
  (3) Planer type
  (4) Special type.

Parts of Milling Machine
The principal parts of a milling machine are:

(1) Base:
- Made up of grey cast iron.
- It serves as a foundation member for all the other parts which rest upon it.
- It carries the column at its one end.
- In some machines, the base is hollow and serves as a reservoir for cutting fluid.

(2) Column: The column is box-shaped, heavily ribbed inside and houses all the driving mechanisms for the spindle and table feed. The front vertical face of the column is accurately machined and is provided with dovetail guideways for supporting the knee.

(3) Knee:
- It is a grey cast iron casting that slides up and down on the vertical ways of the column face.
- The knee houses the feed mechanism of the table and different controls to operate it.

(4) Table:
- The table rests on ways on the saddle and travels longitudinally.
- The top of the table is accurately finished and T slots are provided for clamping the work and other fixtures on it.
- A lead screw under the table engages a nut on the saddle to move the table horizontally by hand or power.
- The longitudinal travel of the table may be limited by fixing trip dogs on the side of the table.

(5) Overhanging arm: The overhanging arm that is mounted on the top of the column extends beyond the column face and serves as a bearing support for the other end of the arbor. (Arbor is an extension of the machine spindle on which milling cutters are securely mounted and rotated.)

(6) Front brace: It is an extra support that is fitted between the knee and the over arm to ensure rigidity to the arbor and the knee.

(7) Spindle: It is located in the upper part of the column and receives power from the motor through belts, gears, clutches and transmits it to the arbor.

The different operations that can be performed in a milling machine are plain milling, face milling, side milling, angular milling, profile milling, end milling, saw milling, gear cutting, helical milling cam milling, thread milling.

1.5 DRILLING MACHINE

Drilling machine is used to make holes in the workpiece. The end cutting tool used for drilling holes in the workpiece is called the drill. The drill is placed in the chuck and when the machine is on the drill rotates. The linear motion is given to the drill towards the workpiece, which is called feed.

Different types of drilling machine used are:
(1) Portable drilling machine
(2) Bench type drilling machine
(3) Sensitive drilling machine
(4) Upright drilling machine
(5) Radial drilling machine
(6) Multiple spindle drilling machine
(7) Deep hole drilling machine
(8) Gang drilling machine
(9) Automatic drilling machine.

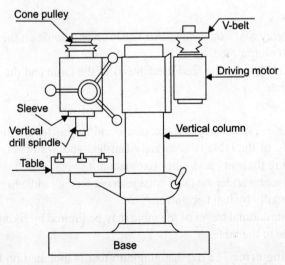

Fig. 1.9: Drilling machine.

Parts of Drilling Machine

The main parts of drilling machine are:

(1) Bed (Base): It is the main support of the column, the driving pulleys, the motor in case of a self-drive, etc. The top surface of the base or bed is machined and provided with T-slot to fix any fixture or big jobs of the table.

(2) Column: It is a vertical member and has to be sufficiently rigid to support the tool post including driving mechanism and the table. The cross-section of column is either round or square. Square section is more rigid than the round one.

(3) Table: The table is mounted around the column having three types of adjustment, radial adjustment about the column, vertical adjustment and circular adjustment about its own axis. After required adjustment of the table, the arm may be fixed in position during operation. The table may be either round or rectangular surface and provided with T slots for fixing vice or workpiece on the table.

(4) Spindle: The spindle is rotated in vertical axis by the top shaft in horizontal axis through two bevel gears. Top shaft rotates from the fast and loose pulley shaft in horizontal axis through two sets of step cone pulleys. This gives different speeds to the spindle shaft. The spindle has two motions one is circular or rotary and the other is vertical motion. Circular or rotary motion gets power from the top shaft through transmission system and vertical motion gets power from the top shaft through spur gears, worm, set of pinions and rack fixed on the hollow cylinder, mounted on spindle guided by bearings. The spindle feeding may be hand or power operated. The lower end of the spindle has a morse taper socket to fit drill bits or drill chucks for parallel shank drill bit. There is a rectangular slot to use drifts for removing drill bits or drill chuck from spindle nose.

Specification

The drilling machine is specified by maximum size of the drill in mm that the machine can operate, maximum dimensions of a job that can mount on a table in square metre, maximum spindle travel in mm, number of spindle speeds and range of spindle speeds in rpm, power input of the machine in H.P. number of automatic spindle feeds available, floor space required in square metre etc.

Operations
The different operations performed on the drilling machine are:
(1) Drilling: Operation of making a circular hole
(2) Boring: Enlarging a hole that has already been drilled.
(3) Reaming: To generate hole of proper size.
(4) Tapping: making internal threads.
(5) Counterboring: Enlarging the entry of a drilled hole to accommodate the bolt head, etc.
(6) Spot facing: For providing smooth seat for bolt head.
(7) Counter sinking: To bevel the top of a drilled hole for making a conical seat.

12 *Mechanical Engineering: Fundamentals*

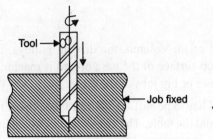

Fig. 1.10: Drilling.

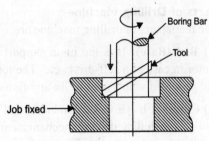

Fig. 1.11: Boring.

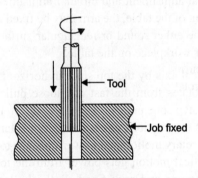

Fig. 1.12: Reaming.

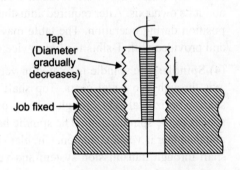

Fig. 1.13: Tapping.

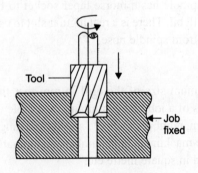

Fig. 1.14: Counterboring.

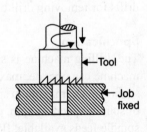

Fig. 1.15: Spot facing.

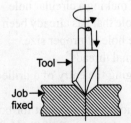

Fig. 1.16: Counter sinking.

1.6 SLOTTER

A slotting machine or slotter can be considered a vertical shaper. The major differene between a shaper and slotter is the direction of the cutting action. The ram carrying the slotting tool reciprocates in vertical guideway of the machine.

Parts of Slotter

Main parts of slotter are:
  (1) Bed
  (2) Column
  (3) Saddle
  (4) Cross slide
  (5) Circular table
  (6) Ram and tool head
  (7) Ram drive
  (8) Feed drive

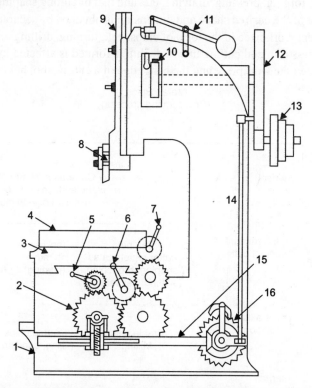

1. Base, 2. Feed gear, 3. Cross slide, 4. Table, 5. Cross feed handle, 6. Longitudinal feed handle, 7. Circular feed handle, 8. Tool, 9. Ram, 10. Crank disc, 11. Lever, 12. Bull gear, 13. Cone pulley, 14. Column, 15. Feed shaft, 16. Pawl actuating crank.

Fig. 1.17: Main parts of a slotter.

Specification

Slotting machine can be specified by maximum length of stroke of the ram in mm, maximum vertical adjustment of the ram in mm, diameter of the table in mm, distance between the centre of tool and the axis of column in mm, maximum longitudinal and cross travel of the table in mm, etc.

Operations

The different operations which can be performed on the slotter are cutting of:
(1) Internal grooves or key ways
(2) Internal gears
(3) Recesses
(4) Concave, circular and convex surfaces.

1.7 INTRODUCTION TO METAL CUTTING

The various working processes fall into two groups—the group of non-cutting shaping, e.g., forging, pressing, drawing, etc. and that of cutting shaping by which finish surface of the desired shape and dimension is obtained by separating a layer from the parent workpiece in the form of chips, e.g., turning, drilling, milling, etc.

The process of metal cutting in which chip is formed is affected by a relative motion between the workpiece and the hard edge of a cutting tool held against the

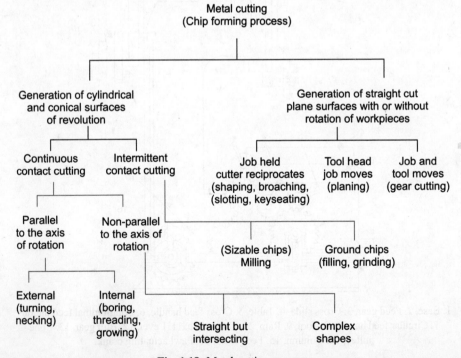

Fig. 1.18: Metal cutting process.

workpiece. Such relative motion is produced by a combination of rotary and translating movements either of the workpiece or of the cutting tool or of both.

Metal Cutting Process

A cutting tool may be used either for cutting apart, as with a knife or for removing chips. Parts are produced by removing metal mostly in the form of small chips.

Chip removal in the metal-cutting process may be performed either by cutting tools having distinct cutting edges or by abrasives used in grinding wheels, abrasive stickes or cloths.

All cutting tools can be divided into two groups. These are:
(1) Single point tools
(2) Multipoint tools

Single point cutting tools are used in lathes and slotting machines. Multipoint cutting tools are merely two or more single-point tools arranged together as a unit, e.g., milling cutter and broaching tools.

Chip Types

In engineering manufacture particularly in metal machining processes hard brittle metals have a very limited use, and ductile metals are mostly used. Chips formed generally are of three types:
(1) The discontinuous or segmental forms
(2) The continuous or ribbon type
(3) The continuous with built-up edge

Discontinuous or segmental chips
- Formed in most brittle materials, e.g., cast iron and bronze.
- These materials rupture during plastic deformation and form chips as separate small pieces.
- As these chips are produced, the cutting edge smoothes over the irregularities and a fairly good finish is obtained.
- Tool life good and low power consumption.

Continuous or ribbon type
- It consists of elements bonded firmly together without being fractured.
- Here the metal flows by means of plastic deformation and gives a continuous ribbon of metal which, under the microscope shows no sign of tear or discontinuity.
- Factors which are favourable for its formation are ductile metals such as mild steel, copper, etc., fine feed, high cutting speed, large rake angle, keen cutting edge, smooth tool face and an efficient lubrication system.

Built-up edge
- It implies the building up of a ridge of metal on the top surface of the tool and above the cutting edge.

- When the cut started in ductile metals, a pile of compressed and highly stressed metal forms at the extreme edge of the tool.
- Formation of false cutting edge on the tool takes place which is generally referred to as the built-up edge.
- Weaker chip metal tears away from the weld as the chip moves along the tool face.
- It is made at low cutting speed, low rake angle, high feed, lack of cutting fluid and large depth of cut.

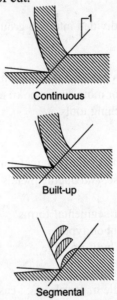

Fig. 1.19: Basic chip forms. 1. Shear plane

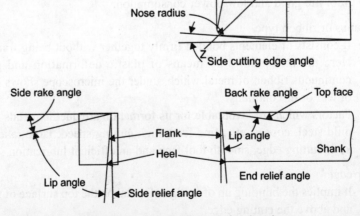

Fig. 1.20: Tool nomenclature and tool angles

1.8 CUTTING TOOL NOMENCLATURE

Cutting tool nomenclature means systematic naming of the various parts and angles of a cutting tool.

Complete nomenclature of the various parts of a single point tool is shown in Fig. 1.20.

The various parts are shank, face, flank, heel, nose, base, back rack, side rake side clearance, end cutting edge, wide cutting edge and lip angle. These elements define the shape of a tool.

Factors affecting tool life.
The life of a tool is affected by many factors such as
   (a) Cutting speed
   (b) Feed
   (c) Depth of cut
   (d) Chip thickness
   (e) Tool geometry
   (f) Material of cutting fluid
   (g) Rigidity of the machine

Types of tool material
The principal cutting materials are:
   (1) Carbon steels
   (2) Medium alloy steels
   (3) High speed steels
   (4) Stellites
   (5) Cemented carbides
   (6) Ceramics
   (7) Diamonds
   (8) Abrasives

Carbon steels
   (i) Carbon percentage is 0.08 to 1.5 per cent.
   (ii) They are low heat and wear resistant.
   (iii) Used only in low cutting speed operations

Medium alloy steels
   [Iron + Carbon + Tungsten or molybdenum or chromium or vanadium]
   They lose their hardness at temperatures from 250° to 350°C.
   High sped steel (HSS)
   (i) HSS operates at cutting speeds 2 to 3 times higher than for carbon steels and retain their hardness up to about 900°C

18 *Mechanical Engineering: Fundamentals*

    (ii) Used as a popular operation of drilling, tapping, hobbing, milling, turning, etc.
    (iii) It may be high tungsten, high molydenum or high cobalt.

Stellites
It is a non-ferrous cast alloy composed of cobalt, chromium and tungsten.

Cemented Carbides
    (i) It is composed principally of carbon mixed with other elements.
    (ii) Carbide suitable for steel machining is 82 per cent tungsten carbide, 10 per cent titanium carbide and 8 per cent cobalt.

Ceramics
Metal cutting tool which uses aluminium oxide is generally referred as ceramics.

Diamond
The diamonds used for cutting tools are industrial diamond which are naturally occurring diamonds containing flaws and therefore of no value as gemstones. Alternatively they can be also artificial.

Abrasive
Abrasive grains in various forms—loose, bonded into wheels and stone and embedded in papers and cloths find wide application in industry.

1.9 CUTTING FLUID

Purpose:
    (1) To cool the tool
    (2) To cool the workpiece
    (3) To lubricate and reduce friction
    (4) To improve surface finish
    (5) To protect the finished surface from corrosion
    (6) To cause chips break up into small parts
    (7) To wash the chips away from the tool.

Cutting fluids used are:
    (1) Water
    (2) Soluble oils
    (3) Straight oils
    (4) Mixed oils
    (5) Chemical additive oil
    (6) Chemical compound
    (7) Solid lubricants

EXERCISE

1. What are the different parts of the lathe explain each one of them.
2. What is compound rest explain
3. Explain how lathes can be specified.
4. Draw the neat stetch of shaper and explain its main parts
5. Define shaper and where it is being used.
6. Differentiate between shaper and planer.
7. What arethe differentoperations that can be performed on planer
8. Draw neat sketch of Milling machine and explain the function of each part.
9. Name the different types of drilling machines
10. Explain how the drilling machines are specified.
11. Define the following in case of drilling
    (a) Counter boring
    (b) Spot facing
    (c) Count sinking
12. Explain the function of saddle in case of slotter
13. Differentiate between shaper and slotter.
14. Explain drilling machine with diagram (MDU Jan 2010)
15. Explain different types of chips in metal cutting? (MDU Jan 2010)
16. Name of different types of cutting fluid used in the machine shop.
17. What is the purpose of using cutting fluid while performing any machine operation.

OBJECTIVE TYPE QUESTIONS

1. Lathe bed is usually made of
   (a) structural steel
   (b) stainless steel
   (c) cast iron
   (d) non-ferrous materials
2. Quick return mechanism is used in
   (a) milling machine
   (b) broaching machine
   (c) grinding machine
   (d) shaper
3. For machining a casting on a lathe, it should be held in
   (a) collect chuck
   (b) magnetic chuck
   (c) three jaw chuck
   (d) four jaw chuck

20 *Mechanical Engineering: Fundamentals*

4. Most machinable metal is one which
   (a) produces discontinuous chips
   (b) permits maximum metal removal per tool grind
   (c) results in maximum length of shear plane.
   (d) all of the above
5. The slowest speed in lathe is adopted for following operation
   (a) normal turning
   (b) thread cutting
   (c) turning big diameter
   (d) taper turning
6. Size of shaper is specified by
   (a) length of stroke
   (b) size of table
   (c) maximum size of tool
   (d) h.p. of motor
7. Size of planer is specified by
   (a) size of table
   (b) stroke length
   (c) size of table and height of cross rail
   (d) no. of tools which operate at a time
8. The feeding of the job in a shaper is done by
   (a) movement of the clipper box
   (b) table movement
   (c) V block
   (d) ram movement
9. Cylindrical parts are held on planer by
   (a) V blocks and arresters
   (b) angle plate
   (c) V-block, T bolts and clamps
   (d) T bolt and clamps
10. The difference between planer and shaper is that in former case
    (a) tool moves over stationary work
    (b) tool moves over reciprocating work.
    (c) both tool and job reciprocate.
    (d) tool is stationary and job reciprocates.
11. which of the follow abrasives is the hardest
    (a) Al_2O_3
    (b) SiC (Silicon Carbide)
    (c) B & C (Boron Carbide)
    (d) diamond
12. The binding material used in cemented carbide tools is
    (a) graphite
    (b) lead

(c) Cobalt
(d) Carbon

13. The type of chip produced when cutting ductile material is
    (a) Continuous
    (b) discontinuous
    (c) With built up edge
    (d) none of the above

14. For drilling operation, the cylindrical job should always be clamped on a
    (a) Collect
    (b) socket
    (c) jaw
    (d) V block

15. The cutting edges of a standard twist drill are called
    (a) flutes
    (b) lips
    (c) wedges
    (d) flanks
    (e) conical points.

16. The angle between the tool face and the ground end surface of a flank is known as a
    (a) lip angle
    (b) rake angle
    (c) clearance angle
    (d) cutting angle

17. Drills are usually made of
    (a) plain high-carbon tool steel
    (b) alloy steel
    (c) high speed steel.
    (d) tungsten carbide

18. The operation of threading a drilled hole is called
    (a) lapping
    (b) reaming
    (c) broaching
    (d) tapping

19. One micron is equal to
    (a) 1 mm
    (b) 0.1 mm
    (c) 0.1 mm
    (d) 0.001 mm

20. Commonly used units of feed in drilling operation are
    (a) mm
    (b) mm/rev
    (c) mm/sec
    (d) mm/mt

21. Chip brakers are provided on cutting tools
    (a) for safety of operator
    (b) to minimise heat generation
    (c) to permit easy access of coolant at tool point
    (d) to permit short segmented chips
22. Tool life is said to be over when
    (a) finish of work becomes too rough
    (b) chips become blue
    (c) chattering starts
    (d) a certain amount of wear or chartering occurs on the flank.
23. Continuous chips are formed when machining
    (a) ductile metal
    (b) brittle material
    (c) heat treated material
    (d) non of the above
24. A quill is a
    (a) steel tube in the head of some machine tools that enclosed the bearings of rotating spindles on which are mounted the cutting tools.
    (b) tool holding device
    (c) work clamping device
    (d) tool used for milling operation
25. The arbor of the milling machine is used to hold
    (a) cutting tool
    (b) spindle
    (c) overarm
    (d) mandrel

ANSWERS

1. c	2. d	3. d	4. b	5. b
6. a	7. c	8. b	9. c	10. d
11. d	12. c	13. a	14. d	15. b
16. a	17. c	18. d	19. d	20. b
21. d	22. d	23. a	24. a	25. a

UNIT 2

Basic Concepts of Thermodynamics

After going through this unit, the reader would be able to understand the following:
(1) Thermodynamic system and its types.
(2) Thermodynamic properties.
(3) Quasi-static process.
(4) Zeroth, I, II and III laws of thermodynamics.
(5) Concept of entropy.

2.1 BASIC CONCEPTS

— Thermodynamics is the science of energy and entropy.
— Thermodynamics is the science that deals with heat and work and the properties of substances that bear a relation to heat and work.
— The basis of thermodynamics is experimental evidence and observation. These observations have been formalized into certain basic laws, which are known as zeroth, first, second and third laws of thermodynamics.

Definitions

Thermodynamic System: It is a three-dimensional region of space or an amount of matter which is under consideration. It is enclosed by an imaginary surface or real surface which may be at rest or in motion, and can change its size or shape. Everything outside the arbitrarily selected boundaries of the system is called surrounding.

Universe: Combination of system and surrounding is called universe.

Open System: The system which can exchange both mass and energy with its surroundings is called an open system. The region of space for observation is described as control volume. The boundary of control volume is called control surface. In such a system, flow type of processes occur. For example, turbine, boiler pump, compressor, condenser, nozzle, diffuser, etc.

Closed System: The system which can exchange energy with its surroundings but not the mass is called as closed system. The quantity of matter thus remains fixed and the system is described as control mass system. The change in shape of the boundary and volume of the system is also permissible.

Isolated System: The system which can neither exchange mass nor the energy which its surrounding is called isolated system. For example: Gas enclosed in insulated box.

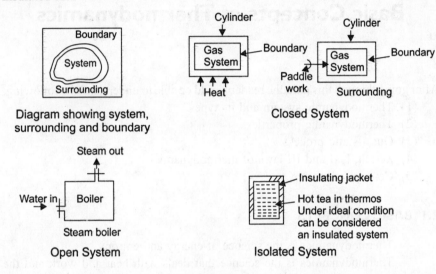

Fig. 2.1

Classical or Macroscopic Thermodynamics: If the analysis of any system is done taking in consideration the average values of different properties not considering the individual moleculer properties, then it is called as classical thermodynamic study.

Microscopic or Statistical Thermodynamics: It is the study where the behaviour of each individual molecule is taken into account for the purpse of analysis. This will further make the study difficult.

2.2 THERMODYNAMIC PROPERTIES

Characteristics with the help of which physical condition may be described are known as properties e.g., volume, temperature, pressure, etc.

It can also be defined as any quantity that solely depends on the state of the system and is independent of the path of the processes by which the system has arrived at the given state. Thermodynamic properties can be divided into two general classses: Intensive and extensive properties.

Extensive Property: Properties which depends on the mass of the system or in other words, value for the whole system is the sum of its values for the various subsystems or parts. Examples: Volume (V), Energy (E).

Intensive Property: These properties have values that are independent of the size or amount of mass of the system. These have fixed value. If a given phase system in equilibrium is divided into n parts, then the value of given intensive property will be the same for each of the subsystems. Examples: Temperature, pressure, density velocity and chemical concentration, etc.

If extensive property like Energy (E) of overall system is divided by the mass (m) of overall system, the resulting property is called specific property. $e = E/m$. A specific property is an intensive property.

State: When all the properties of a system have definite values, then the system is said to exist at a definite state. Properties are the coordinates to describe the state of a system. Any operation in which one or more of the properties of a system changes is called a change of state.

Path: The succession of states passed through during a change of state is called the path of the change of state.

Process: When the path is completely specified, the change of state is called a process, e.g. constant temperature process (isothermal process), constant pressure process (isobaric process).

Cycle: A thermodynamic cycle is defined as a series of state change such that the final state is identical with the initial state.

Phase: A quantity of matter homogeneous throughout in chemical composition and physical stucture is called a phase. Every substance can exist in any one of the three phases:
  (1) Solid
  (2) Liquid
  (3) Gas

Homogeneous System: A system consisting of a single phase is called a homogeneous system.

Heterogeneous System: A system consisting of more than one phase is called a heterogeneous system.

Thermodynamic Equilibrium: If the system is isolated from its surroundings and no change in its macroscopic properties occur, then it is called in thermodynamic equilibrium. A system will be in state of thermodynamic equilibrium if the condition for the following three types of equilibrium are satisfied:
  (a) Mechanical Equilibrium
  (b) Chemical Equilibrium
  (c) Thermal Equilibrium

(a) Mechanical Equilibrium: Absence of an unbalanced force between the system itself and between the system and surroundings, then the system is in mechanical equilibrium.

(b) Chemical Equilibrium: If there is no chemical reaction or transfer of matter from one part of the system to another such as diffusion or solution, the system is said to exist in a state of chemical equilibrium.

(c) Thermal Equilibrium: When the temperature of the system and surrounding are same and transfer of heat is not occurring within the system, then the system is said to be in thermal equilibrium.

Quasi-static Process

A process is said to be quasi-static if it is carried out in such a way that at every instant the system departs only infinitesimally from previous thermodynamic equilibrium state. Thus, it is a succession of equilibrium states. Only a quasi-static process can be reversible and can be represented on a thermodynamic plane. Infinite slowness is the characteristic feature of a quasi-static process.

Reversible Process

When the system changes state in such a way that at any instant during the process, the state point can be located on the thermodynamic plane, then the process is said to be a reversible process, i.e. quasi-static processes are reversible in nature. For a process to be reversible, no friction should be present. Pressure and temperature difference between system and surroundings should be infinitesimally small, i.e. the process must occur with infinitely slowness.

Irreversible Process

If any condition of reversible process is not fulfilled, then the process is called irreversible process.

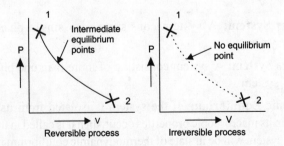

Fig. 2.2

Basic Concepts of Thermodynamics 27

Temperature and Temperature Scale

Temperature scale is referred to a thermometer for measuring inequality of temperature. To construct a thermometer two reference points are chosen, namely ice point and steam point.

Generally, two scales are used, namely Celsius and Fahrenheit. The Celsius scale (often referred to as the centigrade scale) is named after Anders Celsius (1701–44) and Fahrenheit scale is named after Daniel Gabriel Fahrenheit (1686–1736). The lower fixed point (ice point) is the temperature at which pure ice melts. The upper fixed point (steam point) is the temperature at which pure water boils. These are known as freezing point and boiling point respectively. These are 0°C and 100°C in the celsius scale and 32°F and 212°F in the Fahrenheit scale. Following equation is used to derive the relation between these two scales:

$$\frac{C - \theta_{ice}}{\theta_{steam} - \theta_{ice}} = \frac{F - \theta_{ice}}{\theta_{steam} - \theta_{ice}}$$

or

$$\frac{C - 0}{100 - 0} = \frac{F - 32}{212 - 32}$$

or

$$\frac{C}{100} = \frac{F - 32}{180}$$

$$\frac{C}{5} = \frac{F - 32}{9}$$

Likewise we can get relations between different scales:

(1) To convert Celsius to Fahrenheit

$$F = \frac{9}{5}C + 32$$

(2) To convert Fahrenheit to Celsius

$$C = \frac{5}{9}(F - 32)$$

(3) To convert Celsius to Kelvin

$$k = C + 273$$

(4) To convert Fahrenheit to Rankine

$$R = F + 460$$

(5) To convert Kelvin to Rankine

$$R = \frac{9}{5}k$$

(6) To convert Rankine to Kelvin

$$k = \frac{5}{9}R.$$

Comparison to Temperature Scale

	Scale	Fahrenheit °F	Centigrade °C	Kelvin K	Rankine °R
(i)	Steam point	212	100	375.15	671.67
(ii)	Triple point	32.02	.01	273.16	491.69
(iii)	Ice point	32	0	–273.15	491.67
(iv)	Absolute zero	–459.67	–273.15	0	0

Devices used for the measurement of temperature:
(1) Fluid thermometers.
   (a) Mercury in glass—thermometers.
   (b) Beckman thermometer—It can measure temperature rise of usually 6°C with an accuracy of 0.01°C.
   (c) The constant volume gas thermometer.
(2) Temperature gauge using fluids.
   (a) Temperature gauge
(3) Dimetallic strip method.
(4) Pyrometers
   (a) Thermocouples—measured due to Peltier and Thomson effect. The thermocouples used are:
      Chromel (+)   Alumel (−)
      Iron (+)   Constantan (−)
      Copper (+)   Constantan (−)
      Chromel (+)   Copel (−)
   (b) The resistance thermometer—the equation used

$$R = R_0 (1 + \alpha t)$$

where R_0 = original resistance.
        α = coefficient of increase of resistance with temperature t
    (i) The radial pyrometer.
    (ii) The fusion pyrometer.
    (iii) Thermal points.

2.3 ZEROTH LAW OF THERMODYNAMICS

This law gives the idea of temperature.

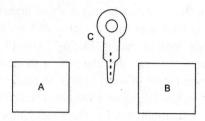

Fig. 2.3: Explaining zeroth law of thermodynamics.

Statement
If two bodies, isolated from the environment, are in thermal equilibrium with a third body, then the two bodies will be in thermal equilibrium with each other.

Illustration
Let A and B be two bodies isolated from each other and C be the third body which acts as a thermometer containing mercury in a glass tube. When the thermometer is steady, it is assumed that the mercury, the glass tube and the body whose temperature is being measured are all at the same temperature and hence are in thermal equilibrium.

If A and B are in thermal equilibrium with C separately and simultaneously, it means showing same temperature reading on C. It also means A and B will also be in thermal equilibrium with each other.

Heat
It is a form of energy which can also be defined as the energy transferred from one system to another system without transfer of mass and movement of boundary, which takes place because of temperature difference between the system and surroundings.

Sign convention heat supplied to a system is taken as positive while heat given out by the system is taken as negative. Word done on the system is taken as negative while work done by the system is taken as positive.

Work
Work is energy in transition. Work is transferred from the system during a given operation if the sole effect external to the system can be reduced to the rise of a weight. Work is done when the point of application of a force moves in the direction of force. The amount of work is equal to the product of force and distance moved in the direction of force.

Sign Convention: The work done on the system is taken to be negative while the work done by the system is taken to be positive.

2.4 INTRODUCTION TO FIRST LAW

The first law of thermodynamics was proposed by Joule in 1851 known as the principle of conservation of energy. Prior to James P Joule (1818–90) heat was considered an invisible fluid flowing from a body of higher calorie to a body of lower calorie and it was known as calorie theory of heat. It was Joule who first established that heat is a form of energy. During 1840–49 he conducted several experiments for the formulation of the first law of thermodynamics. Work was transferred to the measured mass of water kept in an adiabatic vessel by means of a paddle wheel driven by the falling of a weight. The rise in temperature of the water was recorded by a thermometer. Let the initial temperature of the water in the vessel be T_1. The paddle wheel work increased the water temperature to T_2. The cover was then removed and bath was allowed to return to its original temperature T_1. The system thus executed a cycle which consisted of a definite amount of work input W_{1-2} into the system followed by the transfer of an amount of work W_{1-2} into transfer of an amount of heat Q_{2-1} from the system. Joule repeated this experiment for different weights falling through different distances and measured each temperature rise. In each and every case, Joule found that work and heat have a constant relation. This constant of proportionality is called Joule's constant or the mechanical equivalent of heat (J).

$$\text{Work/Heat} = \text{Constant (J)}$$

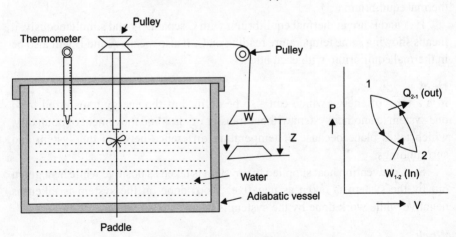

Fig. 2.4: Diagram explaining 1st law.

If the cycle involves several work and heat transfers, the result is the same and can be written as:

$$\Sigma W_{cycle} = J\Sigma Q_{cycle}$$

Statement: 1st law of thermodynamics: If the system undergoes a cyclic change of state, then the algebraic sum of the work delivered to the surroundings is proportional to the algebraic sum of heat taken from the surroundings.

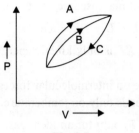

Fig. 2.5

$$\oint \delta W = J \oint \delta Q$$

In S.I. system the heat and work have same unit. Thus,

$$\oint \delta Q = J \oint \delta W$$

$$\oint (\delta Q - \delta W) = 0$$

Internal Energy Is A Property

$$\oint (\delta Q - \delta W) = \int_1^2 (\delta Q - \delta W)_A + \int_2^1 (\delta Q - \delta W)_c = 0 \quad (1)$$

$$\oint (\delta Q - \delta W) = \int_1^2 (\delta Q - \delta W)_B + \int_2^1 (\delta Q - \delta W)_c = 0 \quad (2)$$

From equations (1) and (2)

$$\int_1^2 (\delta Q - \delta W)_A - \int_1^2 (\delta Q - \delta W)_B = 0$$

$$\int_1^2 (\delta Q - \delta W)_A = \int_1^2 (\delta Q - \delta W)_B \quad (3)$$

Thus, it is obvious from equation 3 that quantity $(\delta Q - \delta W)$ is same for both the processes A and B connecting two points 1 and 2. Thus, $(\delta Q - \delta W)$ does not depend on path of process but depends on 1 and 2 (i.e. state points). Therefore, it is a point function and hence it is a property of the system. This property is called overall energy of the system and is represented by E. Thus,

$$\int_1^2 (\delta Q - \delta W) = \int_1^2 dE \quad (4)$$

E = Sum of microscopic and macroscopic modes of energy.
$$E = U + PE + KE$$
PE is bulk potential energy, KE is bulk kinetic energy and U is the total internal energy of the system.

In an ideal gas, there are no intermolecular forces of attraction and repulsion and the internal energy depends only on temperature. Thus,
$$U = f(T) \text{ only for an ideal gas.}$$
A system can possess other forms of energies also, like magnetic energy, electrical energy and surface (tension) energy. In the absence of these forms; the total energy E of a system is given by:
$$E = U + PE + KE \qquad (5)$$
In the absence of motion and gravity, $KE = 0$ and $PE = 0$,
$$E = U$$
Therefore, equation 5 becomes
$$\delta Q - \delta W = dE$$
$\delta Q - \delta W$ is independent of path and depends upon end states only, internal energy U which is equal to $\int (\delta Q - \delta W)$ is also independent of path and hence a property of the system.

Internal Energy: Internal energy can be defined as a form of stored energy in a system in the absence of magnetism, electricity, capillarity, surface tension, motion and gravity.

2.5 LIMITATIONS OF THE FIRST LAW AND INTRODUCTION TO THE SECOND LAW

It is a common experience that while work is very easily converted into heat (by rubbing both hands we generate heat) but heat cannot be easily converted into work. The first law, however, places no restriction on the direction of flow of heat and work. For example, in the case of disc brakes, when it is stopped by friction pad, the pad as well as disc gets heated up, because the KE lost by the disc is converted into heat energy. The first law of thermodynamics would be equally satisfied if the brakes were to cool off and give back its internal energy to disc causing it to resume its rotation again. This may, however, never occur. The action of the brake in stopping the disc by friction is irreversible process. Thus, we can conclude that there exists a directional law which imposes limitation on energy transformation.

Statement of second law also called **Kelvin-Planck statement**. "It is impossible to construct an engine which will work in a complete cycle and produce no other effect except the rising of a weight (the production of work) and the exchange of heat with a single reservoir".

Thus, according to this law, an engine E as shown in figure working on cyclic process (which means that there is no change in the internal energy for the complete cycle) cannot simply receive an amount of heat, say Q from a heat source and deliver work W such that W is equal to Q. Such an engine would satisfy the first law of thermodynamics but violate the second law.

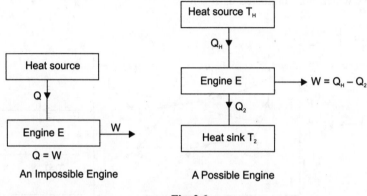

Fig. 2.6

Thus, a practical engine must reject some of the heat which it receives. An engine is a device to convert heat into work whereas a heat pump is a device to pump (transfer) heat from a lower temperature to a higher temperature while operating in a cycle.

Clausius' statement on the second law: "It is impossible to construct a heat pump which will work in a complete cycle and produce no effect except to transfer heat from cold reservoir to a hot reservoir". Or in other words, it is stated as "heat itself cannot flow from a cold to a hot body".

Perpetual motion machine of second kind: A heat engine that exchanges heat with a single body in an equilibrium state and produces work without creating any other effect is called perpetual motion machine of second kind. Obviously, such a machine is impossible. Therefore, based on this concept there is another statement of second law. "Perpetual motion machine of the second kind is an impossibility".

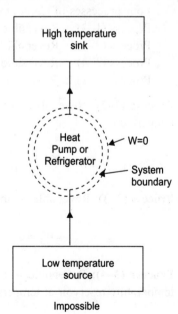

Fig. 2.7: Clausius' statement's Explanation.

2.6 CARNOT CYCLE

In 1824, Nicolas Sadi Carnot, a French Engineer, introduced an imaginary reversible cycle, which is named after him. From the second law of thermodynamics, it has been observed that the efficiency of a heat engine cannot be equal to unity. If the efficiency of the heat engine is less than unity, what is the maximum efficiency of heat engine. The answer to this is given by considering Carnot cycle.

Carnot cycle consists of two reversible isothermal and two reversible adiabatic processes which are connected as follows:

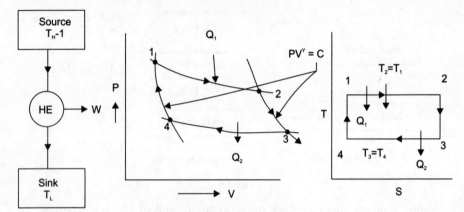

Fig. 2.8: Carnot cycle

Four processes of Carnot cycle are as follows:
Process (1–2) : Reversible isothermal heat addition.
Process (2–3) : Reversible adiabatic expansion.
Process (3–4) : Reversible isothermal heat rejection.
Process (4–1) : Reversible adiabatic compression.

Process (1–2): Reversible isothermal heat addition. Heat is added at constant temperature

$$Q_H = P_1 V_1 \ln \frac{V_2}{V_1} = mRT_H \ln r \qquad r = \frac{V_2}{V_1}$$

Process (2–3): Reversible adiabatic expansion. Work is obtained by expansion.

$$_2W_3 = C_v(T_2 - T_3) = \frac{P_2 V_2 - P_3 V_3}{\gamma - 1}$$

Process (3–4): Reversible isothermal heat rejection. Heat is rejected to a low temperature reservoir at temperature T_L

$$Q_L = P_3 V_3 \ln \frac{V_3}{V_4} = mRT_L \ln r \qquad r = \frac{V_3}{V_4} = \frac{V_2}{V_1}$$

Process (4–1): Reversible adiabatic compression. Work is supplied for compression by an external source in a device.

$$\text{Work done} = \frac{P_4 V_4 - P_1 V_1}{\gamma - 1}$$

Network Done = Heat supplied – Heat rejected

$$Q_H - Q_L = mR(T_H - T_L)\ln r$$

Thermal efficiency of Carnot cycle $= \dfrac{\text{Work done per cycle}}{\text{Heat supplied per cycle}}$

$$= \frac{mR(T_H - T_L)\ln r}{mRT_H \ln r} = \frac{T_H - T_L}{T_H} = 1 - \frac{T_L}{T_H}$$

$$\eta_{\text{carnot}} = \left[1 - \frac{T_L}{T_H}\right].$$

$$\eta_{\text{carnot}} = \frac{T_{\max} - T_{\min}}{T_{\max}}$$

Thus, Carnot efficiency depends only on temperature difference of source and sink and is independent of working substance.

Example
Due to metallurgical limit of boiler tube, piston cylinder and turbine blades maximum temperature of working fluid should not exceed 660°C. Temperature at which heat is rejected from sink should be more than atmospheric temperature. In India, an average atmospheric temperature is considered to be 25°C. Let us consider that the heat rejection temperature is about 45°C (more than 25°C for better heat tansfer). So, maximum value of Carnot efficiency would be about:

$$\eta_{\text{carnot}} = 1 - \frac{45 + 273}{660 + 273} = 1 - \frac{318}{933} = 1 - .34 = .659$$

That is, Carnot efficiency should not exceed 65.9%.

Carnot efficiency gives the maximum value of efficiency which can be acheived by any engine operating between the same temperature limits.

Reversed Carnot Cycle: It is an ideal cycle for refrigerator and heat pumps. The cycle consists of four processes which are as follows:
Process (1–2): Reversible adiabatic expansion in an expanding device.

Process (2–3): Reversible isothermal heat absorption from low temperature source.
Process (3–4): Reversible adiabatic compression in a compressor.
Process (4–1): Reversible isothermal heat rejection to high temperature sink.

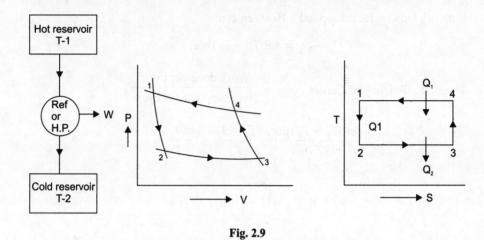

Fig. 2.9

Performance of refrigerator and heat pump:

$$(COP)_{ref} = \frac{\text{Heat abstracted from cold reservoir}}{\text{Work supplied}}$$

$$(COP)_{H.P.} = \frac{\text{Heat rejected to hot sink (surroundings)}}{\text{Work supplied}}$$

$$(COP)_{ref} = \frac{T_{min}}{T_{max} - T_{min}}$$

$$(COP)_{H.P.} = \frac{T_{max}}{T_{max} - T_{min}}$$

Carnot Theorem: No heat engine working between two thermal reservoirs with fixed temperatures can be more efficient than a reversible engine operating between same thermal reservoirs.

The proof is as follows:

Let there be a hot reservoir at temperature T_1 and a cold reservoir at temperature T_2. Imagine a reversible engine R taking 100 kJ from the heat reservoir and converting 60 kJ into work and rejecting 40 kJ to the cold reservoir, the engine efficiency being 60%. If this engine is reversed, it will take 60 kJ to drive it another 40 kJ will be taken from the cold reservoir and 100 kJ will be delivered to the hot reservoir.

Assume an irreversible engine I having efficiency greater than that of the reversible engine say 75% and driving the reversible engine without any losses. Now for the same 60 kJ work done heat taken by irreversible engine = 60/0.75 = 80 kcal from the hot reservoir and heat rejected to the cold reservoir = (80–60) = 20 kJ. In this system, reversible engine delivers (100–80) = 20 kJ heat more to hot reservoir than the irreversible engine takes from it. Also reversible engine takes (40–20) = 20 kJ heat more from the cold reservoir than the irreversible engine delivers to it and this heat is being pumped from the cold reservoir to the hot reservoir without any external power which is contrary to the second law of thermodynamics. Hence, the assumption that irreversible engine is more efficient is wrong.

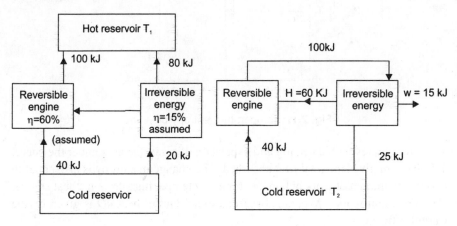

Fig. 2.10

The assumption of higher efficiency for irreversible engine contradicts the second law.

2.7 ENTROPY

Entropy of a substance is a property which increases with the addition of heat. Entropy itself cannot be defined but change of entropy can be defined. Mathematically, in a reversible process, the increase of entropy when multiplied by the absolute temperature, gives the heat received by the fluid from an external source.

The change of entropy with temperature is shown on a diagram known as Temperature entropy (T-S) diagram.

The initial condition of a gas receiving heat is denoted by point 1 while its final state be represented by point 2. Considering any point 0 on the curve, a small addition of heat dQ under reversible conditions increases entropy by dS. If T is the absolute temperature at this instant, by the mathematical definition of entropy, we get $dQ_{rev} = Tds \; ds = \dfrac{dQ_{rev}}{T}$ from figure. Tds is the area under the curve during the change of entropy ds.

$$Q = \int_{T_1}^{T_2} Tds = \text{area under the curve 12}$$

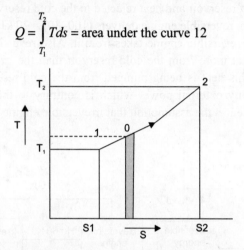

Fig. 2.11: Temperature-entropy diagram

Thus, for any reversible heating and expansion of a gas, the area under the curve, T-S diagram gives the total heat absorbed. The change in entropy is also equal to maximum amount of work obtainable for a unit temperature drop in a heat engine. The maximum work dW obtainable from an amount of heat dQ is given by the Carnot efficiency.

$$dW = dQ \times \dfrac{(T_1 - T_2)}{T_1}$$

If temperature drop (T1–T2) is unity:

$$dW = \dfrac{dQ}{T} = \text{change of entropy } dS$$

Thus, the change of entropy may be regarded as a measure of the rate of availability of heat for transformation into work.

Physical Concept of Entropy

The entropy of a substance is a real physical quantity defining the state of the body and can be easily evaluated for a solid or a perfect gas except for an additive constant representing the entropy at the absolute zero.

Entropy can be viewed as a measure of molecular disorder, or molecular randomness. As a system becomes more disordered, the positions of the molecules becomes less predictible and the entropy increases. Thus, it is not surprising that the entropy of a substance is lowest in the solid phase and highest in the gas phase. In the solid phase, the molecules of a substance continually oscillate about their equilibrium positions, but they cannot move relative to each other and their positions at any instant can be predicted with good certainty. In the gas phase, however, the molecules move about at random, collide with each other, and change direction, making it extremely difficult to predict accurately, the microscopic state of a system at any instant. Associated with this molecular chaos is a high value of entropy.

Heat is, in essence, a form of disorganized energy, and some disorganization (entropy) will flow with heat. As a result, the entropy and the level of molecular disorder or randomness of the hot body will decrease with the entropy and the level of molecular disorder of the cold body will increase. The second law requires that the increase in entropy of the cold body be greater than the decrease in entropy of the hot body and thus, the net entropy of the combined system (the cold body and hot body) increases. That is, the combined system is at a state of greater disorder at the final state.

Clausius Inequality

When any closed system undergoes a cyclic process, the sum of all $(\delta q/T)$ terms at the system boundary for each differential element of the process will always be equal to or less than zero. Thus,

$$\oint \frac{\delta q}{T} \leq \text{for any cyclic process (possible) (Clausius' inequality)}$$

If $\oint \frac{dQ}{T} = 0$, the cycle is reversible $ds = 0$

$\oint \frac{dQ}{T} < 0$, the cycle is irreversible and possible $ds > 0$

$\oint \frac{dQ}{T} > 0$, the cycle is impossible

2.8 THIRD LAW OF THERMODYNAMICS

The third law of thermodynamics often referred to as the Nernst law, provides the basic for the calculation of absolute entropy of the substances.

Third Law Statement
The entropy of all perfect crystalline solid is zero at absolute zero temperature.

40 Mechanical Engineering: Fundamentals

The absolute zero temperature, O.K. or OR, exists in the thermodynamic temperature scale. This correponds to –273.15°C and –459.67°F on the celcius and fahernheit scale respectively.

In practice one can never achieve absolute zero temperature as it is impossible but temperatures very close to it have been attained due to perfect insulation.

Comparison of first and second law of thermodynamics.

First Law	Second Law
1. According to the first law heat and work are of same quality indicating 100% efficiency of a cyclic engine	Work is considered to be a high grade energy where as heat as a low grade energy
2. Results in the definition of the extensive property; internal energy	Results in the definition of the extensive property; entropy
3. States that energy of an isolated system can neither be created nor destroyed	States that entropy of an isolated system cannot be destroyed but it can be created
4. Energy of the universe is constant	The entropy of the universe increases towards a maximum
5. Energy is conserved in every real process	Energy is degraded in every real process

SOLVED PROBLEMS

Problem 1

A freezer is to be maintained at a temperature of 238 K when the ambient temperature is 306 K. In order to maintain the freezer box at 238 K it is necessary to remove heat from it at the rate of 2460 J/sec. What is the maximum possible coefficient of performance of the freezer and what is the minimum power that must be supplied to the freezer.

Solution

The maximum coeffcient of performance is obtained when the freezer operates reversibily.

Max COP of a refrigerator

$$= \frac{Q_2}{Q_1 - Q_2} = \frac{T_2}{T_1 - T_2} = \frac{238}{306} = 3.5 \text{ \textbf{Ans.}}$$

But $Q_2 = 2460$ J/S and $Q_1 - Q_2 = W$

$$\frac{2460}{W} = 3.5 \text{ or } W = \frac{2460}{3.5} = 702.8 \text{ J/S}$$

$$= 702.8 \, W \text{ Ans.}$$

Problem 2
An inventor claims to have developed a cyclic engine which exchanges heat with reservoirs at 130°C and –40°C. It receives onlyu 2100 kJ/min of heat and develops 17.66 kW Is his claim feasible?

Solution
The maximum efficiency of a heat engine can be equal to that of carnot engine

$$\eta_{max} = \frac{T_1 - T_2}{T_1}$$

$$= \frac{(273+130) - (273-40)}{(273+130)} = \frac{433 - 239}{403}$$

$$= 0.422 \text{ or } 42.2\%$$

The claimed efficiency = work output/heat input

$$= \frac{17.66}{2100/60} = 0.50$$

$$= 50.45\%$$

The claimed efficiency is even greater than the carnot engine efficiency which is not possible. Therefore the claim is not feasible.

Problem 3
A heat engine works between hot and cold reservoirs at 556 K and 278 K. the engine reservoirs at 556 K and 278 K. The engine receives 278 kJ/sec of heat. The following results were reported
  (i) 70 kJ/sec of heat was rejected
  (ii) 139 kJ/sec of heat was rejected
  (iii) 208 kJ/sec of heat was rejected
Indicate which of the results shows a reversible cycle, irreversible cycle or impossible cycle.

Solution
We know that

$$\oint \frac{dQ}{T} = 0 \text{ for a reversible cycle}$$

$$\oint \frac{dQ}{T} = 0 \text{ for an irreversible cycle}$$

$$\oint \frac{dQ}{T} > 0 \text{ is impossible}$$

Now taking three cases we have

Since
$$\oint \frac{dQ}{T} = \oint \frac{dQ}{T}_{\text{heat added}} + \oint \frac{dQ}{T}_{\text{heat rejected}}$$

∴ for case (i)
$$\oint \frac{dQ}{T} = \frac{278}{556} - \frac{70}{278} = 0.248$$

This is positive so case (i) is impossible

Case (ii)
$$\oint \frac{dQ}{T} = \frac{278}{556} - \frac{139}{278} = 0$$

∴ the cycle is reversible

Case (iii)
$$\oint \frac{dQ}{T} = \frac{278}{556} - \frac{208}{278} = 0.248$$

This is negative

∴ The cycle is irreversible

Problem 4

Q4(a). In a certain reversible process the rate of heat transfer to the system per unit temperature is given by $\frac{dQ}{dT} = 1.05$ kJ/K. Find the increase in entropy of the system. If its temperature rises from 300°K to 400°K. Also find the change in specific entropy if mass of system in 2 kg.

(b). In a second process between the same and states the temperature rise is obtained by stirring accompanied by a heat addition half as great as in case (a) find the increase in entropy in the case.

Solution

$$\frac{\delta Q}{T} = 1.05 \quad or \quad \delta Q = 1.05 \, dT$$

(a) for reversible process

$$ds = \frac{1.05 \, dT}{T}$$

$$S = 1.05 \int \frac{dT}{T} = 1.05 \, In \frac{T_2}{T_1}$$

$$= 1.05 \, In \frac{400}{300}$$

or $\quad\quad\quad 1.05 \times 0.2876 = 0.3019$ kJ/K

change in specific entropy $= \dfrac{0.3019}{2} = 0.15095$ kJ/k kg.

(b) Between same end states the change of entropy will not depend on the process. Therfore in this case also the increase of entropy is the same, namely 0.30 kJ/K.

EXERCISE

1. Define: (i) property (ii) state (iii) system (iv) control volume (v) process
2. Explain: close system, open system and isolated system.
3. What is the physical significance of zeroth law of thermodynamics. State zeroth law of thermodynamics?
4. What do you understand by first law of thermodynamics and explain the two statements of second law of thermodynamics.
5. What are the limitation of first law of thermodynamics? State second law of thermodynamcis.
6. Explain Clausius' Inequlity of Entropy.

OBJECTIVE TYPE QUESTIONS

1. A closed system is one in which
   (a) mass does not cross boundaries of the system, though energy may do so.
   (b) mass crosses the boundary but not the energy.
   (c) neither mass nor energy crosses the boundaries of the system
   (d) both energy and mass cross the boundaries of the system

2. Intensive property of a system is one whose value.
   (a) depends on the mass of the system like volume
   (b) does not depend on the mass of the system, like temperature pressure etc.
   (c) is not dependent on the path followed but on the state
   (d) is dependent on the path followed and not on the state
3. Heat and work are
   (a) point functions
   (b) system properties
   (c) path functions
   (d) intensive properties
4. Which of the following is the property of a system
   (a) pressure and temperature
   (b) internal energy
   (c) volume and density
   (d) all of the above
5. Work done is zero for the following process
   (a) constant volume
   (b) free expansion
   (c) throttling
   (d) all of the above
6. Entropy change depends upon
   (a) heat transfer
   (b) mass transfer
   (c) change of temperature
   (d) thermodynamic state
7. First law of thermodynamics
   (a) enables to determine change in internal energy of the system
   (b) does not help to predict whether the system will or will not undergo a change
   (c) does not enable to determine change in entropy
   (d) all of the above
8. If a heat engine attains 100% thermal efficiency it violates
   (a) zeroth law of thermodynamics
   (b) first law of thermodynamics
   (c) second law of thermodynamics
   (d) all of the above laws
9. According to Clausius' statement
   (a) heat flow from hot substance to cold substance.
   (b) heat cannot flow from cold substance to hot substance.
   (c) heat can flow from cold substance to hot substance with the aid of external work.
   (d) none of the above

10. In a carnot cycle heat is transferred at
    (a) constant pressure
    (b) constant volume
    (c) constant temperature
    (d) constant enthalpy

11. The value of $\Sigma \dfrac{d\theta}{T}$ for an irreversible process is
    (a) equal to zero
    (b) greater than zero
    (c) less than zero
    (d) unity

12. In case of refrigeration machine C.O.P. will be equal to
    (a) $\dfrac{Q_2}{Q_1 - Q_2}$
    (b) $\dfrac{Q_1}{Q_1 - Q_2}$
    (c) $\dfrac{Q_1 - Q_2}{Q_1}$
    (d) $\dfrac{Q_2 - Q_1}{Q_1}$

13. Efficiency of a carnot engine with $t_1 = 200°C$ $t_2 = 30°C$ is
    (a) 85%
    (b) 36%
    (c) 80%
    (d) 12%
    (e) 15%

14. Internal energy of a substance depends on
    (a) volume
    (b) pressure
    (c) temperature
    (d) entropy

15. Kelvin Plank's law deals with
    (a) conservation of heat
    (b) conservation of work
    (c) conversion of heat into work
    (d) conversion of work into heat

16. A frictionless heat engine can be 00% efficient only if its exhaust temperature is
    (a) below surrounding
    (b) 0°C

(c) 0°K
(d) equal to inlet temperature

17. Calorie is a measure of
    (a) specific heat
    (b) quantity of heat
    (c) thermal capacity
    (d) entropy

18. The value of 1 bar in S.I. units is equal to
    (a) 1 N/m^2
    (b) 1 kN/m^2
    (c) $1 \times 10^4 \text{ N/m}^2$
    (d) $1 \times 10^5 \text{ N/m}^2$

19. Compressed air coming out from a punctured football
    (a) becomes hotter
    (b) becomes cooler
    (c) remains at the same temperature
    (d) attains atmospheric temperature

20. Which of the following cycles has maximum efficiency
    (a) Rankine
    (b) Stirling
    (c) Carnot
    (d) Brayton

21. Which of the following is extensive property?
    (a) entropy
    (b) internal energy
    (c) kinetic energy
    (d) all of the above

22. A process occurs spontaneously if its entropy
    (a) increases
    (b) decreases
    (c) remains same
    (d) becomes zero

23. Entropy is called the property of the system because
    (a) Its derivative is zero for any process.
    (b) It has some value at any two equilibrium states
    (c) It has a single value at each equilibrium state
    (d) It has a constant value at each equilibrium state

24. A reversible engine working between the temperatures limits of 600°K and 12°K receives 50 kJ of heat. The work done by the engine will be
    (a) 50 kJ
    (b) 100 kJ
    (c) 25 kJ
    (4) –25 kJ

25. A closed system receives 50 kJ heat but the internal energy of the system decreases by 25 kJ The work done by the system would be
    (a) 75 kJ
    (b) –75 kJ
    (c) 25 kJ
    (d) –25 kJ

ANSWERS

1. a	2. b	3. c	4. d	5. d
6. a	7. d	8. c	9. c	10. c
11. c	12. a	13. b	14. c	15. c
16. c	17. b	18. d	19. b	20. c
21. d	22. a	23. c	24. c	25. a

UNIT 3

Properties of Steam and Steam Generators (Boilers)

After going through this unit, the reader is required to understand the following:
(1) How to use steam tables.
(2) How to make calculations with Mollier chart.
(3) Methods for finding out the dryness fraction of steam.
(4) Cochran and Babcock and Wilcox boiler.
(5) Boiler mountings.
(6) Boiler accessories.

3.1 PROPERTIES OF STEAM AND BOILERS

Steam is a vapour. It is used for heating and as the working substance in the steam engine and steam turbine plants. Properties of steam were first investigated by Regnault and subsequently by Prof. Callendar. Properties of steam are given in the form of tables and also in form of charts. The datum used for the calculation of properties of steam is 0°C. All values measured above this temperature are considered positive and those measured below are taken as negative. Steam is a pure substance and like any other pure substance, it can be converted into any of the three states i.e. solid, liquid and gas. A system composed of liquid and vapour phases of water is also a pure substance. Even if some liquid is vaporized or some vapour gets condensed during a process, the system will be chemically homogenous and unchanged in chemical composition.

Formation of Steam at Constant Pressure
Assume that a unit mass of steam is generated starting from solid ice at –20°C at 1 atm pressure in a cylinder and piston machine. The distinct regimes of heating are:

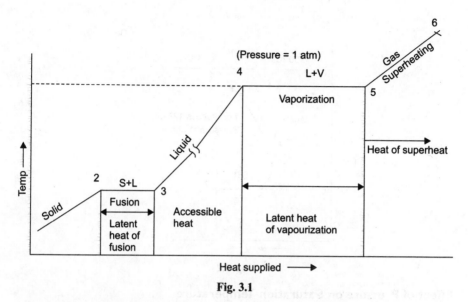

Fig. 3.1

Regime (1–2): The heat given to ice increases its temperature from –20°C to 0°C. The volume of ice increases with the increase in temperature. Point 2 shows the saturated solid condition. At 2, the ice starts to melt.

Regime (2–3): The ice melts into water at constant pressure and temperature. The amount of heat supplied per kg of the substance is called latent heat of fusion. At 3, the melting process ends. There is a sudden decrease in volume at 0°C as the ice starts to melt. It is a peculiar property of water due to breaking of hydrogen bonding.

Regime (3–4): The temperature of water increases on heating from 0°C to 100°C as shown in figure. The volume of water first decreases with the increase in temperature, reaches to its minimum at 4°C and again starts to increase because of thermal expansion. Point 4 shows the saturated liquid condition.

Regime (4–5): The water starts boiling at 5. The liquid starts to get converted into vapour. The boiling ends at point 5. Point 5 shows the saturated vapour condition at 100°C and 1 bar. The amount of heat supplied during the process is called latent heat of vaporization.

Regime (5–6): It shows the superheating of steam above saturated steam point. The volume of vapour increases rapidly and it behaves as a perfect gas. The difference between the superheated temperature and the saturation temperature at a given pressure is called degree of superheat.

Points 2,3,4,5 are known as saturation states. State 2 is saturated solid state while states 3 and 4 are saturated liquid state. State 3 is for fusion and state 4 is for vaporization. State 5 is saturated vapour state. At saturated state, the phase may get changed without change in pressure or temperature.

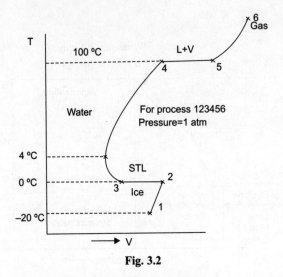

Fig. 3.2

Effect of Pressure on Saturation Temperature

The boiling temperature of water increases with the increase of pressure. The boiling temperature of water at a particular pressure is called saturation temperature and corresponding pressure is known as saturation pressure. The saturation temperature at every pressure is uniquely fixed. The variation of saturation temperature with saturation pressure is shown in the figure.

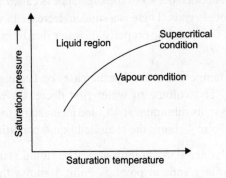

Fig. 3.3

Critical temperature: The critical temperature of water is defined as the temperature at which water evaporates into steam without taking any latent heat. Above this temperature, it is impossible to liquefy the steam. The saturation pressure corresponding to critical temperature is called critical pressure. The critical temperature and critical pressure of water are 374.15°C and 221.20 bar respectively. Above critical point, all differences in the properties of liquid and vapour disappear.

Triple point: At very low pressure, melting point and boiling point coincide together and all the three states–solid, liquid and gas remain in equilibrium. This temperature is called triple point. For water at .006114 bar, the triple point is 273.16°K.

Generation of Steam

The amount of heat required to raise the temperature of ice to the melting point = $h_i = C_{pi}(t_i - t_0)$

where t_0 = initial temperature of ice
t_i = melting point of ice
C_{pi} = specific heat of solid state

Total heat required for melting of ice = $C_{pi}(t_i - t_0) + h_{if}$ where h_{if} = enthalpy of melting = latent heat of fusion. Addition of heat to water from latent point to saturated liquid at saturation temperature is called sensible enthalpy. = $h_f = 1 \times C_{pw}(t_s - t_i)$ where C_{pw} = specific heat of water, t_s = saturation temperature.

Latent heat of evaporation = h_{fg}.

On further heating superheated steam is generated. Total heat added from initial temperature to final temperature of superheated steam (t_{sup}) can be given by:

$$h_{sup} = C_{pi}(t_i - t_0) + h_{if} + h_f + h_{fg} + C_{ps}(t_{sup} - t_s)$$

where C_{ps} = specific heat of superheated steam
t_s = saturation temperaure
Average value of Cps = 2.0934 KJ/Kgk

($t_{sup} - t_s$) is called degree of superheat

Again take an example:
Let us consider 1 kg of water at 0°C (datum temperature of enthalpy) contained in a cylinder filled with a frictionless piston exerting a constant pressure (P) in the cylinder all the time. As heat is supplied the temperature of water continues to increase until saturation temperature is attained. The heat supplied in this process is given by $h_f = C_{pw}(t_s - 0)$.

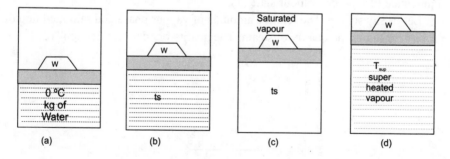

Fig. 3.4

where C_{pw} = specific heat of water at constant pressure. This heat is called enthalpy of saturated liquid. Further addition of heat is used to change the phase from liquid to vapour. This is continued until the entire liquid is converted into vapour. Now the total heat required to convert entire 1 kg of water at 0°C into steam at t_s°C is given by:

$$h_g = h_f + h_{fg}$$

h_{fg} is called latent enthalpy or enthalpy of evaporation and h_g is called enthalpy of saturated vapour. Further addition of heat will increase the temperature of steam above the saturation temperature. One such condition has been shown in figure. The quantity of heat during superheating is given by $C_{ps}(t_{sup} - t_s)$ and total heat of superheated steam

$$= h_{sup} = h_g + C_{ps}(t_{sup} - t_s)$$

Quality of Steam

The steam may exist in three conditions, namely:
(1) Wet
(2) Dry saturated
(3) Superheated

Wet steam: Steam generated in the presence of water with which it is in the state of equilibrium is known as wet steam. The water present in the wet steam is saturated liquid at saturation temperature.

Dry saturated steam: The steam which does not contain any liquid particle but is at saturated temperature is called dry saturated steam.

Superheated steam: If the temperature of the dry steam is increased more than the saturation temperature steam becomes superheated.

The steam ordinarily produced in any boiling vessel is always wet steam as it contains liquid particles in a finely divided liquid droplet, etc. Entire liquid has not taken latent heat. The state of wet steam will be somewhere in between saturated liquid state (2) and dry vapour state (3).

Let us say state 5. The relative amounts of vapour phase and saturated liquid phase determines the quality of steam. The quality or "dryness fraction" of steam

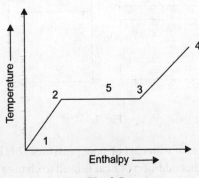

Fig. 3.5

is defined as the ratio of the mass of dry vapour to the total mass of mixture. Thus, dryness fraction

$$x = \frac{\text{Mass of dry vapour in the mixture}}{\text{Mass of the mixture}}$$

For example, if 1 kg of steam contains 0.95 kg of dry vapour, then dryness fraction of steam is 0.95. $(1 - x)$ represents wetness of the steam. The heat required to produce 1 kg of wet steam is given by:

$$h = h_f + x h_{fg}.$$

State 4 represents the condition of superheated steam.

Following advantages occur by using superheated steam in steam power plants:

(i) Increased work capacity without having to increase its pressure.
(ii) Greater heat content of the superheated steam leads to less steam consumption for a given output and that results in saving the cost of the fuel.
(iii) Moisture content of steam leaving the steam turbine/engine can be kept within safe limits. Accordingly, the heat losses due to condensation of steam are avoided to a large extent.
(iv) Superheating is done by utilizing the heat of waste furnace gases which otherwise passes uselessly to the atmosphere.

Thermodynamic Properties of Steam

In the study of various engineering processes and vapour power cycles, thermodynamic properties of steam like specific volume, enthalpy, entropy, internal energy and saturation temperature must be known. The values of these properties are determined either experimentally or otherwise. These properties are available in the form of table and chart. The properties of saturated liquid, saturated vapour and superheated vapour are available. The properties of wet steam are calculated by knowing dryness fraction and pressure of the steam.

The calculation of various properties are given below:

(1) Specific volume

v_f = specific volume of saturated liquid

v_{fg} = specific volume of evaporation

v_g = specific volume of dry steam = $v_f + v_{fg}$

Specific volume of wet steam is given by :

$$v = v_f + x v_{fg} = v_f + x(v_g - v_f) = (1 - x)v_f + x v_g$$

but the specific volume of saturated liquid is very small compared to specific volume of saturated vapour.

Therefore, $v = xv_g$

Specific volume of superheated steam is determined, approximated by the relation

$$v_{sup} = v_g \cdot T_{sup}/T_s$$

because superheated steam behaves as a perfect gas approximately.

(2) Internal energy of steam

Internal energy of steam is the difference of enthalpy and flow work. Therefore, internal energy of saturated liquid is given by:

$$v_f = h_f - P.v_f$$

Internal energy of wet steam

$$v = (h_f + x_{hfg}) - P_{xvg}$$

Internal energy of dry saturated steam

$$= v_g = h_g - P_{vg}$$

Internal energy of superheated steam

$$= v_{sup} = h_{sup} - Pv_{sup} = h_g + C_p(t_{sup} - t_s) - Pv_{sup}$$

(3) Entropy of steam

From definition change of entropy $d_s = dQ/T$

Entropy of saturated liquid

$$d = \int_{273+0}^{273+t_s} C_{pw} \frac{dT}{T} = C_{pw} \ln \frac{273+t_s}{273}$$

$$S_f = C_{pw} \ln \frac{T_s}{273}$$

0°C (273°K) has been taken as datum.

Latent entropy or entropy of evaporation is $S_{fg} = h_{fg}/T_s$ because during evaporation T_s remains constant.

Entropy of wet steam $= S = S_f + xS_{fg}$

Entropy of superheated steam $= S_{sup} = S_g + \int_{T_s}^{T_{sup}} C_{ps} \ln dT$

$$= S_{sup} = S_g + C_{ps} \ln \frac{T_{sup}}{T_s}$$

3.2 STEAM TABLES

The generation of steam at different pressures has been studied experimentally and various properties of steam have been obtained at different conditions. The properties have been listed in tables called steam tables. The steam tables are available for:
1. Saturated water and steam—on pressure basis.
2. Saturated water and steam—on temperature basis.
3. Superheated steam—on pressure and temperature basis for enthalpy, entropy and specific volume.
4. Supercritical steam—on pressure and temperature basis above 221.2 bar and 374.15°C for enthalpy, entropy and specific volume.

Some important points regarding steam tables.
(1) The steam table gives values for 1 kg of water and 1 kg of steam.
(2) The steam table gives values of properties from the triple point of water to the critical point of steam.
(3) For getting values of thermodynamic properties, either saturation pressure or saturation temperature need to be known. Pressure based steam table (i.e. extreme left pressure column is placed) is used when pressure value is known, similarly temperature based steam table is used when temperature value is known.
(4) At low pressure, the volume of saturated liquid is very small as compared to the volume of dry steam and usually the specific volume of liquid is neglected. But at very high pressure the volume of liquid is comparable and should not be neglected.
(5) The specific enthalpy and specific entropy at 0°C are both taken as zero and measurements are made from 0°C onwards.
(6) In computing, the properties of wet steam, it should be noted that only h_{fg} and S_{fg} are affected by dryness fraction but h_f and S_f are not affected by dryness fraction. This means that for steam with dryness fraction x

$$h_{fg} = h_f + xh_{fg}$$
$$S_{fg} = S_f + xS_{fg}$$

(7) The values of properties for superheated steam can be taken from the table for superheated steam or can be calculated by application of ideal gas equation.

Property Table

Property	Wet steam	Dry steam	Superheated steam
Volume	$(1-x)v_f + xv_g$	v_g	$v_g T_{sup}/T_s$
Enthalpy	$h_f + xh_{fg}$	$h_f + h_{fg} = h_g$	$h_g + C_{ps}(t_{sup} - t_s)$
Entropy	$S_f + xS_{fg}$	$S_f + S_{fg} = S_g$	$S_g + C_{ps} \ln(T_{sup}/T_s)$

STEAM PROPERTY CHARTS

In addition to steam table, properties can be represented graphically on a chart. The most convenient charts are the temperature entropy (T–S) and enthalpy-entropy (h–s) charts which were first prepared by Mollier. These charts are discussed below:

(a) Temperature-Entropy Chart (T-S Diagram)

From the diagram, we find the gap between the saturated liquid line and saturated vapour line goes on decreasing with increase of pressure and finally both the lines meet at a point called critical point where latent entropy and latent enthalpy becomes zero. Constant pressure line in the two-phase region is also a constant temperature line. But in superheated region, entropy increases with increase of temperature and therefore, constant pressure line is inclined in superheated region.

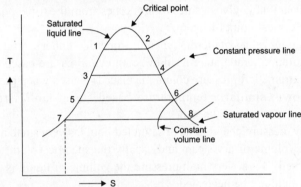

Fig. 3.6

(b) Enthalpy-Entropy Chart (Mollier Chart)

The enthalpy and entropy of the saturated liquid and vapour can be found from steam tables at various pressures. These two lines can be drawn on h–s diagram

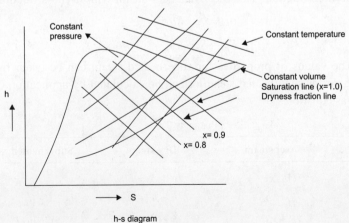

h-s diagram

Fig. 3.7

SOLVED PROBLEMS

Problem 1
Determine the state of the steam, i.e. whether it is wet, dry or superheated in the following cases:
  (i) Pressure 8 bar and specific volume 0.196 m³/kg.
  (ii) Pressure 14 bar and temperature 225°C
  (iii) Pressure 25 bar and 2700 kJ/kg of heat is required to generate steam from water at 0°C.

Solution
(i) The given value of specific volume = 0.196 m³/kg is less than that of taken from steam table at 8 bar, i.e. 0.2402 m³/kg.

$$\text{Dryness fraction } x = \frac{0.196}{0.2402} = \frac{0.196}{0.2402} = 0.815 \text{ Ans. (wet steam)}$$

(ii) The saturated temperature at 14 bar is 195°C which is less than the temperature of given sample. Hence, sample is *superheated* and degree of superheat = (225–195) = 30°C.

(iii) At pressure 25 bar, h_g = 2809.9 kJ/kg and for sample enthalpy is less than it. Therefore, enthalpy of wet steam = $h_f + xh_{fg}$

$$2700 = 961.9 + x1839$$

$$x = 0.945 \text{ Ans.}$$

Problem 2
Evaluate enthalpy, internal energy, volume and entropy of 1 kg of steam having dryness fraction 0.85 and pressure of 20 bar.

Solution
From steam tables at 20 bar, properties are:

t_s = 212.4°C v_f = 0.00177 m³/kg v_g = 0.099549 m³/kg

h_f = 908.6 kJ/kg h_{fg} = 1888.7 kJ/kg h_g = 2797.2 kJ/kg

S_f = 2.447 kJ/kgk S_{fg} = 3.890 kJ/kgk S_g = 6.337 kJ/kg k

Enthalpy = $h_f + xh_{fg}$

= 908.6 + 0.85 × 1888.7 = 2513.995 kJ/kg

Volume $= (1-x)v_f + xv_g \simeq xv_g$

$= (1-0.85) \times 0.00177 + 0.85 \times 0.099549$

$= 0.0848$ m³/kg

Internal energy $= h - Pv$

$= 2513.995 - 20 \times 10^5 \times 10^{-3} \times 0.0848$

$= 2344.395$ kJ/kg

Entropy $= S_f + xS_{fg}$

$= 2.447 + 0.85 \times 3.890$

$= 5.753$ kJ/kgk

Problem 3
In a turbine, steam expands from 20 MPa, 550°C to 0.005 MPa isentropically. Evaluate the work done per kg of steam.

Solution: Equation

20 MPa $= 20 \times 10^6$ Pa

$= 20 \times 10^6$ N/m² $= 200$ bar.

$.005$ MPa $= 0.005 \times 10^6$ Pa

$= .05$ bar

Isentropic expansion process 1–2 is shown in figure:

$S_1 = S_2$

$S_1 = S_g + C_{ps}$ In T_{sup}/T_s

value of C_{ps} is not given. However, average value of $C_{ps} = 2.0934$ kJ/kgk.

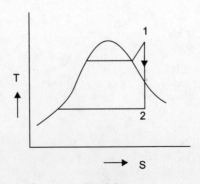

Fig. 3.8

From steam table, saturation temperature T_s = 365.7°C at 200 bar which is less than 550°C. Therefore, steam is superheated. So directly superheated steam table can be used for:

S_1 = 6.325 kJ/kgk and h_1 = 3388.3 kJ/kg at 0.05 bar

$S_2 = S_{f2} - xS_{fg2}$

6.325 = 0.476 + x7.920

x = 0.738

$h_2 = hf_2 + xh_{fg2}$

= 137.8 + 0.738 × 2423.8 = 1926.56

Work done per kg = $(h_1 - h_2)$

= (3388.3 − 1926.56)

= 1461.74 kJ

Problem 4

Steam at 10 bar and 0.9 dryness fraction is available, find the final dryness fraction of steam in each of the following two cases:
  (i) 170 kJ of heat is removed per kg of steam at constant pressure.
  (ii) Steam expands isentropically to a pressure 0.5 bar in a turbine in a flow process. The turbine develops 300 kJ of work per kg of steam.

Solution

At 10 bar and 0.9 dryness,

heat available = $h_f + x_1 h_{fg}$ = 762.6 + 0.9 × 2013.6

= 2574.84 kJ/kg.

If 170 kJ heat is removed, then rest heat = 2574.84 − 170 = 2404.84 kJ/kg

2404.84 = 762.6 + x_2 × 2013.6

x_2 = 0.815 Ans.

(ii) h_1 = 2574.84 kJ/kg

300 = 2574.84 − (340.6 + x_2 × 2305.4) at 0.5 bar

x_2 = 0.83

3.3 METHODS OF DETERMINING THE DRYNESS FRACTION OF STEAM

There are three methods used for determining the dryness fraction of steam:
  (a) Separating Calorimeter

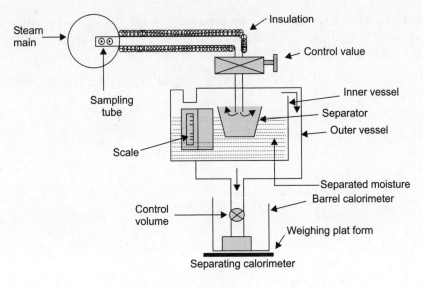

Fig. 3.9

(b) Throttling Calorimeter
(c) Combined Separating and Throttling Calorimeter

Separating Calorimeter

The apparatus used is shown in figure. The supply steam is admitted to the calorimeter from main pipe through a sampling tube. The incoming steam strikes the baffle plate and direction of steam is completely reversed. Therefore, water particles being heavier are separated and collected at the bottom of the vessel. The quantity of separated water is noted from the indicator attached with the inner vessel. The dry steam passes down through the annular space between inner and outer vessel. A barrel calorimeter rests on a weighing platform. The calorimeter receives the dry steam in the form of condensate whose quantity is obtained by weighing machine.

Let m_w = mass of moisture collected in inner vessel in kg.

m_s = mass of steam condensed in barrel calorimeter in kg.

x = dryness fraction of steam.

$$\text{dryness fraction of steam } x = m_s/m_s + m_w$$

Limitations:
(a) The size of the water particles is very small and may be carried away with the steam at high dryness fraction. Therefore, this calorimeter cannot be used for high dryness fraction of steam.
(b) Complete separation of water is not possible.
(c) Method is approximate.

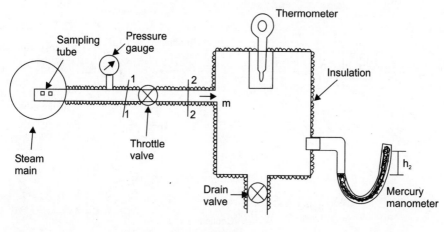

Fig. 3.10

Throttling Calorimeter

The sample of the wet steam enters through the sampling tube as shown in figure:

The sample of the steam is throttled. A pressure gauge and a mercury manometer are provided to measure pressure before and after throttling respectively. The steam after throttling must be superheated. With the help of mercury thermometer, temperature of throttled steam can be measured. During throttling, the enthalpy of steam remains constant. The enthalpy of superheated steam is known if its pressure and temperature are known. If wet steam is throttled to become superheated, then the dryness fraction can be evaluated by equating before and after throttling. The properties of steam are noted from the steam table.

Limitations:
(a) For the measurement of minimum value of dryness fraction, the steam should be superheated by minimum 5°C.
(b) This method cannot be used for low dryness fraction.

Combined Separating and Throttling Calorimeter

The arrangement is shown in figure. Steam from main pipe is collected through sampling tube. First it enters into separating calorimeter where direction is reversed and thereby heavy water particles are collected and separated at the bottom of the separator. The steam with considerable dryness fraction leaves the separator and enters the throttling calorimeter where it is collected and taken to the superheated region. The process is shown on the h–s diagram.

The process 1–2 represents separation of moisture at constant pressure. The dryness fraction x_1 at point 1 is improved to x_2 at point 2. Process 2–3 represents throttling.

Calculation of dryness fraction:

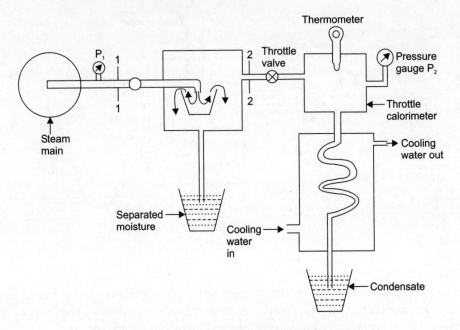

Fig. 3.11: Combined separating and throttling calorimeter

Let m = mass of steam passing through sampling tube in kg.
m_1 = mass of moisture separated in separator in kg.
m_{s1} = mass of steam coming from separating calorimeter in kg.
x = dryness fraction of steam in steam main.
x_1 = dryness fraction of steam leaving the separating calorimeter.
x_2 = dryness fraction of steam leaving the throttling calorimeter.

The dryness fraction of steam after separating calorimeter is given by $x_1 = m_{s1}/m$.

It should be noted that the sample m_{s1} contains moisture m_2 and dry steam m_{s2}. Therefore, the dryness fraction after throttling is given by $x_2 = m_{s2}/m_{s1}$.

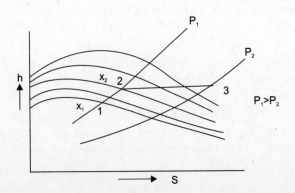

Fig. 3.12: Representation of separating and throttling processes

The total dryness fraction of steam will be the ratio of final dry steam to total initial steam given as:

$x = m_{s2}/m = m_{s2}/m_{s1} \times m_{s1}/m = x_2 \cdot x_1$

$x = x_1 x_2$

Thus, the dryness fraction of steam main is the product of dryness fractions given by separating and throttling calorimeter respectively.

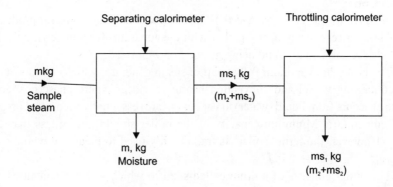

Fig. 3.13: Line diagram for calculation.

Problem 1
A sample of steam from a boiler drum at 3MPa is put through a throttling calorimeter in which the pressure and temperature are found to be 0.1 MPa, 120°C. Find the quality of the sample taken from the boiler.

Solution

$h_f + x h_{fg} = h_g + C_p (t_2 - T_{sup})$ [During throttling enthalpy remains constant]

$$1008 + x \times 1795.73 = 2675 + 2.1 (120 - 100)$$
$$= 2675 + 2.1 (20)$$

$$x = \frac{2675 + 42 - 1008.41}{1795.73}$$

$$= \frac{1708.59}{1795} = 0.951$$

Exercise Problem
Steam flows in a pipeline at 1.5 MPa. After expanding to 0.1 MPa in a throttling calorimeter, the temperature is found to be 120°C. Find the quantity of steam in pipeline. What is the maximum moisture at 1.5 MPa that can be determined with this set-up if at least 5°C of superheat is required after throttling for accurate readings?

3.4 BOILER

Introduction
Steam is the most important vapour used as a working substance in steam engines and turbines working on vapour power cycles. The equipment used for producing steam is called steam generator or boiler. Technically speaking, all steam generators are not boilers but all boilers are steam generators.

Definition
Steam generator (as given by ASME) "A combination of apparatus for producing, furnishing or recovering heat together with the apparatus for transferring the heat so made available to the fluid being heated and vaporized."

The boiler or evaporator is that part of steam generator which consists only of the containing vessel and convection heating surfaces where phase change (or boiling) occurs from liquid (water) to vapour (steam) essentially at constant pressure and temperature. Minimum capacity of the boiler should not be less than 22.4 litres. However, the term 'boiler' is traditionally used to mean the whole steam generator.

The boiler consists of a drum called shell in which water is contained. The thermal energy released due to combustion of fuel (may be solid, liquid or gas) in the combustion chamber is used to evaporate water and to produce steam at a desired pressure and temperature. The steam thus generated is used for:

   (a) **Power generation:** Mechanical work or electrical power may be generated by expanding steam in steam engine or steam turbine.
   (b) **Heating:** The steam is utilized for heating the residential and industrial buildings in winter season and for producing hot water for hot water supply.
   (c) **Industrial Processings:** Steam is used in textile industries, sugar mills and other chemical industries.

Requirements of a Good Boiler
A good boiler should be able to fulfill the following conditions.
   (1) It should be capable of generating the maximum quantity of steam at a required pressure and temperature and quality with minimum fuel consumption.
   (2) It should be light in weight and should not occupy much space.
   (3) It should be safe in working and should conform to the safety regulations laid down in the boiler Act.
   (4) The initial cost, installation cost and maintenance cost of the boiler should be low.
   (5) It should be capable of quick starting and should be able to meet any rapid fluctuation of load.

(6) All parts and components should be easily accessible for inspection and repair.
(7) It should have minimum number of joints and these too should be as far away as possible from the direct flames.
(8) The design and construction of the component parts should be such that they can be easily dismantled and transported.
(9) To make the best use of the heat supplied, the boiler should have proper arrangement of circulation of water and hot gases.
(10) Minimum refractory material should be used.
(11) The heated surface should be entirely free from any deposition of mud or firing particles.

Classification of Boilers

Boilers are classified on the basis of following:
 (i) Tube contents
 (ii) Method of firing
 (iii) Pressure of steam
 (iv) Method of circulation
 (v) Nature of service
 (vi) Position and number of drum
 (vii) Gas passage
 (viii) Nature of draught
 (ix) Heat source
 (x) Once through boiler
 (xi) Fluid used
 (xii) Boiler shell material.

(i) Tube Contents

According to the contents in the tube the boiler is classified as
 (a) Fire tube boiler
 (b) Water tube boiler

(a) Fire Tube Boiler
Example: Cochran, Lancashire, Cornish Locomotive boiler
In the fire tube boiler, the hot gases (flue gases) pass through the tubes and water surrounds them. The products of combustion (hot gases) leaving the furnace pass through fire tubes which are surrounded by water. Heat from hot flue gases is transferred to water which is converted into steam. The spent flue gases are then discharged to atmosphere through chimney.

(b) Water Tube Boiler
Example: Babcock and Wilcox boiler, Stirling boiler

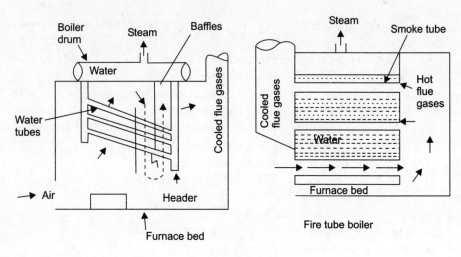

Fig. 3.14

In water tube boiler water flows inside the tube and the hot flue gases flow outside the tubes. A bank of water tubes containing water is connected with a steam water drum by means of two sets of headers. The hot flue gases from the furnaces pass over the tubes and are discharged through the chimney. The water thus absorbs heat from the hot gases and evaporates in the form of steam. The steam thus formed gets accumulated in the steam space.

(ii) Method of Firing
On the basis of method of firing, boilers are classified as:
  (a) **Internally Fired Boiler:** As the name implies, the furnace region is provided inside the boiler shell and is completely surrounded by water cooled surfaces. Examples of the internally fired boiler are Lancashire, Locomotive and Scotch boiler.
  (b) **Externally Fired Boiler:** Furnace is provided outside the boiler. Its furnace region is simple and easy to construct. Example of externally fired boiler is the Babcock and Wilcox boiler.

(iii) Pressure of Steam
Boilers may be classified according to pressure as:
  (a) **Low pressure boiler:** A boiler which generates steam at a pressure below 80 bar is called a low pressure boiler. Examples of low pressure boilers are Cochran, Cornish, Lancashire and Locomotive boilers.
  (b) **High pressure boiler:** A boiler which generates steam at a pressure higher than 80 bar is called a high pressure boiler. Examples of high pressure boilers are Babcock and Wilcox, Lamont, Velox, Benson boilers, etc.

(iv) Method of Circulation of water
  (a) **Natural Circulation:** In these type of boilers circulation of water in the boiler takes place by natural convection current produced by the application of heat. Examples of these are Lancashire, Locomotive, Babcock and Wilcox boilers, etc.
  (b) **Forced Circulation:** In these mechanical means (pumps) are used to increase the circulation. Examples are Lamont, Velox boilers, etc.

(v) Nature of Service
  (a) **Stationary boilers:** Boilers that are used for stationary plants are called stationary or land boilers.
  (b) **Portable boilers:** Boilers that can be readily dismantled and transported from one place to another are called portable boilers.
  (c) **Mobile boilers:** The boilers which are fitted on vehicles that can move from place to place are called mobile boilers. Examples are marine and locomotive boilers.

(vi) Position and Number of Drums
According to position
  (a) Horizontal boiler
  (b) Inclined
  (c) Vertical

According to number of drums
  - Single drum
  - Multi drum

(viii) Nature of Draught
  (a) Natural draught boilers
  (b) Forced draught boilers

(ix) Heat Source
The heat source may be any of the following:
  (a) Combustion of solid liquid or gaseous fuel.
  (b) Electrical or nuclear energy.
  (c) Hot waste gases which are by-products of other chemical processes.

(x) Once through Boiler
In this type of boiler, there is no recirculation of water. The feed water leaves the tube as steam. An example is the Benson boiler. In recirculation boilers only a part of water is evaporated and the remainder is circulated.

(xi) Fluid Used
On the basis of fluid used, the boilers are classified as:
  - Steam boilers: In this water is the working fluid.
  - Mercury boilers: In this mercury is the working fluid.

Difference between Water Tube and Fire Tube Boilers

Fire Tube Boiler	Water Tube Boiler
(i) Hot flue gases flow inside the tubes while water surrounds the tube.	(i) Water flows inside the tubes while the hot flue gases surround the tube.
(ii) It has a large ratio of water content to steam capacity and therefore slow in evaporation.	(ii) It has comparatively small ratio of water content to steam and therefore quick evaporation.
(iii) Failure in water supply is less effective and may not overheat the boiler due to large capacity of water.	(iii) Failure in water supply system brings about a breakdown as the boiler may be overheated due to low capacity of water.
(vi) The stress consideration limits the range of pressure from 17.5 bar to 24.5 bar. Thus, it is suitable for low pressure.	(iv) It can withstand high internal pressure for the same wall thickness and stress. Thus, it is suitable for pressures as high as 200 bar.
(v) It has a small capacity due to smaller heating surface area resulting in slow rate of steam generation. It is not suitable for steam power plants.	(v) It has a wide range of capacity due to larger heating surface area resulting in high rate of steam generation. It is suitable for large power plants.
(vi) Circulation is poor and thus there is every chance of a deposit of impurities on the heated surface. The construction is such that removal of impurities is difficult.	(vi) Circulation is greater thus there are less chances of a deposit of impurities on the heated surface. The construction is such that impurities if deposited from the water are found outside the zone of rapid circulation and can be easily removed.
(vii) An exposition if it occurs becomes a very serious problem in a fire tube boiler because of its large water capacity.	(vii) Because of the small drum size the water is uniformly spread in a large number of tubes, thus the failure of one water tube does not cause any disastrous explosion.
(viii) All parts are not easily accessible for cleaning, inspection or repair.	(viii) All parts are easily accessible for cleaning inspection or repair.
(ix) It is less efficient.	(ix) It is highly efficient.
(x) Though it is suitable for rapid changes in load but large load fluctuations extending over long duration may damage the boiler.	(x) It is suitable for large load fluctuations extending over long durations without danger to the boiler.
(xi) The construction is simple, rigid and compact. Hence the initial cost is less.	(xi) Design is complex hence initial cost is high and it requires periodic examinations.

(xii) Boiler Shell Material

On the basis of materials used for shell the boilers are classified as follows:
    (a) Cast iron boilers
    (b) Steel boilers
    (c) Copper and stainless steel.

Fire Tube Boilers

Cochran Boiler: The Cochran boiler is one of the most popular and best type of vertical, multitubular, fire tube (internally fired) natural circulation boiler. The following particulars relate to a 2.75 diameter shell.

Height of the shell = 5.78 m.

Maximum evaporative capacity = 568 kg/hr of steam from cold feed when burning 36 to 40 kg/hr of coal.

Economical rating = 3/4 of the maximum
Heating surface = 120 m^2
Steam pressure = 6.7 bar.

Construction Details: It consists of a vertical cylindrical shell having a hemispherical top. The shell is made of mild steel plates. The furnace is also hemispherical in shape and it is hydraulically pressed from one plate without a single weld or scum. Thus, the furnace is the strongest structure under compression. The fire grate is arranged in the furnace and the ash-pit is provided below the grate. In the firebox a fire door and a damper are provided. Adjacent to the fire box the boiler has a combustion chamber which is dry-baked and lined with fire bricks. Close to the combustion chamber a number of horizontal smoke tubes are provided. These tubes are of equal length and arranged in a group with wide space in between them and the shell so as to help convection currents. The ends of these smoke tubes are fitted in the smoke box tube plate and combustion chamber tube plates. These tubes are simply pushed in the holes of tube plates and then expanded at the ends to make steam light joints. The smoke box is built of steel plates and is fitted with hinged door which gives an easy access to smoke tubes for cleaning and inspecting. The stack is provided at the top of the smoke box for discharge of the gases to the atmosphere.

The furnace is surrounded by water on all sides except at the opening for the fire door and the combustion chamber. The smoke tubes are also completely surrounded by water.

Working: The hot gases produced from the burning of the fuel on the grate rise up through the flue pipe and reaches the combustion chamber. The flue gases from the combustion chamber passes through the fire tubes and the smoke box and finally are discharged through the chimney. The flue gases during their travel from fire box to the chimney give heat to the surrounding water to generate steam.

The circulation of water in the shell is shown by arrows. The water courses down by the cooler wall of the shell and rise up past the fire tubes by natural circulation due to convection current.

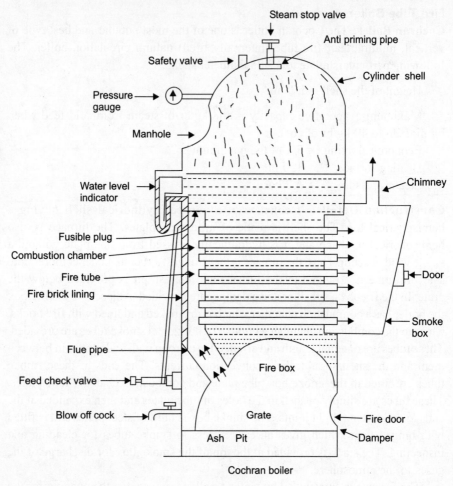

Fig. 3.15

LANCASHIRE BOILER

A Lancashire boiler is fire tube boiler. Its normal working pressure range is 15 bar and steaming capacity is about 8000 kg/hr. Its size varies from about 8 metres to 9 metres in length and from 2 to 3.5 metres in diameter.

Constructional details: It consists of the following parts.
 (1) **Feed check valve:** Feed water is supplied to the boiler under pressure, feed check valve stops its escaping back.
 (2) **Pressure gauge:** It is used for measuring the pressure of steam.
 (3) **Water level gauge:** It indicates the level of water in the boiler.
 (4) **Dead weight safety valve:** It is for safety against pressure in excess of the rated pressure.

Properties of Steam and Steam Generators (Boilers) 71

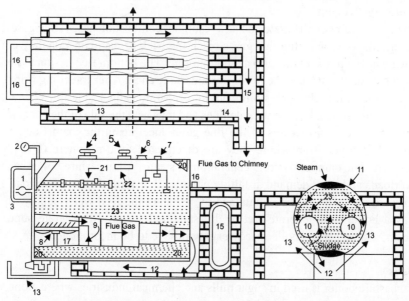

1. Feed check valve, 2. Pressure gauge, 3. Water level gauge, 4. Dead weight safety valve, 5. Steam stop valve, 6. Main hole, 7. Low water high steam safety valve, 8. Fire grate, 9. Fire bridge, 10. Flue tubes, 11. Boiler shell, 12. Bottom flue, 13. Side flue, 14. Dampers, 15. Main flue, 16. Doors, 17. Ashpit, 18. Blow off cock, 19. Blow off pit, 20. Gusset stays, 21. Perforated feed pipe, 22. Anti primming pipe, 23 Fusible plug.

Fig. 3.16

(5) **Steam stop valve:** It opens and closes the supply of steam for using.
(6) **Manhole:** It is for cleaning and inspection of the drum.
(7) **Fire grate:** The grate in which the solid fuel is burnt.
(8) **Fire bridge:** This is used for deflecting the gases of combustion upwards.
(9) **Flue tubes:** These are for the first pass of the flow of the flue gases. The flue tubes are tapered being larger in diameter at the front and smaller at the back.
(10) **Boiler shell:** It is used for containing water and steam.
(11) **Damper:** It is used for controlling the flow of flue gases. These are iron doors which slide up and down in the grooves by rope and pulley.
(12) **Blow off cock:** Blow off cock is provided at the bottom of the shell by a bent pipe to blow off sludge at intervals. Since the water shell is not perfectly horizontal, a few degrees tilting towards the front enables all the sediments to accumulate in front near the blow off cock.
(13) **Anti priming devices:** It is for separating out suspended moisture and allowing as far as possible the dry steam through the stop valve.
(14) **Fusible plug:** It is for safety against exposing the fuel tube to excessive heat when the water level falls too low.

Working Process

The fuel is burnt on the grate and as a result hot combustion gases (flue gases) are produced. As soon as the flue gases reach the back of the main flue they deflect downwards and travel through the bottom flue. The bottom flue is situated below the water shell and heats the lower portion of the shell. After travelling from back to the front, these flue gases bifurcate into separate paths inside flues. Now they travel from front to back in the side flues and thus heat the side of the water shells.

After these two steams of the flue gases meet again the main flue passing through damper from where they are discharged to atmosphere through the chimney. The damper controls generation of steam. A Lancashire boiler can take overloads and meet demand at falling pressure. For short duration overloads, the pressure of steam can be maintained by reducing the feed water supply. In this way a boiler will generate wet or dry saturated steam. To generate superheat steam, a superheater (boiler accessory) will have to be incorporated.

Uses

Lancashire boiler is used in sugar mills and chemical industries where along with power, steam is also required for process work.

Locomotive Boiler

The locomotive boiler is a horizontal, multitubular, natural circulation, artificial draft, internally fired, fire tube type of portable boiler. The unit is so designed that it is capable of meeting the sudden and fluctuating demands of steam which may be imposed because of variation of power and speed. In addition to railways road rollers and haulage engines, the locomotive boilers have been used in agricultural fields, saw mill plants and stationary power service where semi-portability is desired.

The principal parts of the unit are:

(1) **Fire box:** This forms a combined grate and combustion chamber. The fire box is water cooked on all the three sides except the bottom.
(2) **Mutitubular barrel:** It contains an envelope of water in which fire tubes are immersed.
(3) **Smoke box:** It is equipped with a very short chimney.

Working

The coal is burnt in the fire box and produces the hot flue gases. These flue gases rising from the grate are deflected upwards by a fire bridge and so that it comes into contact with the walls and roof of the firebox. Due to the motion of the locomotive, a strong draught is created and the atmospheric air rushes into the fire-box through the dampers. The function of the dampers is to control the quantity of air entering in the fire box. The ash of the coal burnt on the grate falls into the ash-pit. The hot flue gases pass from the furnace box to the smoke box through horizontal smoke tubes. A large door in the front of the smoke box gives access to

Properties of Steam and Steam Generators (Boilers) 73

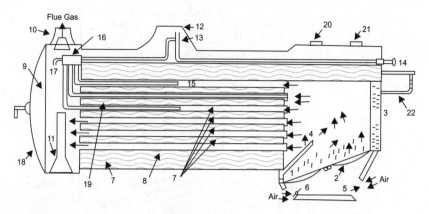

1. Firebox	2. Grate	3. Fire hole
4. Fire bridge arch	5. Ash pit	6. Damper
7. Fine Tubes	8. Barrel or shell	9. Smoke box
10. Chimney (short)	11. Exhaust steam pipe	12. Steam dome
13. Regulator	14. Lever	15. Superheater tubes
16. Superheater header	17. Superheater exit pipe	18. Smoke box door
19. Feed check valve	20. Safety valve	21. Whistle
22. Water gauge		

Fig. 3.17: Locomotive boiler.

it and the tubes for examination and cleaning purposes. The hot gases from the smoke box are discharged to the atmosphere through a short chimney.

During the travel of hot gases from the grate to the chimney, they give heat to the water and generate steam. The generated steam is collected in the steam dome. The function of steam dome is to increase the steam release capacity and to increase the distance of steam from water line which reduces priming. The driver operates the regulator by turning a lever which leads the dry saturated steam to the engine for expanding and doing work. To get superheated steam, the steam is diverted to superheated heater with the help of a regulator and lever arrangement and then to superheater tubes. They start from the superheater header and are laid inside the large diameter fire tubes which are placed at the highest of the boiler shell. The hot flue gases passing through the fire tube supply heat to the superheater tube that heat the steam inside superheater tubes which get superheated steam.

Babcock and Wilcox Boiler

Babcock and Wilcox boiler is a water tube high pressure boiler. This is used for large generation rate at high pressure. The boiler has three main components– steam and water drum, water tubes and furnace.

It is a longitudinal drum, externally fired water tube natural circulation type of stationary boiler. Since the boiler is externally fired, it is suitable to all types of fuels and for hand and stroker firing. Evaporative capacity in such boiler ranges from 20,000 to 40,000 kg/hr of steam and operating pressures from 11.5 to 17.5

74 *Mechanical Engineering: Fundamentals*

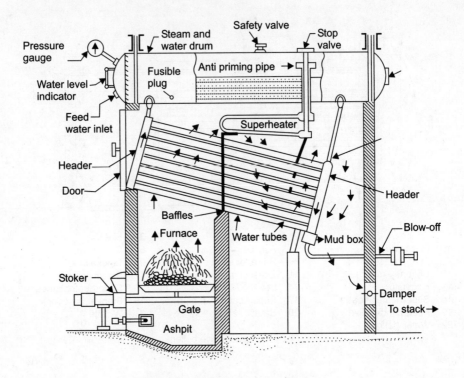

Fig. 3.18: Babcock and Wilcox boiler

bar are quite common. But the operating pressures may be as high as 42 bar. It is suitable for small size thermal power plants and other industrial works.

Construction: It consists of a high pressure-drum mounted at the top. The drum of the boiler is made of welded steel or single course joined by longitudinal butt strap. The heads of the drum are forced by hydraulic press and are dished to a radius equal to the diameter of the drum. From each end of the drum, connections are made with the upper header and down take header. A large number of water tubes connect the uptake and down take headers. The water tubes are inclined 5 to 15 degrees to promote water circulation. The water tubes are straight, solid drawn steel tubes about 10 cm in diameter and are expanded into the bored holes of the headers to ensure proper fixing. The headers have a serpentine (sinusoidal) form. This serpentine form of headers arranges the water tubes such that they are staggered and this exposes the complete heating surface to flue gases. The heating surface of the unit is the outer surface of the tubes and half of the cylindrical surface of the water drum which is exposed to flue gases. A mud box is attached to the bottom of the downtake header.

The whole of the assembly of water tubes is hung along with the drum from steel girder frame by steel rods called slings in a room made of masonry work lined with fire bricks. Below the uptake header the furnace of the boiler is arranged. The coal is fed to chain grate stoker. There is a bridge wall deflector which deflects the products of combustion upwards. Two baffles are also arranged which provide three passes of the flue gases. A damper is placed at the inlet of the chimney to regulate the draught.

Working Process

The hot combustion gases produced by burning of fuel on the grate rise upwards and are deflected by the bridge wall deflection to pass over the front portion of the water tubes and drum. In this way they complete the first pass and provision of baffles deflect the gases downwards, so that they complete the second pass. Again due to the provision of baffles they rise upwards and complete the third pass and finally come out through the chimney. During their travel they give heat to water and steam is formed. The circulation of water in boiler is due to natural circulation set up by convection currents.

Feed water is supplied by a feed water inlet pipe. The hottest water and steam rise from the tubes to the uptake header and then through the riser it enters the boiler drum. The steam vapours escape through the water to upper half of the drum. The cold water flows from the drum to the down take header and thus the cycle is completed. A set of superheater coils are provided to superheat the steam, which enters these tubes from the steam space in the boiler shell, through the saturated steam box. The superheated steam is taken to the steam stop valve through the steam pipe, which is connected to superheated steam box.

BOILER MOUNTINGS AND ACCESSORIES

For safe and efficient operations, boilers are equipped with two categories of components such as:
    (a) Boiler mountings.
    (b) Boiler accessories

(a) Boiler mountings: Boiler mountings are those machine components which are mounted over the body of the boiler itself for the safety of the boiler and for complete control of the process of steam generation. These mountings form an integral part of the boiler. According to the Indian boiler regulations, the following mountings should be fitted on the boiler.
    (i) Two safety valves
    (ii) Two water level indicators
    (iii) Pressure gauge
    (iv) Fusible plug
    (v) Steam stop valve
    (vi) Feed check valve
    (vii) Blow off cock

(viii) Inspector's test gauge
(ix) Man and mudholes.

(b) Boiler accessories are those machine components which are installed either inside or outside the boiler to increase the efficiency of the plant and or to help in the proper working of the plant.

The following accessories are generally used in the boiler.
(i) Air pre-heater
(ii) Economiser
(iii) Superheater
(iv) Feed pump
(v) Steam trap
(vi) Steam separator
(vii) Pressure reducing valve

Boiler Mountings
Safety Valve: Safety valves are needed to blow off the steam when the pressure of the steam in the boiler exceeds the working pressure. These are placed on the top of the boiler. There are four types of safety valves.
(1) Dead weight safety valve
(2) Spring loaded safety valve
(3) Lever safety valve
(4) High steam and low water safety valve.

Dead weight safety valve: In this valve, the steam pressure in the upward direction is balanced by the downward force of the dead weight acting on the valve. In this the weight in form of cylindrical cast iron discs is placed on the valve. The valve is made up of gun metal and it rests on a gun metal seat secured on the top of a vertical cast iron pipe bolted to the mounting block, which is riveted to the top of the boiler shell. At normal conditions the upward force exerted by steam in the boiler is balanced by the downward force equivalent to the load on the valve. When this load is greater than the force due to steam pressure acting on the valve, the steam will not escape, but in case it is less than the force due to steam pressure the valve is lifted up from its seat and the steam will escape to the enclosed discharge pipe which is connected to the discharge casing, from where it is directed outside the boiler house.

Uses
(1) Stationary boilers but not for marine portable boiler because of its great weight and reduction in effectivity when the ship is in midst of heavy waves.
(2) Used for low pressure boiler.
(3) It cannot be easily tampered.

Spring Loaded Safety Valve

Here the dead weight is replaced by a spring. It consists of cast iron body which has two valve chests for the flow of steam. The flange of the body is bolted to a mounting block on the boiler shell. There are two valves made of gun metal of the same size and they have their seats in the upper ends of the hollow valve chests. The valves are held down on their seat against the steam pressure by applying load on the valve through spring and lever arrangement. The lever has two pivots one is pinned to the lever while the other is forged on the lever. The pivots rest on the centres of the valves. The lever is pulled downwards by the spring and one half of this force is shared by each of the valves because the upper end of the spring is fixed to the lever midway between the valves. This valve has certain advantages which are listed below:

(1) The heavy weights are eliminated.
(2) Easy maintenance and examination.
(3) Suitable for marine and locomotive engines.

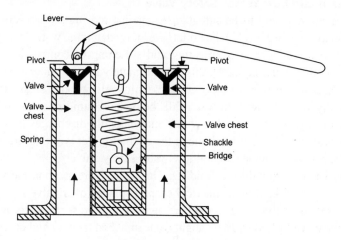

Fig. 3.19: Spring loaded safety valve

Lever Safety Valve

It consists of a cast iron body, the flange of which is bolted to the boiler mountings so as to keep it in communication with steam in the boiler.

It has a gun metal valve and valve sheet. The valve is held by a mild steel or wrought iron lever fulcrumed at one end and loaded at the other end by an external weight. The thrust is applied to the valve through a strut placed very near to the fulcrum and hinged to the lever.

The position of the weight on the lever can be fixed according to choice which will be determined according to steam pressure below the valve. The weight

78 *Mechanical Engineering: Fundamentals*

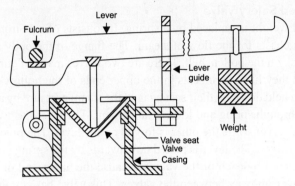

Fig. 3.20: Lever safety valve

is firmly screwed to the lever by a pin and locked so that not unauthorised person can displace it. The weight is applied at the longer arm of the lever so a small weight can give a larger thrust at the valve.

High Steam and Low Water Safety Valve

The boiler has not only to be safeguarded against too high a pressure but also against too low a water level in the boiler. We know that if the water level becomes too low, there is danger of overheating and softening of combustion chamber, furnace and other boiler parts. All this may ultimately lead to boiler explosion.

In this case we use a safety device which allows the steam to escape automatically, once the level of water falls below a certain limit. Along with that we use high-pressure safety valve which automatically discharges the steam out when the pressure of steam rises above the normal boiler working pressure.

Figure shows a high steam and low water safety valve.

It consists of two valves V_1 and V_2. The valve V_2 rests upon the valve seat and valve V_1 which is of hemispherical shape is placed over valve V_2 which acts as a valve seat for valve V_1. To safeguard against high pressure it acts as a simple lever safety valve loaded by two weights, one attached to lever and other to valve V_1. The valve V_2 is attached to external lever. The lever is hinged at its one end and a weight is at the other end. A short pivot attached on the lever is placed on the valve which keeps it in position under normal working pressure. If the pressure exceeds the normal limit, the valve V_2 opens and the steam escapes through the passage between the valve seat and the valve V_2.

The low water safety arrangement is inside the boiler. A lever is hinged from the boiler shell. At one end of the lever is a weight W_1 and at the other end of the lever a large float is provided. The hemispherical valve V_1 is connected with a spindle, the lower end of which also carries weight W_2. The knife edge provided on the lever touches a collar on the spindle under normal pressure. Now when the water level falls below a certain level the float is uncovered causing an increase in its weight according to Archimedes' principle which causes a swing in the lever.

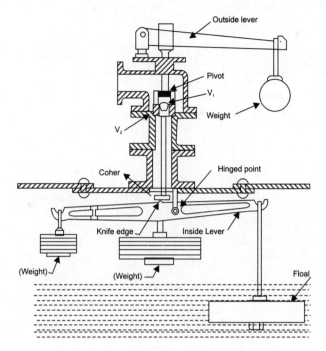

Fig. 3.21: High steam low water safety valve.

This swing pushes the spindle upwards through knife edge collar arrangement and the valve V_1 opens, steam escapes between the passage between the valves V_1 and V_2. This produces a large hissing sound which will draw attention of the boiler attendant about the low water level in the boiler.

Water Level Indicator

The function of the water level indicator is to show the level of water in the boiler. The upper end of the valve opens in the steam space while the lower end opens in the water. The valve consists of a strong glass tube. The ends of the tube pass through stuffing boxes formed in the hollow castings. These castings are flanged and bolted to the boiler. It has three cocks, two of them control the passages between the boiler and the glass tube while the third one (the drain cock) remain closed. Two balls are provided, one is to stop the flow of water and another to stop the flow of steam. Both the balls move to the end passage as shown by dotted (ball) to stop the flow of water and steam when the water gauge glass tube is broken.

Pressure Gauge

It shows the pressure of the steam formed at any instant. The pressure generated should be nearly constant and should change with the fluctuations of load. Figure shows the details of the Bourdon's pressure gauge. The gauge is actuated by the

80 *Mechanical Engineering: Fundamentals*

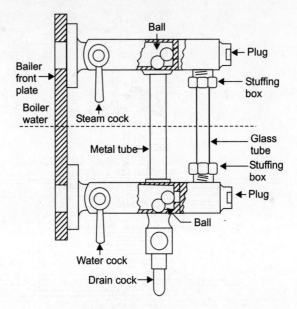

Fig. 3.22: Water level indicator

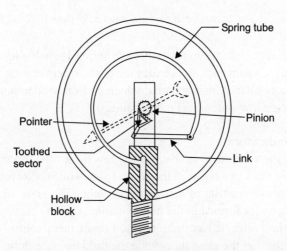

Fig. 3.23: Pressure gauge

Bourdon's tube. One end of this solid drawn elliptic section is connected to the steam space and the other end is plugged and is connected to the clock-work mechanism of the gauge. When pressure is applied to the interior of the Bourdon tube, the Bourden tube tends to straighten and move outwards. In doing so it causes pressure on the link which operates the toothed sector. Design of link mechanism is done in such a way that a slight movement of link is magnified and the pointer gives a deflection which is read on the calibrated scale.

Steam Stop Valve or Junction Valve

The function of the steam stop valve is to stop or allow the flow of steam from the boiler to the steam pipe or from the steam pipe to the turbine (or engine). Conventionally the smaller sizes are called stop valves and large sizes are called junction valves. When this valve is mounted on the topmost portion of the boiler drum, it is customary to call it a junction valve. When it is mounted in the steam pipe leading to the steam turbine or engine it is customary to call it a steam stop valve.

It consists of a valve chest made of cast iron which has two flanges at right angles. One flange is bolted to the mounted block at the highest point of the steam space and second flange is bolted to the steam pipe. A valve seat made of gun metal is screwed on the valve chest by means of logs. A gun metal valve seat is made in the chest. A valve made up of gun metal rests on the valve seat. The valve is connected to the spindle by a nut lower end of which comes in contact with a collar on the lower end of the spindle. By this construction the spindle can rotate freely in the valve but at the same time it carries the valve with it when raised or lowered. The spindle has a hand wheel at the top end by which it is rotated. The spindle passes out of a gland and stuffing box formed in the cover of the body.

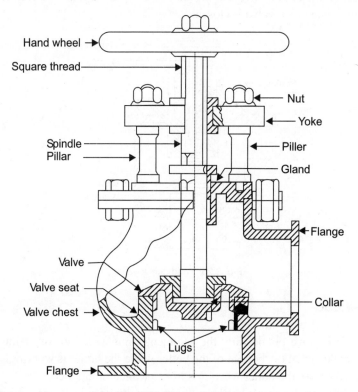

Fig. 3.24: Steam stop valve

The upper portion of the spindle has screw threads which passes through a nut in a yoke. The yoke is carried by two pillars fixed to the body.

Feed Check Valve

This valve is fitted to the boiler slightly below the working level of the water. The function of the feed check valve is to allow the feed water under pressure to pass into the boiler and to prevent simultaneously any water escaping back from the boiler in the event of failure of the feed pump.

The feed check valve consists of a check valve whose lift is controlled by the lower end of a spindle. The valve rests on its seat and is operated by the difference of pressure of water acting on its top and bottom side. The valve can be kept in a closed position by pressing it down by the spindle which can be lowered or raised by the hand wheel. Thus, the valve also carries out the function of a feed valve. This type of valve has one disadvantage that the check valve cannot be inspected or cleaned while the boiler is working.

An another design in which a separate check valve is placed over and above the check valve and the feed valve is connected to the spindle. The feed valve is placed near the boiler. Therefore, when it is closed, the boiler water pressure cannot act on the check valve and the check valve can be removed for inspection or cleaning when feed valve is closed.

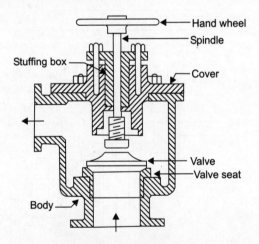

Fig. 3.25: Feed check valve

Blow off Cock

The function of blow off cock is:
  (i) To remove periodically the sediments (mud, scale or other impurities) collected at the bottom of the boiler while the boiler is working.
  (ii) To empty the boiler when it is to be cleaned or emptied.
  (iii) To lower the water level rapidly in case the level becomes too high.

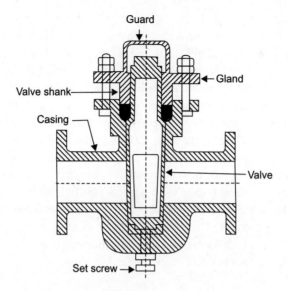

Fig. 3.26: Blow off cock

The blow off cock consists of a conical, hollow gun metal plug type valve which fits accurately into a corresponding hole in the casing. The plug valve has a hole which when brought in line with the hole in the casing by turning the plug value water will flow out of the boiler. The flow of water can be stopped by turning the plug such that its solid part comes in line with the hole in the casing .

The plug valve shank projects out of the stuffing box gland. The stuffing box is provided to prevent the leakage of water at the shank. The gland can be tightened by the nuts provided. A set screw is provided under the plug to force it off its seat, if jammed. The casing has two flanges. The blow off cock is connected to boiler by one flange and the other flange is connected to the pipe carrying the blow off water out of the boiler house.

Manhole
The function of manhole is to provide an opening through which a man can enter a boiler for cleaning and inspection purposes. Generally, it is oval in shape and a cover is provided at the top. The size of the manhole is usually 40 cm × 30 cm.

Fusible Plug
The function of the fusible plug is to extinguish the fire in the furnace of the boiler. When the water level in the boiler falls below the safe limit thereby preventing the boiler from explosion as a result of overheating of the firebox crown plate or fire tubes.

It consists of a hollow gun metal body screwed into the firebox crown plate. This body has hexagonal plane for fixing it in position with the help of a spanner.

A gun metal plug A is screwed into the upper portion of the gun metal body with the help of a hexagonal flange provided in plug A. There is another solid plug B made of copper with a conical top and rounded bottom which is kept in the hole of plug A and held firmly by putting fusible metal between them. The fusible metal may be of tin or lead which has low fusing point. Under normal working conditions the fusible plug is submerged in water, fusible metal is protected from direct contact with water or steam or furnace gases being placed between the gun metal plug A and the copper plug B. Thus, the temperature of the fusible metal is below its melting point.

When the water level falls below a certain limit in the boiler the fusible plug is uncovered out of water and is exposed to steam. This overheats the plug, with the result that the fusible metal melts and the copper plug B released and rests with the gun metal body by the ribs. This opens a way between steam space and furnace thus water and steam rush to the firebox and extinguish the fire. Before starting the boiler again, the operator fixes a new fuse.

Boiler Accessories

The boiler accessories are provided to increase the efficiency of a boiler. The following accessories are commonly attached to boilers:

    (a) Air preheater
    (b) Economiser
    (c) Superheater
    (d) Feed pump
    (e) Injector
    (f) Steam trap
    (g) Steam separator.
    (h) Pressure reducing valve.

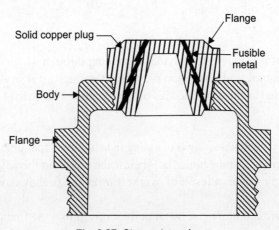

Fig. 3.27: Steam stop valve

Air Preheater

Air preheater is the waste heat recovery device which is placed in the path of the waste flue gases going to the chimney thereby abstracting heat from the flue gases and transferring it to the air before its use for economical combustion: hence the name is preheater. It is placed near the chimney and above the economiser.

The preheating of air facilitates the burning of poor grades of fuel, thus permitting a reduction in excess air and so improving the efficiency. The increase in overall efficiency of the boiler varies from 2 to 10 per cent.

Usually there are three types of preheater.
 (i) Tubular
 (ii) Plate type
 (iii) Regenerative type

Economizer

The function of the economizer is to recover some of the heat from the heat carried away in the flue gases up the chimney or stack and to utilize it for heating the feed water supplied to the boiler. It is placed in the path of the flue gases in between the exit from the boiler and entry into the chimney.

Advantages

(1) The economiser reduces the temperature differences in the boiler and prevents the formation of stagnation pockets of cold water. Thus, it helps to reduce the thermal stress created in the boiler tubes. It also helps to improve internal circulation of water substance. For every 6°C rise in feed water temperature the gain is near about 1%.
(2) It reduces the fuel consumption in boiler furnace.
(3) The economiser helps to reduce the scale formation in boiler tubes, as most of the impurities which create temporary hardness are deposited on the inside of economiser tubes.

Working

The feed water from the feed pump on its way to the boiler enters the bottom heater. From the bottom header water passes inside the vertical tubes and reaches the upper header. From the upper header the water enters the boiler through the feed check valve. There are two stop valves for water. One is connected to the bottom header to stop or allow water to enter the economiser and the other is connected to the top header to allow or stop the water from going to the boiler from economiser. The water while passing inside the vertical tubes gains heat from the hot flue gases which passes over the vertical tube.

Soot that is deposited over the vertical tube is scraped by scrapers which are kept slowly moving up and down to clean the surfaces.

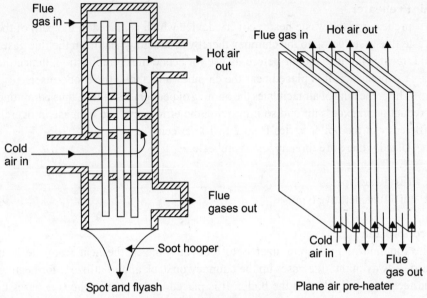

Fig. 3.28: Tubular air pre-heater

Superheaters

In superheater the temperature of steam is increased above saturation temperature keeping its pressure constant. This superheated steam improves the efficiency of boiler by reducing the steam consumption by producing more work per kg of steam, by reducing the condensation losses, and by eliminating the erosion of steam turbine blades. The wet steam flows inside the tubes and the hot flue gases flows over the tubes. By this way the wet steam takes heat from the flue gases and first get dried at the same temperature and pressure then its temperature is raised above the saturation temperature at the same pressure. According to the mode of heat reception, the superheaters are classified as:

  (i) Convective superheaters
 (ii) Radiant superheaters
(iii) Combination superheaters.

Feed Pumps

The function of the feed pump is to pump the feed water to the boiler. The three types of feed pump commonly used are reciprocating, rotary and injector pumps.

The rotary pumps are generally high speed centrifugal type and are driven either by small steam turbine or electric motor. A rotary pump is used when a large quantity of feed water is to be supplied to the boiler. In the reciprocating pump, the pumps are continuously run by the steam from the same boiler to which the water is to be fed.

Properties of Steam and Steam Generators (Boilers) 87

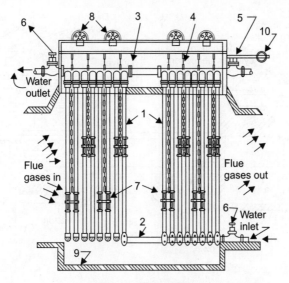

Green's Economizer

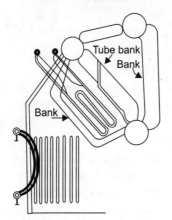

Location of convection and radiant wave

Fig. 3.29

Duplex Feed Pump

It is a double acting reciprocating type feed pump. It consists of two pumps mounted side by side. One is a water pump and other is a steam pump. Both the cylinders are connected to its own piston rod which is finally connected to a common cross head so that the steam pump also serves as driver of the water pump due to the expansion of steam in the steam cylinder.

Steam Injector

An injector is a device which is used to deliver feed water into the boiler under pressure. It consists of a group of nozzles so arranged that the steam expanding in these nozzles imparts its kinetic energy to a mass of water.

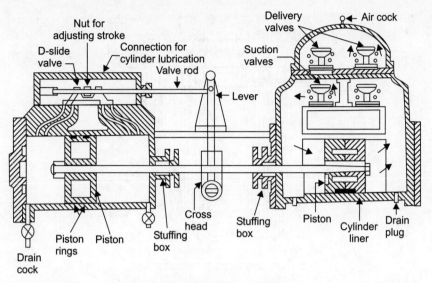

Fig. 3.30: Duplex feed pump

Construction and Working

The steam coming through the steam pipe operated by a valve enters the steam chamber, the outlet of which is in the form of a converging nozzle called a mixing tube. Inside the steam chamber there is a converging nozzle whose annular opening may be adjusted by a valve attached to the spindle operated by a hand wheel. In the nozzle section the velocity of the steam increases forming a jet of steam while the pressure decreases. Due to decrease in pressure of steam at the exit of the converging nozzle than that of feed water tank the feed water enters through the pipe. The resulting jet of water mixes with water in the mixing tube. The steam condenses and a strong vacuum is created in the mixing tube and the mixing chamber. Because of this, more water rushes from the feed tank to the mixing chamber. The velocity at the end of the mixing tube is the greatest and in the mixing chamber itself there is only high velocity jet of water. This jet of water enters the diverging cone where the kinetic energy of water is converted into pressure energy. The increase in pressure of water is sufficient to force the water into the boiler through feed check valve.

Steam Trap

The function of steam trap is to drain off water resulting from partial condensation of steam from steam pipes and jackets without allowing the steam to escape through it.

There are two types of steam traps:
(1) Bucket or float type
(2) Thermal expansion type.

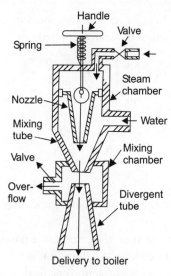

Fig. 3.31: Steam injector

Steam Separator

The function of a steam separator is to separate suspended water particles carried by steam on its way from the boiler to the engine or turbine. It is installed in the main steam pipe very near to the engine or turbine.

Pressure Reducing Valve

The function of a pressure reducing valve is to maintain constant pressure on its delivery side of the valve irrespective of fluctuating demand of steam from the boiler.

Antipriming Pipe

An antipriming pipe is a cast iron box which is fitted in the steam space of the boiler shell and under the mounting block on which the steam stop valve is to be bolted. When the steam with water particles passes through the perforations made in the upper half of the antipriming pipe, the heavier water particles separate out and are collected at the bottom of the pipe. The water thus collected is later on drained to the boiler through the holes which are made at the ends of the pipe.

SOLVED PROBLEMS

Problem 1

During a test carried to measure the dryness fraction of steam by separating calorimeter, the following observations were noted.

Solution

The mass of water separated = 0.85 kg/min.

The mass of steam passed out as recorded by measuring gauge = 10 kg/min.
Determine the dryness fraction of steam supplied.
since we have.

Dryness fraction of steam $x = \dfrac{m_s}{m_s + m_w}$

$$= \dfrac{10}{10 + 0.85}$$

Problem 2

In a throttling calorimeter, the pressure before and after the throttling are 14 bar and 1 bar respestively.

(a) Determine the dryness fraction of steam before passing through the throttling calorimeter, if the temperature after the throttling is 120°C.

(b) What is the value of the dryness fraction can be determined for same pressures before and after throttling.

Take Cp for steam to be 2.1 kJ/kgk.

Solution
Given that

(a) pressure before the throttling is 14 bar (state 1), pressure and temperature after throttling are 1 bar 120°C (state 2).

State (1) from steam table.

$$h_{f_1} = 830.0 \text{ kJ/kg}$$

$$h_{fg_1} = 1957.7 \text{ kJ/kg}$$

state (2) from steam table.

$$hg_2 = 2675.5 \text{ kJ/kg}$$

$$h_{f_1} + x h_{fg_1} = h_{g_2} + C_{ps}(T_2 - T_{sat})$$

$$830.0 + x \times 1957.7 = 2675.5 + 2.1\,(140 - 99.6)$$

$$x = 0.986$$

(b) Minimum dryness fraction that can be determined exist when no super heating takes place.

$$h_f + x' hg_1 = hg_2$$

$$830.0 + x' \times 1957.7 = 2675.5$$

$$x' = 0.9426$$

Problem 3

In a test with a separating and throttling calorimeter the following observations were made.

The absolute presure of steam, before throttling 8 bar.

The pressure and temperature of steam after throttling are 754 mm of Hg absolute and 120°C.

1.5 kg of water is trapped at the separator.

10.5 kg of condensed water is collected from the condenser.

Determine the dryness fraction of steam in the main take Cp for steam to be 2.1 kJ/kg k and barometric reading to be 750 mm of Hg.

Solution

The dryness fraction measured by separating calorimeter is

$$x_1 = \frac{m_s}{m_s + m_w} = \frac{10.5}{10.5 + 1.5} = 0.875$$

Absolute pressure after throttling = 754 mm of Hg.

$$= 754 \times 10^{-3} \times 9.8 \times 13.6 \times 10^3$$

$$= 1.005 \text{ bar.}$$

for throttling calorimeter.

$$h_{f_1} + x_2 h_{g_2} + C_{ps}(T_2 - T_{sat})$$

$$721.11 + x_2 (2048.0) = 2675.68 + 2.1 (120 - 99.64)$$

$$x_2 = 0.975$$

∴ Drynes fraction of steam in the main $x = x_1 \cdot x_2$

$$x = 0.875 \times 0.975 = 0.853$$

Problem 4

1.0 kg of wet steam of quality 0.7 at 0.3 MPa. pressure is heated at constant pressure till the temperature rises to 300°C, calculate the amount of energy added to heat.

Solution: At point 1

$$x = 0.7$$

$$p_1 = 0.3 \text{ MPa}$$

92 *Mechanical Engineering: Fundamentals*

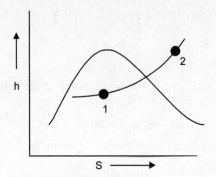

from steam table.

$$hf = 561.47 \text{ kJ/kg}.$$

$$h_{fg} = 2163.8 \text{ kJ/kg}.$$

$$h_1 = 561.47 + 0.7$$

$$h_1 = 2163.8 \text{ kJ/kg}.$$

h_2 at 0.3 MPa and 300°C from steam tables = 3069.3 kJ/kg.

$$h_2 = 3060 \text{ kJ/kg}$$

Heat added = $h_2 - h_1$ = 3069.3 − 2076.13 = 993.17 kJ/kg

EXERCISE

1. Differentiate between
   (a) Wet steam and superheated steam.
   (b) Throttling and isentropic expansion.
   (c) Fire tube and water tube boilers.
2. Explain the working of.
   (a) Separating calorimeter.
   (b) Throttling calorimeter.
   (c) Separating and throttling calorimeter
   (d) Cochran boiler.
   (e) Babcock and Wilcox boiler.
3. What do you mean by quatity of steam.
4. (a) What is a steam generator? Draw a neat sketch of a steam, generator and label its various parts.
   (b) Explain function of the following.

(1) Feed check value
(2) Fusible plug
(3) Economiser
(4) Superheater
5. What do you understand by mountings? Explain the function of all mountings used in boilers.
6. What do you understand by accessories? Explain all accessories used in boiler.

OBJECTIVE TYPE QUESTIONS

1. Heating of dry steam above saturation temperature is known as
   (a) enthalpy
   (b) superheating
   (c) super saturation
   (d) latent heat
2. Superheating of steam is done at
   (a) Constant volume
   (b) Constant temperature
   (c) Constant pressure
   (d) Constant entropy
3. The saturation temperature of steam with increase in pressure increases
   (a) linearly
   (b) rapidly first and than slowly
   (c) slowly first and then rapidly
   (d) inversely
4. One kg of steam sample contains 0.8 kg dry steam; its dryness fraction is
   (a) 0.2
   (b) 0.8
   (c) 1.0
   (d) 0.6
5. If x_1 and x_2 be the dryness fractions obtained in separating calorimeter of the steam will be.
   (a) $x_1 x_2$
   (b) $x_1 + x_2$
   (c) $\dfrac{x_1 + x_2}{2}$
   (d) $x_1 - x_2$
6. If a steam sample is nearly dry condition, then its dryness fraction can be most accurately determined by
   (a) throttling calorimeter
   (b) separating calorimeter
   (c) combined separating and throttling calorimeter
   (d) none of the above

7. A wet vapour can be completely specified by
   (a) temperature only
   (b) dryness fraction only
   (c) pressure only
   (d) pressure and dryness fraction only
8. On a Mollier chart, the constant pressure lines
   (a) diverge from left to right
   (b) diverge from right to left
   (c) are equally spaced throughout
   (d) first rise up and than fall
9. If x is the weight of dry steam and y is the weight of water in suspension then dryness fraction is equal to

   (a) $\dfrac{x}{x+y}$

   (b) $\dfrac{y}{x+y}$

   (c) $\dfrac{x}{x-y}$

   (d) $\dfrac{y}{x-y}$

10. Fire tube boiler are those in which
    (a) flue gases pass through tubes and water around it.
    (b) water passes through the tubes and flue gases around it.
    (c) forced circulation takes place
    (d) tubes are laid vertically
11. Which of the following is a fire tube boiler
    (a) locomotive boiler
    (b) Babcock and Wilcox boiler
    (c) Stirling boiler
    (d) all of the above
12. Which of the following is a water tube boiler
    (a) Locomotive boiler
    (b) Cochran boiler
    (c) Cornish boiler
    (d) Babcock and Wilcox boiler
13. One kg steam sample contains 0.4 kg water vapour. Its dryness fraction is
    (a) 0.4
    (b) 0.6
    (c) 0.4/1.5
    (d) 0.4 × 0.6

14. The water tubes in a Babcock and Wilcox boiler are
    (a) horizontal
    (b) vertical
    (c) inclined
    (d) horizontal and inclined
15. Which of the following boilers is best suited to meet fluctuating demands
    (a) Babcock and Wilcox
    (b) locomotive
    (c) Lancashire
    (d) Cochran
16. A fusible plug is fitted in small boilers in order to
    (a) avoid excessive built-up of pressure
    (b) avoid explosion
    (c) extinguish fire if the water level in the boiler falls below alarming level.
    (d) Control steam dome
17. The high pressure boiler is one producing steam at a pressure more than
    (a) atmospheric pressure
    (b) 5 kg/cm^2
    (c) 75–80 kg/cm^2
    (d) 40 kg cm^2
18. One kilowatt-hour energy is equivalent to
    (a) 1000 J
    (b) 250 kJ
    (c) 3600 kJ
    (d) 3600 kW/sec
19. The basic purpose of drum in a boiler is to
    (a) serve as storage of steam
    (b) serve as storage of feed water
    (c) remove salts from water
    (d) separate steam from water
20. A safety valve in a locomotive starts leaking. The leaking medium will be
    (a) water
    (b) dry steam
    (c) wet steam
    (d) superheated steam
21. Fusible plug for boiler is made of fusible metal containing tin, lead and
    (a) Bismath
    (b) copper
    (c) aluminium
    (d) nickel
22. The Conomiser is used in boilers to
    (a) Increase thermal efficiency of boiler
    (b) Conomise on fuel.

(c) increase flue gas temperature
(d) to heat feed water by bled steam
23. An economiser in a boiler
    (a) increases steam pressure
    (b) increases steam flow
    (c) decreases flue consumption
    (d) decreases steam pressure
24. The blow down cock in boiler is used for
    (a) regulating drum level by blowing unwanted water
    (b) emptying the boiler in case of shut down
    (c) To remove sludge or sediments from drum.
    (d) none of the above
25. Formation of scale on a boiler tube
    (a) protects it
    (b) increases its life
    (c) decreases its life
    (d) life is unaffected

ANSWERS

1. b	2. c	3. b	4. b	5. a
6. a	7. d	8. a	9. a	10. a
11. a	12. d	13. a	14. c	15. b
16. c	17. c	18. c	19. d	20. d
21. a	22. a	23. c	24. c	25. c

UNIT 4

Refrigeration and Air-Conditioning

4.1 INTRODUCTION

American Society of Refrigerating Engineers defines refrigeration as "the science of providing and maintaining temperatures below that of surrounding atmosphere". In different types of the refrigeration system, some physical property of matter is used for producing cold.

The different methods of refrigeration are:

(1) Ice refrigeration
This type is used in hotels for keeping the drinks cold. It consists of an insulated cabinet equipped with tray or tank at the top, for holding blocks of ice pieces. Shelves for food are located below the ice compartment which helps in cooling the food on the shelves below. Air returns from the bottom of the cabinet up, the sides and back of the box cabinet which is warmer, flows over the ice and again flows down over the shelves to be cooled. This type of system is called direct contact refrigeration system as shown in Fig. 4.1.

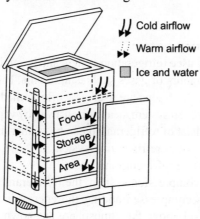

Fig. 4.1: Direct contact refrigeration system.

(2) Evaporative refrigeration

Heat is absorbed when a liquid evaporates. Evaporation of moisture from the skin surface of a man helps to keep him cool. Similar example is that of the desert bags which are used to keep drinking water cool. Another example is of earthen pots where water moves out through microholes and evaporates making the remaining water cool.

(3) Refrigeration by the expansion of air

According to the first law of thermodynamics applied to a closed system

$$Q = W + dU$$

for reversible adiabatic process $Q = 0$ therefore

$$W = - dU$$

This means the reduction in temperature. The temperature of the gas can be reduced by an adiabatic expansion of the gas. The principle was used in the Bell-Coleman air refrigeration system.

(4) Refrigeration by throttling of the gas

In adiabatic throttling process enthalpy remains constant and as the enthalpy is only a function of temperature, temperature of the perfect gas remains constant after and before throttling. However, with actual gases, the temperature of the gas after throttling may either increase, decrease or remain constant.

The term which indicates the magnitude and sign of the change in temperature is called the Joule-Thomson coefficient and it is given by

$$\mu = \left(\frac{\partial T}{\partial P} \right)_h$$

which is the change in temperature with respect to the pressure during constant enthalpy process.

(5) Vapour refrigeration system

It is the most commonly used refrigeration system and is employed in the household refrigerators. In this vapours of gases such as ammonia, carbon dioxide, sulphur dioxide, hydrocarbons like freons are used as working medium.

In a vapour refrigeration system, heat carried away by vapour in the refrigerator is in the form of latent heat of refrigerant. It can be further divided into two types:
  (1) Vapour compression refrigeration system.
  (2) Vapour absorption refrigeration system.

In case of vapour compression system, the refrigerant vapour is sucked into the compressor and is compressed adding the energy in the form of work to increase its thermal level above the atmosphere as shown in Fig. 4.2. The high

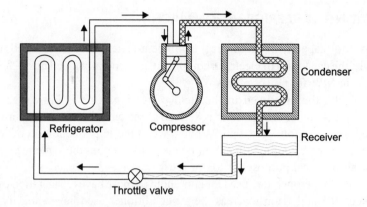

Fig. 4.2: Vapour compression refrigeration system.

pressured fluid is cooled in condenser at constant pressure. A throttling valve is attached after condenser where expansion of highly pressured fluid takes place at constant enthalpy. There is no work or heat transfer to long place. The fluid is then passed in evaporator where it takes or sucks heat from the environment and gets heated up. Generally change of phase occurs from liquid to gas in this section. The gaseous fluid then goes in the compressor to start the next cycle.

In case of absorption refrigeration system, a working fluid (refrigerant) is taken which has high affiliation to dissolve in water as shown in Fig. 4.3. The refrigerant which is commonly used in absorption refrigeration system is ammonia. The liquid, strong in ammonia, is heated further by the application of external heat to generate vapour and its temperature is increased above atmospheric temperature.

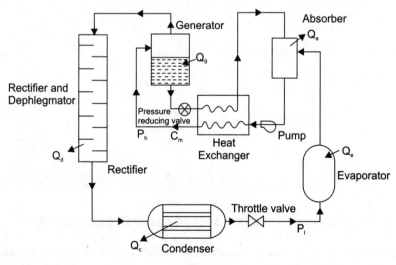

Fig. 4.3

4.2 RATING OF REFRIGERATION MACHINE

The power or capacity of mechanical equipment is generally given in HP or kW. Similarly, the capacities of electrical equipment are generally given in kW for small units and in MW for power plant.

Similarly, the unit used in the field of refrigeration is known as ton of refrigeration. A ton of refrigeration is defined as the quantity of heat required to be removed from one ton ice within 24 hours when the initial condition of water is 0°C, because same cooling effect will be given by melting the same ice.

For all practical purposes, 1 ton of refrigeration = 3.5 kJ/sec is taken.

Usually, the refrigeration used for household purpose have refrigeration capacity in between 0.5 and 1 ton whereas the air conditioner used have capacity in the range of 1.5 to 5 tons cinema hall: air-conditioning plant had capacity 200 tons.

C.O.P.—Coefficient of Performance

In refrigerator $T_1 < T_2$ where T_2 is atmospheric temperature and for doing so most economically, the maximum Q must be taken from sink with the minimum amount of W so that the performance of the refrigerator is taken into account by a ratio Q/W and it is known as coefficient of performance (C.O.P.). Refer Fig. 4.4.

$$\text{C.O.P.} = \frac{Q}{W}$$

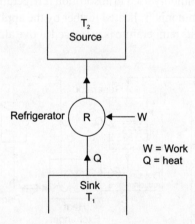

Fig. 4.4: C.O.P.

4.3 SIMPLE REFRIGERATION VAPOUR COMPRESSION CYCLE

The liquid freon-12 is under atmospheric pressure 1.013 bar and the saturation temperature corresponding to this pressure is approximately –29.8°C. Thus, the liquid freon-12 evaporates at this low temperature and latent heat of vaporisation is

Refrigeration and Air-Conditioning 101

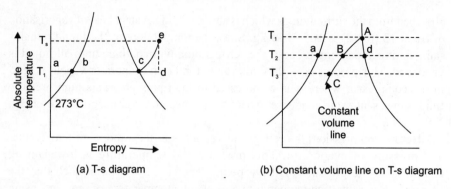

(a) T-s diagram

(b) Constant volume line on T-s diagram

Fig. 4.5: Simple refrigeration vapour compression cycle.

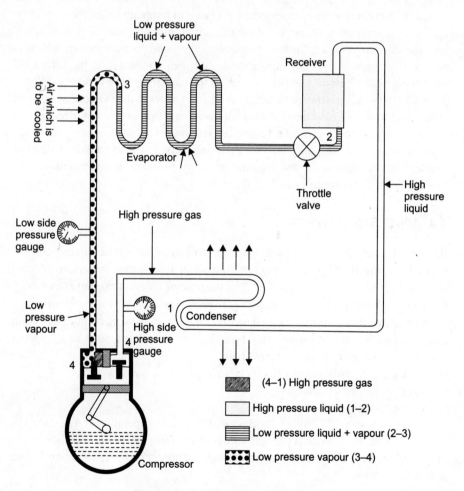

Fig. 4.6: Vapour-compression refrigerator.

absorbed from the surroundings which is above 0°C. The latent heat of vaporisation of freon-12 (R–12) is 165.91 kJ/kg. Since the temperature of the liquid remains same during vaporising process, the refrigeration will continue until all liquid is vaporised. This process take place in evaporator. From evaporator vapour is send in the compressor where it is compressed to high pressure. This high pressure fluid is then sent to the condenser where it loses heat to the atmosphere. Refer Fig. 4.5 and 4.6.

Before compression, the refrigerant vapour is at the vaporising temperature and pressure of evaporator. This pressure and temperature is low. During compression the pressure of the vapour is increased to a point such that the corresponding saturation temperature is above the temperature of the condensing medium used. At the same time, mechanical work is done on the vapour in compressing it to higher pressure, the internal energy and the enthalpy of the vapour is increased with corresponding increase in temperature of the vapour.

After compression the high temperature, high pressure vapour is discharged into the condenser where heat is rejected to the condensing medium which is normally air or water. Thus, vapour cools and then condenses at the saturation temperature corresponding to the pressure in the condenser. The condensed liquid passes on to the liquid storage vessel from where it flows to the evaporator through expansion valve. The function of the expansion valve is to allow the liquid refrigerant under high pressure to pass at a controlled rate into the low pressure part of the system. Some of the liquid evaporates passing through the expansion value but greater portion vaporises in the evaporation at low pressure (low temperature). In this way next cycle repeats itself.

4.4 AIR-CONDITIONING

Air-conditioning signifies the automatic control of an atmospheric environment either for comfort of human beings or animals or for the proper performance of some industrial or scientific process. The purpose of air conditioning is to supply sufficient volume of clean air containing a specific amount of water vapour and at a temperature capable of maintaining predetermined atmospheric conditions within a selected enclosure. The space may be a small compartment, such as a research test cabinet, shopping mall, a cinema hall.

Broadly air-conditioning can be classified into two types:
(a) Comfort air-conditioning
(b) Industrial air-conditioning

Industrial air conditioning provides air at required temperature and humidity to perform a specific industrial process successfully. Here the design conditions are not based on the feelings of the human beings but purely on the requirement of the industrial process.

Comfort Air-Conditioning		
Summer Air-Conditioning In this main objective is to reduce the sensible heat and the water vapour content of air by cooling and dehumidifying.	**Winter Air-Conditioning** In this the problem encountered is to increase the sensible heat and water vapour content of the air by heating and humidification.	**Year Round Air-Conditioning** This type of air-conditioning assures the control of temperature and humidity of air in an enclosed space throughout the year when the atmospheric conditions change as per the season

Psychrometric Properties

Psychrometry is the science of study of properties of mixture of air and water vapour.

Dry air: It is considered a mixture of nitrogen and oxygen neglecting the small percentage of other gases, i.e., 79% nitrogen, 21% of oxygen. Molecular weight: of dry air is taken as 29.

Moist air: It is a mixture of dry air and water vapour.

Water vapour: The mixture of air and water vapour at a given temperature is said to be saturated when it contains maximum amount of water vapour that it can hold. The water vapour present in air is known as moisture and its quantity in air is an important factor in all air-conditioning systems.

Dry bulb temperature: The temperature of air measured by ordinary thermometer is known as dry bulb temperature.

Wet bulb temperature: The temperature measured by the thermometer when its bulb is covered with wet cloth and is exposed to a current of moving air is known as wet bulb temperature.

Dew bulb temperature: The temperature of the air is reduced by continuously cooling, then the water vapour in the air will start condensing at a particular temperature. The temperature at which the condensing starts is known as Dew-point temperature.

Specific humidity: It is the mass of water vapour present per kg of dry air. It is given in grams per kg of dry air.

Absolute humidity: The weight of water vapour present in unit volume of air is known as absolute humidity.

Degree of saturation: It is defined as the ratio of mass of water vapour associated with unit mass of dry air to mass of water vapour associated with unit mass of dry air saturated at the same temperature.

Relative humidity: It is defined as the ratio of actual mass of water vapour in a given volume to the mass of water vapour if the air is saturated at the same temperature.

Sensible heat of air: The quantity of heat which can be measured by measuring the dry bulb temperature of the air is known as sensible heat.

4.5 PSYCHROMETRIC CHART

A chart which shows the interrelation of all the important properties is known as psychrometric chart. This chart for a given barometric pressure is very important to illustrate the different psychrometric processes. Refer Fig. 4.7.

In this chart dry bulb temperature is taken along abscissa and specific humidity as ordinate to the right-hand side of the chart. All other properties are shown by different lines on the chart. The different lines shown on this chart are constant specific volume lines, constant enthalpy lines, constant relative humidity lines and constant wet-bulb temperature lines.

The various pscychrometric processes such as sensible heating, sensible cooling, humidification, dehumidification, chemical dehumidification, humidification by steam injection can be easily represented on this chart and thus it saves lot of labour in making calculations.

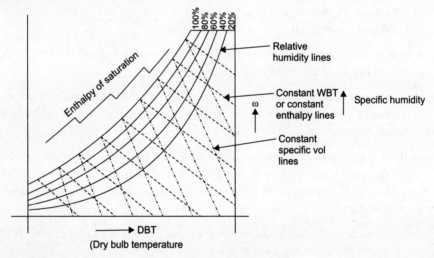

Fig. 4.7: Psychrometric chart.

Human Comforts

Requirements of comfort air-conditioning.

The comfort feeling of the people in an air-conditioned plant depends upon the following factors:
 (1) Supply of oxygen and removal of carbon dioxide
 (2) Removal of body heat dissipated by the occupants.
 (3) Removal of body moisture dissipated by the occupants.

(4) To provide sufficient air movement and air distribution in occupant space.
(5) To maintain the purity of air by removing odour and dust.

The real comfort feeling cannot be achieved unless above-mentioned factors are properly controlled. The above-mentioned factors are discussed below.

(1) Oxygen supply

Each person requires nearly 0.65 cu.m of O_2 per hour under normal conditions and produces 0.2 cu.m of CO_2. The percentage of CO_2 in the atmosphere is nearly 0.6% and it is necessary to maintain this percentage for proper functioning of respiratory system. When the percentage of CO_2 in air exceeds above 2% the partial pressure of oxygen will be reduced to a value such that the breathing becomes more difficult. Extreme discomfort exists when the percentage of CO_2 reaches 6% and unconsciousness occurs at 10% of CO_2. Hence, the quantity of CO_2 should not exceed the minimum in the air-conditioned space.

(2) Heat Removal

If a space of 6 cu.m is provided to each person and if there is no transfer of heat and air from the outside source, then the space temperature will rise through 0.15°C for each kJ of heat added to the space and rise in temperature of 48°C per hour would result as man's body dissipates 320 kJ of heat per hour.

In a good ventilation system, sufficient circulation of air is there so that excessive rise in temperature of air in air-conditioned space does not take place.

(3) Moisture removal

The body ability to dispose of heat by evaporation to atmosphere decreases as the air humidity increases. High humidity of air reduces the freshness of air in an enclosed space in addition to the difficulty in disposing of body heat. The ventilation system must be capable to maintain the relative humidity below 70%.

(4) Air Motion

Air velocity in the air-conditioned space should not be more than 6 to 9 m/min at 20°C and 9 to 15 m/min at 22°C. The following table gives the comfortable ranges of air velocity and humidity with respect to room air temperature.

Room Air Temp. °C	Velocity m/sec		R.H. %	
	Minimum	Maximum	Minimum	Maximum
20	0.04	0.12	35	65
21	0.04	0.14	35	65
22	0.05	0.17	35	65
23	0.07	0.21	35	65
24	0.09	0.24	35	65
25	0.12	0.32	35	65
26	0.16	0.40	35	65

Air distribution is defined as a uniform supply of air to an air conditioned system. Air motion without proper air distribution is responsible for local cooling sensation known as draft.

(5) Purity of Air

The quality of air in regard to odour, dust, toxic gases and bacteria is considered for defining the purity of air. Now days to get purified air the plasma air purifier is being installed in most Air conditioning system.

4.6 PLASMA AIR PURIFIER

In the fourth state of matter called plasma, high voltages break down disease-causing germs and viruses into harmless ions, leaving the air clean and healthy.

With the use of this technology, it is not necessary to change the filter for a life-time. The principle of this is explained in Fig. 4.8 which is self explanatory.

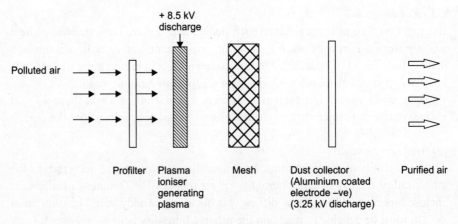

Fig. 4.8: Plasa Air Purifier—Principle.

4.7 COMFORT CHARTS

Thermodynamically the ideal human comfort exists when the rate of heat production becomes equal to the rate of heat loss. This equilibrium condition is maintained when proper conditions of temperature, humidity, air velocity and purity are maintained in the air-conditioned space. The feeling of comfort to individual depends upon factors as habits of eating, types of clothes used, duration of stay age and sex and rate of activity, etc. This feeling also differs from individual to individual, so it is difficult to set up standards for ideal comfort feeling.

A scientific method is introduced to measure the comfort feeling of human beings by introducing a concept of effective temperature.

Effective temperature: It is a measure of feeling warmth or cold to the human body in response to the air temperature, moisture content and air motion.

A chart is prepared showing the percentage of people feeling comfort at different effective temperatures. The data concerning the percentage of people feeling comfort at different effective temperatures is collected.

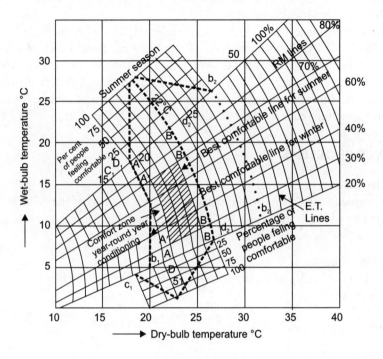

Fig. 4.9

Desirable inside design conditions for different buildings (Summer conditions)

Type of Building	DBT(°C)	R.H.
Living room	20–30	55 to 65%
Bedrooms	15–20	55 to 65%
Halls and corridors	15–20	55 to 65%
Wards	24	60%
Operating theatres	20°C	65%
	26–5°C	50%
	20–23	60–65%
	15–20	55–60%
	12–16	50–60%
	16–19	60–65%
Heavy work	15	50%
Light work	18	55%
Sedentary work	22	60%
	19–23	60%
	25–28	50–60%

The chart which gives different percentage of people (feeling comfort at different effective temperatures) is known as comfort chart as shown in Fig. 9.9.

On this chart area A'A"B"B' shown is used for year round air-conditioning design. So in this chart from the opinion of the people area is found out where most of the people feel comfortable.

Here first psychometric chart is drawn taking DBT on x-axis and WBT on y-axis. The 100% RH line is drawn as DBT and WBT are same at each point on this line then the other RH lines are drawn. Then the effective temperature lines are drawn. The percentage of people feeling comfortable at different effective temperatures for summer and winter conditions are drawn in the graph and then comfort zone for year round air-condition is drawn in the chart.

SOLVED PROBLEMS

Problem 1
Ice is formed at 0°C from water at 15°C. The temperature of the brine is –8°C. Find the kg of ice formed per 1 kWh. Assume that the refrigeration cycle used is perfect reversible carnot cycle latent heat of ice = 336 kJ/kg.

Solution
Heat taken out per kg of water = $4.2 \times (15 - 0) + 336 = 399$ kJ/kg
one kWH = 3600 kJ

C.O.P. of carnot cycle $\dfrac{T_1}{T_2 - T_1} = \dfrac{-8 + 273}{15 - (-8)} = \dfrac{265}{23} = 11.52$

Assume m_c is the ice formed per kWh

$$C.O.P. = 11.52 = \dfrac{m_c \times 399}{3600}$$

$m_c = 103$ kg of ice

Problem 2
A carnot refrigerator extracts 410 kJ of heat per minute from a cold room which is maintained at –16°C and it is discharged to atmosphere which is at 30°C. Find an ideal kW-capacity of motor required to run the unit.

Solution

$T_1 = -16 + 273 = 257$

$T_2 = 30 + 273 = 303$ K

for the reversed Carnot cycle

$$\dfrac{Q_1}{T_1} = \dfrac{Q_2}{T_2} \quad \therefore Q_1 \times \dfrac{T_2}{T_1}$$

$$\therefore Q_2 = 410 \times \frac{303}{257} = 484.38 \text{ kJ/min}.$$

w(workdone) = $(Q_2 - Q_1)$ = 483.38 – 410

= 73.38 kJ/min

= 1.22 kJ/sec

kW required = 1.22 kW.

Problem 3
The capacity of a refrigerator is 500 tons when working between –4°C and +20°C. Find the mass of ice produced within 24 hours when water is supplied at 10°C also find the minimum kW required. Assume the cycle as carnot cycle and latent heat of ice = 336 kJ/kg.

Solution
Heat extraction capacity of the plant

= 500 × 3.5 = 1750 kJ/sec.

Heat removed from water = 4.2 (10 – 0) + 336 = 378 kJ/kg.

\ Mass of ice formed in 24 hr = $\dfrac{\text{Heat extraction capacity of plant per hour}}{\text{Heat extracted from each ton of ice}}$

$$= \frac{1750 \times 3600 \times 24}{378 \times 1000} = 400$$

C.O.P. of the cycle = $\dfrac{T_1}{T_2 - T_1} = \dfrac{-4 + 273}{20 - (-4)} = \dfrac{269}{24} = 11.20$

C.O.P. = $\dfrac{\text{Refrigeration effect/sec}}{\text{workdone/sec}} =$

work done per sec = $\dfrac{\text{Refrigeration/sec}}{\text{C.O.P.}} = \dfrac{1750}{11.20} = 156.25$ kJ

kW required = 156.25 kJ.

Problem 4
A cold storage plant is required to store 15 tons of fish.
The temperature of fish when supplied = 30°C
Storage temperature of fish required = –10°C
Specific heat of fish above freezing point = 2.95 kJ/kg°C
Specific heat of fish below freezing point = 1.25 kJ/kg °C

Freezing point of fish = $-3°C$
Latent heat of fish = 230 kJ/kg
If the cooling is achieved within 10 hours find.
(a) Capacity of the refrigerating plant
(b) Carnot cycle C.O.P. between this temperature range.
(c) If the actual C.O.P. is 1/4 of the carnot C.O.P. find the kW required to run the plant.

Solution

Heat removed from each kg of fish in 10 hours.

= 2.95 [30 – (–3)] + 230 + 1.25 [–3–(–10)] =

327.35 + 8.75 = 336.1

Heat removed by the plant per sec. = $\dfrac{336.1 \times 15 \times 10^3}{10 \times 3600}$ = 140.04 kJ.

Capacity of the plant = $\dfrac{140.04}{3.5}$ = 40.01 tons

C.O.P. of carnot cycle = $\dfrac{T_1}{T_2 - T_1} = \dfrac{263}{303 - 263} = \dfrac{263}{40}$ = 6.575

Actual C.O.P. = $\dfrac{6.575}{4}$ = 1.64

Work done/sec = $\dfrac{140.04}{1.64}$ = 85.39

kW required = 85.39 kW

EXERCISE

1. Define the following terms
   (a) Refrigerating effect
   (b) Ton of refrigeration
   (c) C.O.P.
   (d) Refrigerator
   (e) Heat pump
2. Explain how a refrigerant produces cooling effect?
3. Explain the working principle of vapour compression refrigeration system with a neat diagram.
4. What are the different requirements of comfort air-conditioning explain.
5. Explain comfort chart?

6. Explain the difference between dry bulb temperature and wet bulb temperature.
7. What is a psychrometric chart explain?
8. Explain the uses of psychrometric chart?

OBJECTIVE TYPE QUESTIONS

1. Which of the following cycles uses air as the refrigerant.
    (a) ericsson
    (b) stirling
    (c) carnot
    (d) Bell-Coleman
2. Ammonia-absorption refrigeration cycle requires
    (a) very little work input
    (b) maximum work input
    (c) nearly same work input as for vapour compression cycle.
    (d) zero work input
3. In vapour compression cycle the condition of refrigerant is saturated liquid.
    (a) after passing through the condenser.
    (b) before passing through the condenser
    (c) after passing through the expansion
    (d) before entering the expansion value
4. In vapour compression cycle, the condition of refrigerant is very wet vapour.
    (a) after passing through the condenser.
    (b) before passing through the condenser
    (c) after passing through the expansion or throttle value.
    (d) before entering the expansion value.
5. In vapour compression cycle, the condition of refrigerant is high pressure saturated liquid.
    (a) before passing through the condenser.
    (b) after passing through the expansion or throttle value.
    (c) before entering the expansion value.
    (d) before entering the compressor.
6. In vapour compression cycle the condition of refrigerant is superheated vapour
    (a) after passing through the condenser
    (b) before passing through the condenser.
    (c) after passing through the expansion or throttle value.
    (d) before entering the expansion value.
7. In vapour compression cycle the condition of refrigerant is dry saturated vapour.
    (a) before passing through the condenser.
    (b) after passing through the expansion or throttle value.
    (c) before entering the expansion value.
    (d) before entering the compressor

8. One ton of refrigeration is equal to the refrigeration effect corresponding to melting of 1000 kg of ice
   (a) in 1 hour
   (b) in 1 minute
   (c) in 24 hours
   (d) in 12 hours
9. One ton refrigeration corresponds to
   (a) 50 kcal/min
   (b) 50 kcal/hr
   (c) 80 kcal/min
   (d) 80 kcal/hr
10. In S.I. unit, one ton of refrigeration is equal to
    (a) 210 k
    (b) 21 kJ/min
    (c) 420 kJ/min
    (d) 840 kJ/min
11. The vapour compression refrigerator employs the following cycle.
    (a) Rankine
    (b) Carnot
    (c) Reversed Rankine
    (d) Reversed Carnot
12. The refrigerant for a refrigerator should have
    (a) high sensible heat
    (b) high total heat
    (c) high latent heat
    (d) low latent heat
13. Rating of a domestic refrigerator is of the order of
    (a) 0.1 ton
    (b) 5 ton
    (c) 10 tons
    (d) 40 tons
14. The C.O.P. of a domestic refrigerator is
    (a) less than 1
    (b) more than 1
    (c) equal to
    (d) None of the above
15. If T_1 and T_2 be the highest and lowest absolute temperatures encountered in a refrigeration cycle working on a reversed carnot cycle, then C.O.P. is equal to
    (a) $\dfrac{T_1}{T_1 - T_2}$

(b) $\dfrac{T_2}{T_1-T_2}$

(c) $\dfrac{T_1-T_2}{T_2}$

(d) $\dfrac{T_1-T_2}{T_2}$

16. In vapour compression refrigeration system, refrigerant occurs as liquid between.
    (a) condenser and expansion value
    (b) compressor and evaporator
    (c) expansion value and evaporator
    (d) Compressor and condenser
17. Aqua ammonia is used as refrigerant in the following type of refrigeration system
    (a) compression
    (b) direct
    (c) indirect
    (d) absorption
18. Most of the domestic refrigerators work on the following refrigeration system
    (a) vapour compression
    (b) vapour absorption
    (c) carnot cycle
    (d) electrolux refrigerator
19. Psychrometric Chart
    (a) is seldom used for air-conditioning design
    (b) provides plots for moist air-conditioner's
    (c) enables to determine wet bulb, and dew point temperature
    (d) is a chart for conversion of British system into metric system
20. On psychrometric chart, dry bulb temperature lines are
    (a) horizontal
    (b) vertical
    (c) curved
    (d) none
21. The comfort conditions in air-conditioning system are defined by
    (a) 22°C dry bulb temperature (DBT) and 60% relative RH.
    (b) 25°C DBT and 100% RH
    (c) 20°C DBT and 75% RH
    (d) 15°C DBT and 80% RH
22. On psychrometric chart, relative humidity lines are
    (a) horizontal
    (b) vertical

(c) straight inclined sloping downward to the right
(d) curved

23. Dust and other impurities in the air are removed by
   (a) air waster
   (b) electrostatic precipitation
   (c) adhesive impregnated filters
   (d) All of the above.

24. Dew point is
   (a) the temperature at which condensation of steam in saturated air well start.
   (b) the lowest attainable temperature for a mixture of air and steam
   (c) dependent on pressure of air
   (d) none of the above

25. Chemical formula of foreon –12 is
   (a) CCl_2F_2
   (b) CCl_2F_3
   (c) CCl_3F_2
   (d) CCl_3F_3

ANSWERS

1. d	2. a	3. a	4. c	5. c
6. b	7. d	8. c	9. a	10. a
11. d	12. c	13. a	14. b	15. b
16. c	17. d	18. a	19. b	20. b
21. a	22. d	23. d	24. a	25. a

UNIT 5

Hydraulic Turbines and Pumps

5.1 INTRODUCTION

Hydraulic machines are defined as those machines which convert either hydraulic energy (energy possessed by water) into mechanical energy (which is further converted into electrical energy) or mechanical energy into hydraulic energy. The hydraulic machines which convert the hydraulic energy into mechanical energy are called turbines while the hydraulic machines which convert the mechanical energy into hydraulic energy are called pumps. Turbines consist of mainly the study of Pelton turbine, Francis turbine and Kaplan turbine while pumps consist of the study of centrifugal pump and reciprocating pump.

Turbines

Turbines are defined as the hydraulic machines which convert hydraulic energy into mechanical energy. This mechanical energy is used in running an electric generator which is directly coupled to the shaft of the turbine. Thus, mechanical energy is converted into electrical energy. The electric power which is obtained from the hydraulic energy (energy of water) is known as hydroelectric power. At present the generation of hydroelectric power is the cheapest as compared by the power generated by other sources such as oil, coal, etc.

Figure 5.1 shows a general layout of a hydroelectric power plant which consists of:
  (1) A dam constructed across a river to store water.
  (2) Pipes of large diameters called penstocks which carry water under pressure from the storage reservoir to the turbines. These pipes are made of steel or reinforced concrete.
  (3) Turbines having different types of vanes fitted to the wheels.
  (4) Tail race which is a channel which carries water away from the turbines after the water has worked on the turbines. The surface of water in the tailrace is also known as tail race.

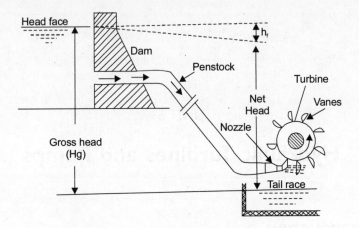

Fig. 5.1: Layout of a hydroelectric power plant

5.2 CLASSIFICATION

Following are the important ways for the classification of hydraulic turbines.

(1) According to energy available at the inlet of turbine.
  (a) Impulse turbine—Pelton turbine.
  (b) Reaction turbine—Francis turbine, Kaplan turbine

Impulse turbine works on the principle of impulse-momentum equation. The change in momentum of water produces impulse on the blades of turbine, which acts as a torque for rotation of turbine shaft.

$$\underbrace{F dt}_{\text{Impulse}} = \underbrace{d(mv)}_{\text{Change in momentum}}$$

Reaction turbine produces back thrust or reaction on the blade since the pressure at outlet of the turbine is less than that of inlet. But in case of Impulse turbine pressure remains constant (same) at inlet and high velocity of jet impinges on the bucket.

(2) According to the direction of flow of water.
  (a) Tangential flow turbine.
  (b) Radial flow turbine
  (c) Axial flow.
  (d) Mixed flow turbine.

(a) *Tangential flow turbine:* If water flows along the tangent of the runner, the turbine is known as tangential flow turbine. Example Pelton turbine.

(b) *Radial flow turbine:* If the water flows in radial direction inward or outward, the turbine is known as radial flow turbine.

(c) *Axial flow turbine:* If water flows through the runner along the axis of the turbine, it is known as axial flow turbine. Example Kaplan turbine.

(d) *Mixed flow turbine:* If water flows through the runner in radial direction but leaves in axial direction, then it is known as mixed flow turbine. Ex: Francis turbine.

(3) According to head.
    (a) High head turbine; H > 180 m (Pelton turbine)
    (b) Medium head turbine: 60 < H < 150 m (Francis turbine)
    (c) Low head

(4) According to the specific speed of turbine.
    (a) Low specific speed turbine (10 to 25) – Pelton turbine.
    (c) Medium specific speed turbine (60–300) – Francis turbine
    (c) High specific speed turbine (300–1000) Kaplan turbine.

5.3 TERMINOLOGY USED FOR TURBINES

(1) *Gross head:* The difference between head of water in a reservoir level (head race level) and tail race level when no water is flowing is known as gross head (Hg).

(2) *Net Head:* It is also called effective head and is defined as the head available at the inlet of the turbine. When water is flowing from head race to the turbine, a loss of head due to friction between the water and penstock occurs. Though there are other losses also such as loss due to bend, pipe fittings, loss at the entrance of penstock, etc., yet they have small magnitude as compared to head loss due to friction. If h_f is the head loss due to friction between penstock and water then net head on turbine is given by.

$$H_{net} = H_g - h_f$$

where h_f = Head loss due to friction $h_f = \dfrac{4fLV^2}{2gD}$

    f = Coefficient of fiction.
    D = Diameter of penstock.
    L = Length of penstock
    V = Velocity of water

(3) *Hydraulic efficiency*:

$$\eta_h = \frac{\text{Power delivered to runner}}{\text{Power at inlet of the turbine}}$$

Power at inlet of the turbine is more than the power supplied to the runner due to losses during flow through the runner.

Difference between impulse and reaction turbine.

Impulse turbine	Reaction Turbine
(1) All the available hydraulic energy is converted into kinetic energy prior to impingement on turbine blades, i.e., buckets.	A part of the available energy is converted into kinetic energy, rest of the energy remains in the form of pressure energy.
(2) The jet strikes the buckets at atmospheric pressure and the pressure remains the same throughout the action of water on the runner through the velocity of jet changes.	At the inlet to the turbine the pressure of water is very high, but both the pressure and velocity change while it passes through the runner.
(3) No water-tight casing. The casing only prevents splashing, and guides water to tail race.	Water-tight casing because the pressure inside the turbine can be less than atmospheric pressure.
(4) Water enters the turbine only in the forms of jet. The jets may be one or more than one.	Water is admitted over the entire circumference of the runner.
(5) The turbine does not run full and the air has a free access to the buckets.	Water completely fills the passage between the blades and casing throughout the operation of the turbine.
(6) The turbine unit is installed above tail race and no draft tube is required.	The turbine is connected to the tail race through a draft tube, so it may be installed above or below the tail race.

(4) *Mechanical efficiency:*

$$\eta_{mech} = \frac{\text{Power at the shaft of the turbine}}{\text{Power delivered by water to the runner}}$$

(5) *Volumetric efficiency*:

$$\eta_v = \frac{\text{Volume of water actually striking the runner}}{\text{Volume of water supplied to the turbine}}$$

The volume of the water striking the runner of a turbine is slightly less than the volume of the water supplied to the turbine. Some of the volume of the water is discharged to the tail race without striking the runner of the turbine.

(6) *Overall Efficiency (η_0):*

$$\eta_0 = \frac{\text{Power available at the shaft of the turbine}}{\text{Power supplied at the inlet of the turbine}}$$
$$= \eta_m \times \eta_h$$

5.4 PELTON WHEEL CONSTRUCTIONAL DETAILS

Pelton Turbine

Pelton turbine is a high head, tangential flow, low specific speed turbine. It is named after L.A. Pelton—an American Engineer.

It is suitable for high head. It is efficient and reliable when employed under these conditions. It is less efficient for low head because for a given power if the head is reduced, the rate of flow has to be increased. Increased flow will require bigger jet diameter, hence bigger runner diameter. On the other hand, jet velocity and peripheral velocity of runner will be reduced. Thus, turbine becomes bulky and slow running for low heads.

Parts of Pelton Turbine

Main parts of Pelton turbine:
(1) Penstock
(2) Runner with buckets
(3) Nozzle and spear mechanism
(4) Casing
(5) Braking jet.

Penstock: The penstock is a steel or concrete conduit which is used to bring the water from reservoir to the nozzle. It is equipped with control valves and screens at inlet. The control valves regulate the flow from reservoir to the turbine. The screens are also called Trash racks which prevent the debris and foreign materials from entering into the penstock.

Runner with buckets: It consists of a circular disc on the periphery of which a number of buckets evenly spaced are fixed. The shape of the buckets is of a double hemispherical cup or bowl. Each bucket is divided into two symmetrical

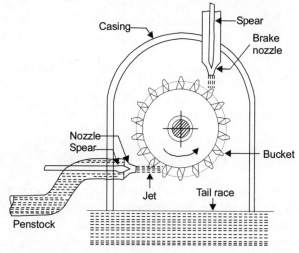

Fig. 5.2: Pelton wheel

parts by a dividing wall which is known as splitter. The splitter helps the jet to be divided without shock into two parts moving sideways in opposite direction. The rear of the bucket is so shaped that water should not interfere during the passage the bucket is proceeding in order of rotation. The jet should be deflected backward at an angle of 160° to 165°, the materials used for bucket may be cast iron bronze or stainless steel.

Nozzle and spear mechanism: The pressure head of water at the inlet of nozzle is converted in the kinetic energy and the velocity of water is $v = \sqrt{2gH}$ where H is the net head. The high velocity water strikes the bucket which deviate from the initial direction. Due to the change in momentum, impulse is produced on the bucket which rotates the runner and produce shaft power. Spear is used to control the flow of water from the nozzle which in turn control the rotational speed of turbine shaft. The function of the spear is to change the flow area of nozzle by forward or backward movement.

If spear moves in forward direction, the flow area will decrease and if it moves in backward direction the flow area, will increase. But due to sudden decrease in flow area or increase in flow area, water causes high pressure in penstock which is dangerous. Therefore, to prevent the high pressure generation in the penstock, a deflector is used in front of nozzle, which deflects the direction of flow of water hence the shaft speed is reduced.

Casing: The function of casing is to prevent the splashing of water and to discharge water to tail race. It also works as safeguard for the wheel. It has no a hydraulic importance.

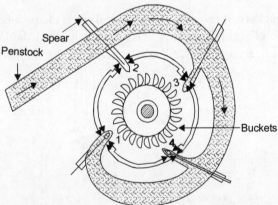

Fig. 5.3: Pelton wheel (Multi nozzle type)

Braking jet: A small nozzle is provided to direct the jet of water on the back of the vanes, to stop the runner, when the nozzle is completely closed by moving the spear in the forward direction. The amount of water striking the runner reduces to zero but the runner goes on revolving due to inertia, so a braking jet is needed to stop its motion.

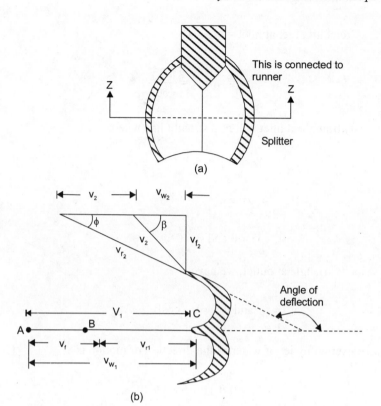

Fig. 5.4: Shape of Bucket

Velocity Triangle and Work Done

The jet of water from the nozzle strikes the bucket at the splitter which splits up the jet into two parts. These parts of the jet, glide over the inner surfaces and come out at the outer edge. Figure shows the section of the bucket at Z-Z. The splitter is the inlet tip and outer edge of the bucket is the outlet tip of the bucket.

The inlet velocity triangle is drawn at the splitter and outlet velocity triangle is drawn at the outer edge of the bucket.

H = Net head acting on the Pelton wheel

$$= H_g - h_f$$

where H_g = Gross head and $h_f = \dfrac{4f l v^2}{D^* \times 2g}$

where D^* = Dia of Penstock. N = Speed of wheel in rpm.

D = Diameter of wheel d = Diameter of jet.

Then v_1 = Velocity of jet at inlet = $\sqrt{2gH}$

$$v = v_1 = v_2 = \frac{\pi DN}{60}$$

The velocity triangle at inlet will be a straight line where

$$v_{r_1} = v_1 - u_1 = v_1 - u$$

$$v_{w_1} = v_1$$

$$\alpha = 0 \text{ and } \theta = 0$$

From velocity triangle at outlet, we have

$$v_{r_2} = v_{r_1} \text{ and } v_{w_2} = v_{r_2} \cos\beta_2 - u_2$$

The force exerted by jet of water in the direction of motion is

$$F_x = \rho a v_i (v_{w1} + v_{w2})$$

The work done by the jet on the runner per second

$$= F_x \cdot U$$

Work done per second per unit weight of water striking

$$= \frac{\rho_a v_1 [v_{w1} + v_{w2}] \cdot U}{(\rho a\, v_1) g}$$

$$= \frac{1}{g}[v_{w1} + v_{w2}]U$$

The kinetic energy of the jet at the inlet = $\frac{1}{2} \cdot mv_1^2$

$\therefore$ Kinetic energy supplied by jet per second = $\frac{1}{2}(\rho_a v_1) \cdot v_1^2$

v_1, v_2 = Absolute velocity of jet at inlet and outlet.
u_1, u_2 = Speed of bucket at inlet and outlet.
v_{r1}, v_{r2} = Relative velocity of jet at inlet and outlet
v_{f1}, v_{f2} = Velocity of flow at inlet and outlet.
v_{w1}, v_{w2} = Velocity of whirl at inlet and outlet
α_1 = Angle between direction of jet and direction of motion of the plane
β_1 = Angle made by relative velocity v_{r1} with the direction of motion at inlet.
α_2 = Angle made by velocity v_2 with the direction of motion of the bucket at outlet.
β_2 = Angle made by relative velocity v_{r2} with the direction of motion of the bucket at outlet

$$a = \text{Area of jet} = \frac{\pi d^2}{4}$$

Now hydraulic efficiency $(\eta_h) = \dfrac{\text{Work done per second}}{\text{K-E supplied per second}}$

$$\eta_h = \frac{\rho_a v_1 [v_{w1} + v_{w2}] \cdot u}{\frac{1}{2}(\rho a v_1) \cdot v_1^2}$$

$$= \frac{2[v_{w1} + v_{w2}] \cdot u}{v_1^2}$$

$$= \frac{2[v_1 + (v_1 - u)\cos\beta_2 - u] \cdot u}{v_1^2}$$

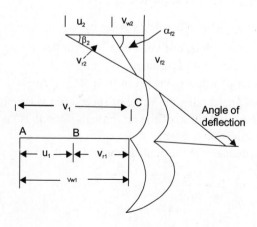

Angle of deflection

$$= \frac{2[v_1 - u + (v_1 - u)\cos\beta_2].u}{v_1^2}$$

$$\eta_h = \frac{2(v_1 - u)[1 + \cos\beta_2].u}{v_1^2} \qquad (1)$$

Maximum hydraulic efficiency

The efficiency will be maximum for a given value of v_1 when

$$\frac{d(\eta_h)}{du} = 0$$

$$\frac{d}{du}\left[\frac{2u(v_1 - u)(1 + \cos\beta_2)}{v_1^2}\right] = 0$$

$$\frac{1 + \cos\beta_2}{v_1^2} \frac{d}{du}(2uv_1 - 2u^2) = 0$$

$$\frac{d}{du}(2uv_1 - 2u^2) = 0 \qquad \left(\because \frac{1 + \cos\beta_2}{v_1^2} \neq 0\right)$$

$$2v_1 - 4u = 0$$

$$u = \frac{v_1}{2} \qquad (2)$$

Putting equation (2) in equation (1)

$$(\eta_h)_{max} = \left(\frac{1 + \cos\beta_2}{2}\right)$$

Points to be remembered. (Pelton wheel).

(i) The velocity of the jet at inlet is given by $v_1 = c_v\sqrt{2gH}$
   where c_v = coefficient of velocity = 0.98 or 0.99.
   H = net head of turbine
(ii) The velocity of wheel (u) is given by $u = C_v\sqrt{2gH}$.
   where C_v = speed ratio, the value of speed ratio varies from 0.43 to 0.48.
(iii) The angle of deflection of the jet through buckets is taken at 165° if no angle of deflection is given.
(iv) The mean diameter or the pitch diameter D of the Pelton wheel is given by

$$u = \frac{\pi DN}{60} \quad \text{or} \quad D = \frac{60u}{\pi N}.$$

(v) **Jet ratio:** It is defined as the ratio of the pitch diameter (D) of the Pelton wheel to the diameter of the jet (d). It is denoted by m.

$$m = \frac{D}{d} (= 12 \text{ for most cases})$$

(vi) Number of buckets on a runner is given by

$$z = 15 + \frac{D}{2d} = 15 + 0.5\, m$$

m = Jet ratio.

(vii) Number of jets is obtained by dividing the total rate of flow through the turbine by the rate of flow of water through a single jet.

5.5 FRANCIS TURBINE CONSTRUCTIONAL DETAILS

The inward flow reaction turbine having radial discharge at outlet is known as Francis turbine, after the name of J.B. Francis an American engineer who in the beginning designed inward radial flow reaction turbine. The flow of water in the turbine may be radially from outward to inward or from inward to outward, but in both the cases the flow passing through the runner has velocity component in a plane normal to the exis of runner. In modern Francis turbine water enters the runner of the turbine in the radial direction at outlet and leaves in the axial direction at the inlet of the runner. Thus, the modern Francis turbine is a mixed flow type of turbine in which water enters radially at its outer periphery but leaves axially at its centre.

Parts of Francis Turbine
(1) Penstock
(2) Casing
(3) Guide mechanism
(4) Runner
(5) Draft tube

(1) **Penstock:** Similar to the penstock mentioned for pelton wheel except that the diameter of the pipe is more because more quantity of water is required in this case.

(2) **Casing:** The casing and the runner are always full of water. The casing is of spiral shape in which area of cross-section goes on decreasing gradually. The casing completely surrounds the runner of the turbine. It is having a spiral shape so that water may enter the runner at constant velocity throughout the circumference of the runner. The casing is made of concrete, cast steel or plate steel.

(3) **Guide Mechanism:** The guide mechanism consists of guide vanes fixed on a stationary circular wheel all around the runner of the turbine. It regulates the

discharge according to the load of the turbine. The guide vanes are of airfoil section to have a smooth flow free of eddies. They allow the water to strike the vanes fixed on the runner without stock at inlet. Also the width between two adjacent vanes of guide mechanism can be altered so that the amount of water striking the runner can be varied. The movement of the guide wheel is regulated by a regulating shaft which in turn is operated by a governor through a servo mechanism. These are made up of cast steel.

(4) **Runner:** It is a circular wheel on which a series of radial curved vanes are fixed. The surface of the vanes are made very smooth. The radial curved vanes are so shaped that the water enters and leaves the runner without shock. The runners are made of cast steel. Cast iron or stainless steel. The runner is keyed to the shaft which is coupled to the generator shaft.

(5) **Draft Tube:** The pressure at the exit of the runner of a reaction turbine is generally less than atmospheric pressure. The water at exit cannot be directly

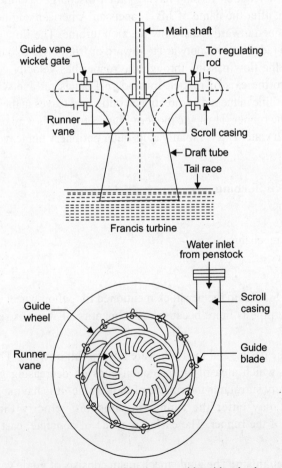

Fig. 5.5: Francis turbine runner with guide wheel

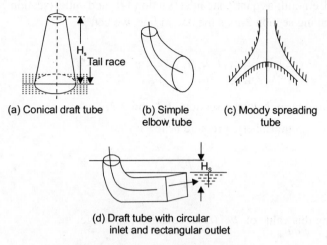

Fig. 5.6: Various shapes of draft tube

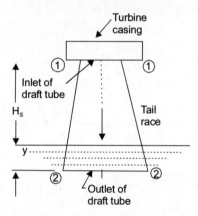

Fig. 5.7: Diagram for explaining the theory of draft tube

discharged to the tail race. A tube or pipe of gradually increasing area is used for discharging water from the exit of the turbine to the tail race. This tube of increasing area is called draft tube.

In addition to serve a passage for water discharge draft tube has the following two purposes also.

 (i) The turbine may be placed above the tail race and hence turbine may be inspected properly.
 (ii) The kinetic energy $v_2^2/2g$ rejected at the outlet of the turbine is converted into useful pressure energy.

Draft Tube Theory

Let H_a = Vertical height of draft tube above the tail race
 H_b = Distance of bottom of draft tube from tail race

Applying Bernoulli's equation to inlet (section 1-1) and outlet (section 2-2) of draft tube and taking section 2-2 as the datum line, we get.

$$\frac{p_1}{\rho g} + \frac{v_1^2}{2g} + (H_a + H_b) = \frac{p_2}{\rho g} + \frac{v_2^2}{2g} + 0 + h_f \qquad (1)$$

h_f = Loss of energy between sections 1-1 and 2-2.

But $p_2/\rho g$ = Atmospheric pressure head $+y$

$$= \frac{p_a}{\rho g} + y$$

Substituting this value of $p_2/\rho g$ in equation (1), we get

$$\frac{p_1}{\rho g} + \frac{v_1^2}{2g} + (H_a + H_b) = \frac{p_a}{\rho g} + y + \frac{v_2^2}{2g} + h_f$$

but $y = H_b$

$$\frac{p_1}{\rho g} + \frac{v_1^2}{2g} + H_a = \frac{p_a}{\rho g} + \frac{v_2^2}{2g} + h_f$$

$$\frac{p_1}{\rho g} = \frac{p_a}{\rho g} + \frac{v_2^2}{2g} + h_f - \frac{v_1^2}{2g} - H_a \qquad (2)$$

$$= \frac{p_0}{\rho g} - H_a - \left(\frac{v_1^2}{2g} - \frac{v_2^2}{2g} - h_f\right)$$

from the equation it is seems that $p_1/\rho g$ is less than atmospheric pressure.

Working of Francis Turbine

Since the modern Francis turbine is mixed flow type, the water enters radially at the outer periphery of the runner and leves axially. This turbines operate under medium heads and requires medium quantity of water. The water first passes through guide vanes, which in turn direct the water at the proper angle so that the water enters the runner vanes without shock and glides properly. As the water passes through the runner and proceeds towards outlet. Some part of the head acting on the turbine is transformed into kinetic head and rest remains pressure head. There is a difference of pressure between guide vanes and the runner is called reaction pressure. The reaction pressure is responsible for the motion of the runner. Since the pressure at inlet of the turbine is much more than the pressure at the outlet, the water in the turbine must flow in a closed conduit and the runner should always be full at water. After passing through the runner the water is discharged to tail race through a draft tube of suitable shape and size.

Work done for Francis Turbine

The velocity triangles can be drawn at both the edges to calculate the workoutput.

$$= \rho a v_1 [v_{w1} \cdot u_1 \pm v_{w2} \cdot u_2]$$

$$= \rho Q [v_{w1} \cdot u_2 \pm v_{w2} \cdot u_2]$$

(2) Work done per second per unit weight of water striking.

$$= \frac{\rho a v_1 (v_{w1} + v_{w2}) \cdot u}{(\rho a v_1) \cdot g}$$

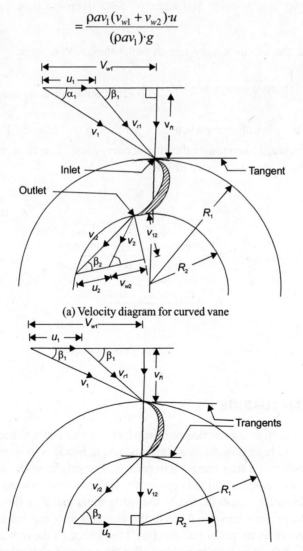

(a) Velocity diagram for curved vane

(b) Velocity diagram for radial curved vane of Francis turbine ($V_{w2} = 0$)

Fig. 5.8

$$= \frac{1}{g}(v_{w1} + v_{w2}) \cdot u$$

v_{w1}, v_{w2} = Velocity of whirl at inlet and outlet
u_1, u_2 = Tangential velocity of wheel at inlet and outlet
D_1, D_2 = Inner and outer diameter of the runner
N = Speed in r.p.m.
Q = Discharge
v_1 = Absolute velocity of water at inlet.

When $\alpha_2 < 90$ then +ve sign and when $\alpha_2 > 90°$ then –ve sign.

If $\alpha_2 = 90°$ then $v_{w2} = 0$.

To minimize the kinetic energy loss at outlet the absolute velocity v_2 should be minimum $\left(\because \text{kinetic energy head} = \dfrac{v_2^2}{2g} \right)$. From outlet velocity triangle it can be observed that v_2 is minimum when $\alpha_2 = 90°$, $v_{w2} = 0$. Thus, the Francis turbine has radial discharge to minimize the kinetic energy loss. Therefore, work done per second per unit weight of water striking

$$= \frac{1}{g} v_{w1} \cdot u_1$$

$$\text{Hydraulic efficiency} = \frac{\text{Power produced per unit mass}}{\text{Water power per unit mass}}$$

$$\eta_h = \frac{(v_{w1} \cdot u_1 \pm v_{w2} \cdot u_2)}{gH}$$

$$\eta_h = \frac{v_{w1} \cdot u_1}{gH} \quad (\because v_{w2} = 0 \text{ for Francis turbine}).$$

5.6 KAPLAN TURBINE

The propeller turbine is a reaction turbine which is particularly suited for low head (upto 30 m) and high flow rate installations, i.e., at barrages in rivers. The unit is like the propeller of a ship operating in reverse. The ship propeller rotates, thrusts the water away behind it and thus causes the ship to move forward. Water enters the turbine laterally, gets deflected by the guide vanes and then flows through the propeller. When the vanes are fixed to the hub and they are not adjustable, the turbine is known as propeller turbine. But if the vanes on the hub are adjustable, the turbine is known as Kaplan turbine after the name of V. Kaplan, an Austrian engineer.

Construction

Main components are:
- (1) Scroll casing
- (2) Guide vane mechanism
- (3) Hub with vanes or runner of turbine
- (4) Draft tube.

(1) Scroll Casing: The water enters the spiral casing in which area of cross-section decreases continuously. The reduction in area is proportional to the decreasing volume of water to be handled. It maintains a constant velocity of water along its path. It is made of cast iron on rolled steel.

(2) Guide Vane Mechanism: The Kaplan turbine has double regulation which comprises the movement of guide vanes and rotation of runner blades. The mechanism employs two servomotors; one controls the guide vanes and second operates on the runner vanes. The governing is done by the governors (servomotors) from the inside of the hollow shaft of the turbine runner and the movement of piston is employed to twist the blades through suitable linkages. The double regulation ensures a balanced and most satisfactory relationship between the relative positions of the guide and working vanes. Both the servomotors are synchronised; they are activated simultaneously and a high efficiency is maintained at all loads.

(3) Hub with Vanes: The runner is in the form of boss. It is an extension of bottom end of the shaft into a bigger diameter. On the periphery of the boss are mounted equidistant 3 to 6 vanes made of stainless steel. Thus it will have less constant surface as compared to Francis (16 to 24 blades) with water and as such a low value of frictional resistance.

(4) Draft Tube: It performs the same function as in case of Francis turbine.

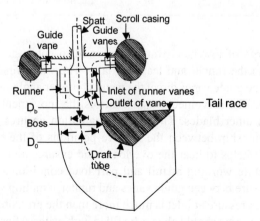

Fig. 5.9: Main components of Kaplan turbine

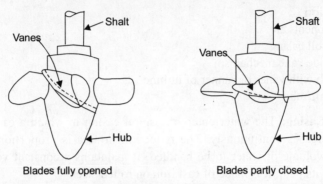

Fig. 5.10: Kaplan turbine runner

Difference between impulse and reaction turbine

Francis turbine	Kaplan turbine
(1) The flow of water is radially or mixed flow type.	(1) It is purely axial flow turbine.
(2) The shaft may be located horizontally or vertically.	(2) The shaft of runner is always vertical.
(3) The runner vanes are not adjustable.	(3) The runner vanes are adjustable
(4) The number of vanes is large (16-24 blades).	(4) The number of vanes is small (3-8 blades).
(5) It works with medium discharge at medium head.	(5) It works with high discharge with low heads.
(6) Specific speed ranges from 50 to 250.	(6) Specific speed ranges from 250 to 850
(7) The correct adjustment of guide vanes and moving blades is obtained at full load conditions.	(7) The correct adjustment is possible at all load condition by turning the moving blades.
(8) It requires one servomotor for adjustment of guide vanes.	(8) It requires servomotors for adjustment of vanes and turning the runner blades.
(9) The servomotor is placed outside the rotor shaft.	(9) The servomotors are placed inside the hollow shaft of turbine runner.

Working

The high discharge of low head water is allowed to enter the scroll casing. It enters axially into the runner and leaves axially. The water first passes through guide vanes and the guide vanes direct the water at proper angle to avoid the shock waves formation. The servomechanism controls the movement of guide vanes and rotation of runner blades. The double regulation ensures a balanced and satisfactory relationship between the relative positions of the guide vanes and working vanes. It helps to bear the overloads and ensures high efficiency at all gate openings while working at full and part load conditions. Because of the difference in pressure between guide vanes and runner, a motion will be present in the runner. As the pressure at inlet is much more than the pressure at outlet a draft tube is must. The turbine should always be filled with water. After passing through the runner, the water is discharged to tail race through draft tube.

5.7 SPECIFIC SPEED

Specific speed is the speed of a geometrically similar turbine (i.e., a turbine identical in shape, dimensions, blade angles and gate openings, etc.) which would develop unit power when working under a unit head.
The derivation of which can be done in the following manner.

Overall efficiency (η_0) of the turbine $= \dfrac{\text{shaft power}}{\text{water power}}$

$$\eta_0 = \dfrac{\text{Power available at the shaft of turbine}}{\text{Power supplied at the inlet of turbine}}$$

$$\eta_0 = \dfrac{P}{\left(\dfrac{W.Q.H.}{1000}\right)} \tag{1}$$

P = Shaft power
Q = Discharge through turbine
W = Weight density of water = $\rho.g$.
H = Head under which the turbine is working
s = Density of power
g = Acceleration due to gravity.
From equation (1), we observe that

$$P = \eta_0 WQ.H.$$

$$P \;\alpha\; QH \qquad (\because \eta_0 \text{ and } w \text{ are constant})$$

Now let
D = Diameter of actual turbine
N = Speed of actual turbine
u = Tangential velocity.
V = Absolute velocity of water
N_s = Specific speed of the turbine

The absolute velocity, tangential velocity and the head on the turbine are related as

$$u \;\alpha\; v \tag{3}$$

$$v \;\alpha\; \sqrt{H} \tag{4}$$

From equations (3) and (4)

$$u \;\alpha\; \sqrt{H} \tag{5}$$

$$u = \dfrac{\pi DN}{60} \tag{6}$$

134   *Mechanical Engineering: Fundamentals*

From equations (5) and (6)

$$\sqrt{H} \alpha\ DN$$

$$D \alpha\ \frac{\sqrt{H}}{N} \quad (7)$$

Since the discharge through the turbine is given by

$$Q = \text{Area} \times \text{velocity} \quad (8)$$

$$\text{Area} \ \alpha\ (B \times D)$$

$$B \ \alpha\ D$$

$$\text{Area} \ \alpha\ D^2 \quad (9)$$

$$\text{Velocity} \ \alpha \sqrt{H} \quad (10)$$

From equations (8), (9) and (10)

$$Q \ \alpha\ D^2 \times \sqrt{H}$$

Using equation (7)

$$Q \ \alpha \left(\frac{\sqrt{H}}{N}\right)^2 \cdot \left(\sqrt{H}\right)$$

$$Q \ \alpha\ \frac{H^{3/2}}{N^2}$$

From equation (2) and (11) we get

$$P \ \alpha\ \frac{H^{3/2}}{N^2} \cdot H$$

$$P \ \alpha\ \frac{H^{5/2}}{N^2} \quad (12)$$

$$P = K \cdot \frac{H^{5/2}}{N^2} \quad \text{where } K \text{ is constant of proportionality.}$$

By definition when $P = 1, H = 1, N = N_s$.

From equation (12)

$$1 = \frac{K \cdot 1^{5/2}}{N_s^2} \qquad (13)$$

$$N_s^2 = K$$

From equation (12) and (13)

$$P = N_s^2 \cdot \frac{H^{5/2}}{N^2}$$

$$N_s^2 = \frac{N^2 \cdot P}{H^{5/2}}$$

$$N_s = \frac{N\sqrt{P}}{H^{5/4}}$$

where P is in kW and H is metres.

Significance of Specific Speed

Specific speed plays an important role in the selection of the turbine. A turbine with high specific speed needs low installation cost but at very high specific speed the cost will increase on account of high mechanical strength required. Also, the increase in specific speed of the turbine is accompanied by decrease in overall efficiency of the turbine and greater depth of excavation of draft tube. The turbine will have lower specific speed at high heads. The type of turbine for different specific speed is given.

Specific speed S.I. Units	Type of turbine
8.5 to 30	Pelton wheel with single jet.
30 to 51	Pelton wheel with two or more jets.
51 to 255	Francis turbine (head < 370 m).
255 to 860	Kaplan or propeller turbine.

5.8 UNIT QUANTITIES

A turbine operates most efficiently at its design point, i.e., at a particular combination of head, discharge, speed and power output. In practice these variables seldom remain constant. To fulfil this purpose the results are expressed in terms of quantities which may be obtained when the head on the turbine is reduced to unity. Each parameter is determined for a unit head on the turbine and expressed as a function of head assuming that the efficiency of the turbine remains unchanged.

This is being studied and this helps in:
(1) Predicting the behaviour of a turbine working under different conditions.
(2) Make comparison between the performance of turbines of the same type but of different sizes.
(3) Compare the performance of turbines of different types.
(4) Facilitate, comparison, corelation and use of experimental data.

The following are the three important unit quantities which must be studied:
(1) Unit speed
(2) Unit power
(3) Unit discharge.

(1) Unit Speed: It is defined as the speed of a turbine working under unit head (i.e. under head of 1 m). It is denoted by N_u. The expression for unit speed (N_u) is obtained as.
N = speed of turbine under head H.
H = Head under which a turbine is working.
u = Tangential velocity.
The tangential velocity, absolute velocity of water and head on the turbine are related as

$$u \propto v \qquad \text{where } v \propto \sqrt{H}$$

$$\propto \sqrt{H} \qquad \qquad \text{(i)}$$

Also tangential velocity (u) is given by

$$u = \frac{\pi DN}{60}$$

where D is the diameter of turbine.

For a given turbine, the diameter (D) is constant

$$u \propto N$$

or $\qquad\qquad N \propto u$

or $\qquad\qquad N \propto \sqrt{H} \qquad\qquad (\because \text{from (1)})$

(2) Unit Power: It can be defined as the power developed by a turbine, working under a unit head.
Let H = Head under which the turbine is working.
P = Power developed by the turbine under a head H.
P_u = Unit power
Q = Discharge through turbine under head H.

By definition

$$\eta_0 = \frac{\text{Power at shaft of turbine}}{\text{Water power}} = \frac{p}{\left(\dfrac{\rho g\, QH}{100}\right)}$$

$$P = \frac{\eta_0 \rho g\, Q\, H}{100}$$

$$p \propto Q.H$$

$$Q \propto \sqrt{H} \qquad (1)$$

$$p \propto H \cdot \sqrt{H} \qquad (2)$$

$$p \propto H^{3/2}$$

$$p = K_2 \cdot H^{3/2} \qquad (3)$$

H = 1 m.

$$p = p_u \qquad (4)$$

$$p_u = K_2\, 1^{3/2} = K_2 = p_u.$$

From equation 4

$$p = p_u\, H^{3/2}$$

$$p_u = \frac{p}{H^{3/2}}$$

(3) Unit Discharge: It is defined as the discharge passing through a turbine which is working under a unit head (i.e., 1 m). It is denoted by the symbol Q_u.

$$Q = \text{Area} \times \text{velocity}$$

But, the area of flow of a turbine remains unchanged.

$$Q \propto \text{velocity}$$

$$\text{Velocity} \propto \sqrt{H}$$

$$Q \propto \sqrt{H} \qquad (1)$$

$$Q = K_3 \cdot \sqrt{H} \qquad (2)$$

At $H = 1$ m

$$Q = Q_u$$

$$Q_u = K_3 \cdot \sqrt{1} \Rightarrow K_3 = Q_u$$

$$Q = Q_u \times \sqrt{H}$$

$$Q_u = \frac{Q}{\sqrt{H}}$$

Thus, the behaviour of a turbine can be estimated by using unit quantities as

$$N_u = \frac{N_1}{\sqrt{H_1}} = \frac{N_2}{\sqrt{H_2}}$$

$$P_u = \frac{P_1}{H_1^{3/2}} = \frac{P_2}{H_2^{3/2}}$$

$$Q_u = \frac{Q_1}{H_1^{1/2}} = \frac{Q_2}{H_2^{3/2}}$$

5.9 TURBINE SELECTION

Three major criterion for selection of water turbines are as follows:
(1) Turbines operating on part load.
(2) Available head
(3) Specific speed

(1) Turbines Operating on Part Load: Efficiency of turbine decreases at part load condition. For the turbine to have maximum efficiency turbines with variable vane are to be used for variable load condition. Deriaz turbines are found to be suitable for part load and overload condition while for low heads Kaplan turbines may be used.

(2) Available Head: This is the most important criteria for the selection of turbine. The table shows the type of turbine selection for the available head.

Available head	Turbine to be selected
2–5 metre	Bulb turbine.
5–15 metre	Propeller turbine
15–30 metre	Kaplan turbine
30–60 metre	Francies/propeller/Kaplan
60–150 metre	Francis turbine
150–350 metre	Either Pelton wheel or Francis turbine
$H > 350$ metre	Pelton wheel.

(3) Specific Speed: Synchronous speed of the generator decides the specific speed of a turbine according to which the turbine is selected. Also the performance of a turbine can be predicted by knowing the specific speed of the turbine.

Specific speed	Type of turbine
S.I. Units	Selected
8–30	Pelton wheel with single jet
30–50	Pelton wheel with two or more jets
51–330	Francis turbine
250–1000	Kaplan or propeller turbine

5.10 PUMPS

The hydraulic machines which convert the mechanical energy into hydraulic energy are called pumps. This is being done to raise the fluid to the higher levels. The hydraulic energy is in the form of pressure energy.

According to the principle of operation and design, pumps are classified as follows.
(1) Centrifugal pumps
(2) Positive displacement pumps

5.11 CENTRIFUGAL PUMPS

The centrifugal pump works on the principle of forced vortex flow which means that when a certain mass of liquid is rotated by an external torque, the rise in pressure head of the rotating liquid takes place. This rise in pressure head at any point of the rotating liquid is proportional to the square of tangential velocity of the liquid at that point. (i.e., rise in pressure head $= \frac{v^2}{2g}$ or $\frac{w^2 r^2}{2g}$). Thus, at the outlet of the impeller where radius is more, the rise in pressure head will be more and the liquid will be discharged at the outlet with high pressure head. In turn the liquid can be lifted to a high level.

Main parts of a centrifugal pump
(1) Suction pipe with a foot valve and a strainer.
(2) Casing
(3) Impeller
(4) Delivery pipe

(1) Suction pipe with a foot valve and a strainer: Figure shows suction pipe. At the lower end of the suction pipe a foot valve (non-return valve on one way

140 Mechanical Engineering: Fundamentals

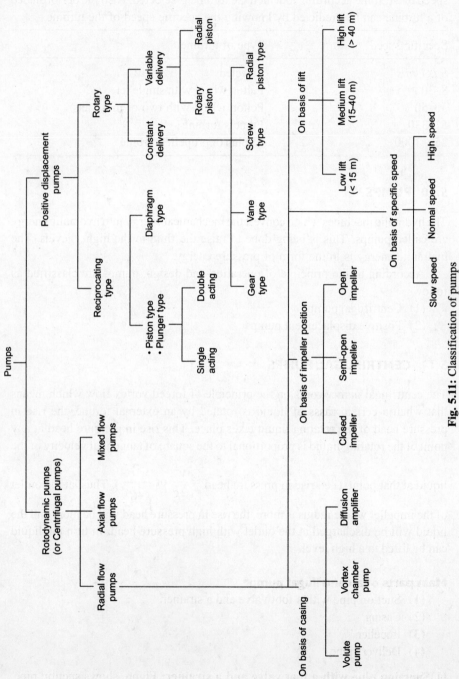

Fig. 5.11: Classification of pumps

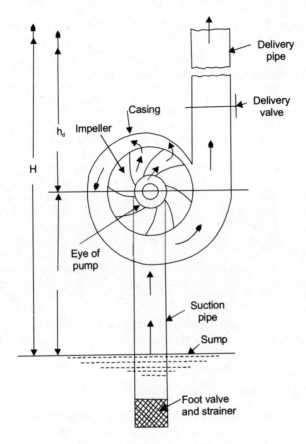

Fig. 5.12: Main parts of a centrifugal pump

valve) is fitted. The foot valve opens only in the upward direction. A strainer for removing the large size particles is attached below it.

(2) Casing: It is an airtight passage surrounding the impeller and is designed in such a way that the kinetic energy of the water discharged at the outlet of the impeller is convered into pressure energy before the water leaves the casing and enters the delivery pipe. The following three types of the casings are commonly adopted.
    (a) Volute casing
    (b) Vortex casing
    (c) Casing with guide blades

(3) Impeller: The impeller is that rotating part of pump which is mounted on a shaft which is further connected to the shaft of an electric motor. It consists of a series of backward curved vanes.

(4) Delivery pipe: The pipe which delivers the water to the required height from the pump is called delivery pipe.

Comparison of centrifugal pumps and reciprocating pumps

Property	Centrifugal pump	Reciprocating pump
Discharge capacity	Very high	Low
Viscosity of liquid	For lifting highly viscous liquids	Pure water or liquids free of impurities
Speed	Can operate at very high speeds	Limited due to separation and cavitation
Maintenance cost	Low	High (value to be replaced)
Initial cost	Low	More (3 to 5 times that of centrifugal pumps)
Size	Compact, occupies less floor space	Occupies more space (5–7 times that of centrifugal pumps)
Weight and installation cost	Small and easy to install	More and installation is difficult

Two important ways of classification of pumps are as follows.

(1) On the basis of direction of flow of liquid:
  (a) Radial flow pumps
  (b) Axial flow pumps
  (c) Mixed flow pumps

(a) Radial flow pumps: In these pumps, the flow of liquid through the impeller is completely in radial direction. The head is developed by the action of centrifugal force. These pumps are useful when low discharge is required at high head.

(b) Axial flow pumps: Flow of liquid is in the axial direction. No centrifugal action in their operation. The head is developed by the propelling action as the liquid enters and discharges axially. Very large quantities of liquids at low head are delivered by these pumps.

(c) Mixed flow pumps: The head developed partially by action of centrifugal force and partially by axial propulsion. In these pumps, the direction of flow of liquid is changed from pure radial to a combination of axial and radial flows. More application of these pumps are found in irrigation. These are most suitable when medium discharge is needed at medium head.

(2) On the basis of type of casing
On this basis major three categories are there.
  (a) Volute casing
  (b) Vortex casing
  (c) Diffuser pump

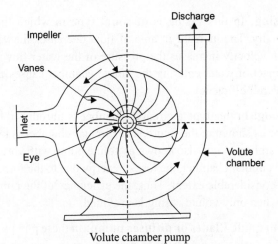

Volute chamber pump

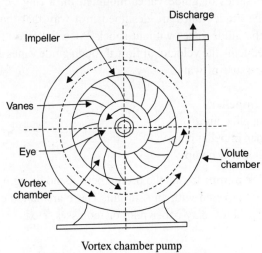

Vortex chamber pump

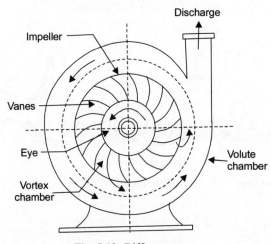

Fig. 5.13: Diffuser pump

(a) Volute casing: In this casing is of spiral type in which the area of flow increases gradually. The increase in area of flow decreases the velocity of flow. The decrease in velocity increases the pressure of the water flowing through the casing. The impact of water entering the volute chamber causes eddy losses and hence lower overall efficiency.

(b) Vortex casing: In this type a circular chamber is introduced between casing and the impeller as shown in the figure. The circular chamber is known as vortex or whirlpool chamber, so it is also known as volute pump with vortex chamber. By introducing the circular chamber, the loss of energy due to the formation of eddies is reduced to a considerable extent. Thus, the efficiency of the pump is more than the efficiency when only volute casing is provided.

(c) Casing with guide Blades or diffuser pump: In these pumps, the impeller is surrounded by a series of guide vanes mounted on a ring which is known as diffuser ring. The guide vanes are designed in a way that the water from the impeller enters the guide vanes without shock. Also the area of the guide vanes increases, thus reducing the velocity of flow through guide vanes and consequently increasing the pressure of water.

(3) Position of Impeller.
    (a) Open impeller pump
    (b) Semi-open impeller pump
    (c) Closed impeller pump

(a) Open impeller pump: In this the design permits no clogging and because of which they are not covered with any shroud. These are suitable for liquids with suspended sand, pebbles, clay paper pulp or molasses, etc.

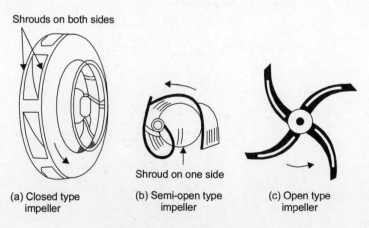

Fig. 5.14: Different types of impellers

(b) Semi-open impeller pump: On one side the shroud is provided in these pumps. These are useful for viscous liquids and mixture of solid and liquid particles. They are also called non-clog pump.

The number of vanes on impeller is less and their height is more in order to give satisfactory service. The material of impeller blades is selected on the basis of chemical nature of the liquid to be pumped.

(c) Closed impeller pump: On both sides of the impeller shrouds are present in these pumps. The criteria for selection of the material of the impeller is the liquid which is to be handled. These pumps are useful for non-viscous liquids. The liquid must be free from external impurities to avoid clogging. For water the impeller may be made up of cast iron but for chemicals it should be stainless steel or coated with some non-reactive material.

(4) On the basis of head that can be provided by the pump.
  (a) Low head pump (upto 15 m)
  (b) Medium head pump (upto 16 m to 41 m)
  (c) High head pump (above 41 m)

5.12 RECIPROCATING PUMP

Classification of Reciprocating Pump
  (a) Single acting pump.
  (b) Double acting pump

(a) Single acting pump: (i) Liquid is in contact with only one side of piston. (ii) It has one suction and one delivery pipe connected to cylinder. (iii) The suction is caused by vacuum while delivery is caused by high pressure built up in cylinder. (iv) The increased pressure causes the suction valve to close and the discharge valve to open and forces the liquid out of the cylinder in discharge pipe.

(b) Double acting pumps (Important features): (i) Liquid is present on both the sides of piston. (ii) Two suction and two delivery valves are present on one cylinder. (iii) During each stroke suction takes place on one side of the piston while delivery on other side. (iv) Uniform discharge because of the continuity of suction and delivery strokes.

5.13 ROTARY PUMPS

Classification of Rotary Pumps
These pumps can further be classified into two categories:
  (a) Constant delivery pumps
  (b) Variable delivery pumps

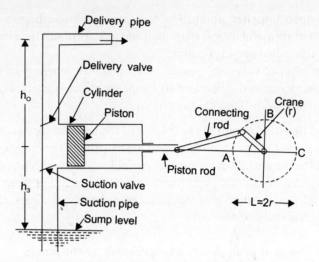

Fig. 5.15: Main parts of a reciprocating pump.

1. A cylinder with a piston, piston rod, connecting rod and a crank,
2. Suction pipe,
3. Delivery pipe,
4. Suction valve, and
5. Delivery valves.

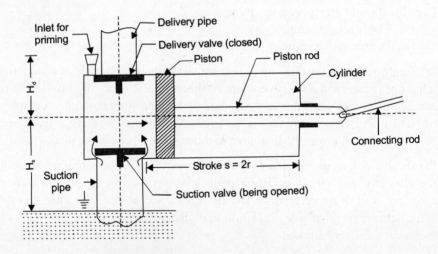

Fig. 5.16: Reciprocating single acting piston pump

(a) Constant delivery pumps: They have a continuous discharge of liquid at uniform rate of flow. In this category the different types of pump are:
  (i) Gear pumps
  (ii) Screw pump
  (iii) Vane pump
  (iv) Radial piston pumps

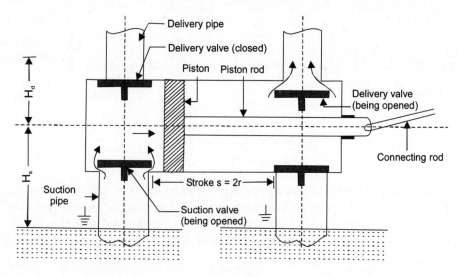

Fig. 5.17

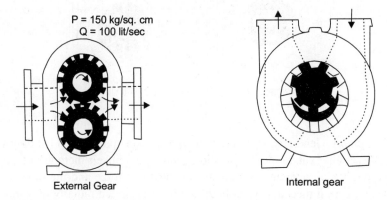

Rotary pump – External gear pump.　　　Rotary pump – Internal gear pump.

Fig. 5.18

(i) Gear pumps: These pumps are used for machine tool drives, for combined hydraulic control system of water turbines and governing of steam turbines, etc. The gear pump consists of intermeshing spur wheels. One gear is keyed to the driving shaft of a motor and the other revolves idly. The liquid enclosed in the spaces between the teeth and the casing is carried around between the gears from the suction part to the discharge part. The liquid cannot slip back into the inlet side due to the meshing of gears.

(ii) Screw Pump: In these pumps there are usually one, two or three numbers of threaded screws which rotate in a stationary housing. They have a true helical screw rotor rotating eccentrically in a double internal helical threaded stator or housing. As the rotor revolves within the stator because of its peculiar shape, the

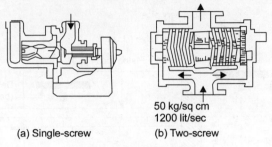

(a) Single-screw   (b) Two-screw

Fig. 5.19: Rotary pumps

same rotor axis travels also in a circular path and because of the helical shape of the stator this movement enables the rotor surface to come in contact with the entire surface of the stator in a forward direction making the fluid trapped between the voids of the stator and rotor to move forward giving the pump positive displacement and uniform discharge.

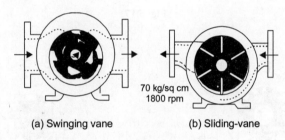

(a) Swinging vane   (b) Sliding-vane

Fig. 5.20

(iii) Vane pump: Generally, there are two types of vane pump (i) (1) sliding vane type (2) Swinging vane type. In the sliding vane pump itself there are two types namely, one which operates as a result of centrifugal force and the other which operates on spring mechanism. In both, the rotor is slotted radially or slantingly and the vanes either slide freely or are pressed against the housing by springs. The rotor is fitted on a shaft which is eccentric to the housing. As the rotor rotates in the former, centrifugal forces cause the vanes to be held in contact against the housing. Inlet ports are so arranged that they are located where the vanes move outwards.

In swinging vane types, a series of hinged vanes swing out with the rotation of the rotor and work similarly as in the sliding. Vane type vane pumps can be used for large discharges with speeds upto 1800 rpm with pressure upto 70 kg/sq.cm and with efficiency up to 80%.

(iv) Radial piston pumps: In the radial piton type, pistons are fixed radially and the cylinder block and piston assembly revolve by means of a drive shaft about a central valve spindle or pintle. Oil is supplied to the pistons through bored holes in the spindle on one half of the circumference. Similarly, discharge takes place in the other half of the circumference through similar ports in the spindle. The pistons

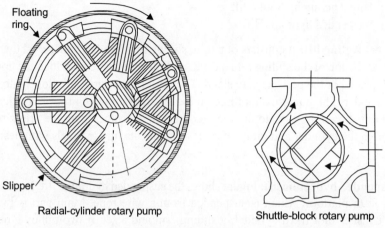

Radial-cylinder rotary pump   Shuttle-block rotary pump

Fig. 5.21

are mounted in the cylinder bores in the cylinder rotor which contains a reaction ring at the periphery against which the pistons come in contact at the outer circumference.

5.14 HYDRAULIC LIFT

The hydraulic lift is a device used for carrying passenger or goods from one floor to another in multi-storeyed building. The hydraulic lifts are of two types.

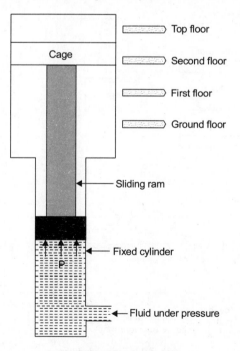

Fig. 5.22: Direct acting lift

150   *Mechanical Engineering: Fundamentals*

(1) Direct acting hydraulic lift
(2) Suspended hydraulic lift

(1) Direct Acting lift: It consists of a ram, sliding in a fixed cylinder as shown in figure. At the top of the sliding ram a cage or platform is there on which the goods may be placed or the persons may stand. The liquid under pressure flows into the fixed cylinder. The liquid exerts force on the sliding ram which moves vertically up and thus raises the cage to the required height. To come down the pressure is released in the cylinder and ram comes down again staying in levels with desired floors.

(2) Suspended hydraulic lift: Figure shows the suspended hydraulic lift. It consists of a cage or platform which is suspended from a wire rope. Main part is jigger which consists of a fixed cylinder, a sliding ram and a set of two pulley blocks. One of the pulley blocks is movable and the other is a fixed one. The movable pulley is connected to the sliding ram. One end of the wire rope is fixed and other end is taken around all the pulleys of the movable and fixed blocks and finally over the guide pulley. At the other end cage is suspended. The raising or lowering of platform is done by the jigger.

The water under high pressure is forced into the cylinder which pushes the ram outwards and the set of pulleys attached to the ram will move with it, while the pulley attached to the cylinder remains fixed. The lift cage move upwards

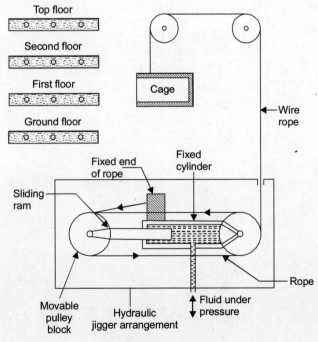

Fig. 5.23: Suspended lift

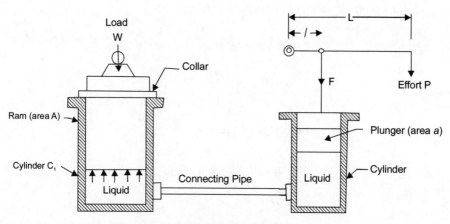

Fig. 5.24: Hydraulic jack

between guide rails. To bring the cage down the fluid is taken out and the tension in the rope moves the ram into the cylinder.

This increases the free length of rope.

5.15 HYDRAULIC JACK

The function of hydraulic jack is to lift heavier load by applying a much smaller effort. It is a close circuit unit in which oil pressure can support a dead weight. Figure above shows the hydraulic jack.

The operation of a hydraulic jack is based on Pascal's law which states that intensity of pressure is transmitted equally in all directions through a mass of fluid at rest. A force F applied to the plunger produces an intensity of pressure p_1 which is transmitted equally in all directions through the liquid. If the plunger and ram are at the same level and if their weights are neglected then pressure intensity p_2 acting on the ram must equal p_1

Now
$$p_1 = \frac{F}{a} \text{ and } p_2 = \frac{W}{A}$$

$$\frac{F}{a} = \frac{W}{A}$$

$$W = F\left(\frac{A}{a}\right)$$

Formulae indicate that a heavy load W gets lifted up by ram when a small force F is applied to the ram.

Mechanical advantage of hydraulic press is increased by applying forces on the plunger by means of a lever.

Taking moments about the fulcrum of lever.

$$F \times l = P \times L; \quad F = \frac{PL}{l}$$

$$\frac{W}{P} = \frac{A}{a} \times \frac{L}{l}$$

The ratio L/l is known as leverage of press

$\frac{A}{a} \times \frac{L}{l}$ is called velocity ratio

EXERCISE

1. What is a hydraulic turbine? How can we classify them.
2. Describe the main components of a pelton wheel with the help of a neat sketch.
3. Explain pelton wheel turbine with the help of a neat sketch. (MRIU Dec 2009)
4. Differentiate between impulse turbine and reaction turbine.
5. Define the following:
    (a) Specific speed of turbine.
    (b) Priming of a pump.
    (c) Unit power.
    (d) Selection of turbine.
    (e) Volute casing of pump.
    (f) Vortex casing of pump.
    (g) Rotary pump.
    (h) Vane pump.
    (i) Hydraulic lift.
    (j) Hydraulic lift.
6. Explain the working principle of reciprocating pumps with the help of a line sketch naming all the main parts.
7. What are the advantages of centrifugal pumps over reciprocating pumps.
8. What is the difference between radial, axial and mixed flow pumps.
9. Explain the construction and working of the following with the help of neat sketch.
    (a) Kaplan turbine
    (b) Francis turbine
    (c) Suspended hydraulic lift
    (d) Hydraulic jack
    (e) Centrifugal pump.

OBJECTIVE TYPE QUESTIONS

1. Reciprocating pumps are no more to be seen in industrial applications (in comparison to centrifugal pumps) because of
   (a) high initial and maintenance cost
   (b) lower discharge
   (c) lower speed of operation
   (d) all of the above
2. In a centrifugal pump casing, the flow of water leaving is
   (a) radial
   (b) centrifugal
   (c) rectilinear
   (d) Vortex
3. Centrifugal pump is started with its delivery value
   (a) kept fully closed
   (b) kept fully open
   (c) irrespective of any position
   (d) kept 50% open
4. One horse power is equal to
   (a) 102 watts
   (b) 75 watts
   (c) 550 watts
   (d) 735 watts
5. Low specific speed of turbine implies that it is a
   (a) propeller turbine
   (b) Francis turbine
   (c) impulse turbine
   (d) any one of the above
6. Medium specific speed of turbine implies it is
   (a) propeller turbine
   (b) Francis turbine
   (c) impulse tube
   (d) any one of the above
7. High specific speed of turbine implies it is
   (a) propeller turbine
   (b) Francis turbine
   (c) impulse turbine
   (d) any one of the above
8. The specific speed of turbine is defined as the speed of a unit
   (a) of such a size that it delivers unit discharge at unit head
   (b) of such a size that it delivers unit discharge at unit power
   (c) of such a size that it requires unit power per unit head.
   (d) of such a size that it produces unit horse power with unit head.

9. A double acting reciprocating pump compared to single acting pump will have nearly
    (a) double efficiency
    (b) double head
    (c) double flow
    (d) double weight
10. For pumping viscous oil which pump will be used
    (a) centrifugal pump
    (b) reciprocating pump
    (c) turbine pump
    (d) screw pump
11. Casing of a centrifugal pump is designed so as to minimise
    (a) frictional loss
    (b) cavitation
    (c) static head
    (d) loss of kinetic energy
12. An impulse turbine
    (a) operates submerged
    (b) requires draft tube
    (c) is most suited for low head application
    (d) operates by initial complete conversion to kinetic energy
13. A pelton wheels is
    (a) axial flow impulse turbine
    (b) radial flow impulse turbine
    (c) inward flow impulse turbine
    (d) outward flow impulse turbine
14. Pelton wheels are used for minimum of following heads
    (a) 20 m
    (b) 100 m
    (c) 125 m
    (d) 180 m or above
15. The ratio of width of bucket for a pelton wheel to the diameter of jet is of the order of
    (a) 2
    (b) 3
    (c) 4
    (d) 5
16. The ratio of depth of bucket for a pelton wheel to the diameter of jet is of the order of
    (a) 1
    (b) 1.2
    (c) 1.5
    (d) 1.8

17. If D is diameter of pelton wheel and d is the diameter of the jet then number of buckets on the periphery of a pelton wheel is equal to
    (a) $\dfrac{D}{2d}$
    (b) $\dfrac{D}{2d}+10$
    (c) $\dfrac{D}{2d}+15$
    (d) $\dfrac{D}{2d}+20$
18. Impulse turbine is used for
    (a) low head
    (b) high head
    (c) medium head
    (d) high flow
19. In reaction turbine
    (a) The vanes are partly filled
    (b) Total energy of fluid is converted to kinetic energy in the runner.
    (c) It is exposed to atmosphere.
    (d) It is not exposed to atmosphere
20. Francis turbine is best suited for
    (a) medium head application from 24 to 180 m
    (b) low head installation upto 30 m
    (c) high head installation above 180 m
    (d) all types of head
21. The ratio of power produced by the turbine to the energy actually supplied by the turbine is called.
    (a) mechanical efficiency
    (b) hydraulic efficiency
    (c) overall efficiency
    (d) turbine efficiency
22. The ratio of actual work available at the turbine to energy imparted to the wheel is called
    (a) mechanical efficiency
    (b) hydraulic efficiency
    (c) overall efficiency
    (d) turbine efficiency
23. The ratio of the work done on the wheel to the energy (or head of water) actually supplied to the turbine is called.
    (a) mechanical efficiency
    (b) hydraulic efficiency

(c) overall efficiency
(d) turbine efficiency
24. Francis, Kaplan and propeller turbines fall under the category of
    (a) Impulse turbine
    (b) Reaction turbines
    (c) Axial flow turbines
    (d) mixed flow turbines
25. Reaction turbines are used for
    (a) low head
    (b) high head
    (c) high head and low discharge
    (d) low head and high discharge

ANSWERS

1. a	2. d	3. a	4. d	5. c
6. b	7. a	8. d	9. c	10. d
11. a	12. d	13. a	14. d	15. d
16. b	17. c	18. b	19. d	20. a
21. c	22. a	23. b	24. b	25. d

UNIT 6

Power Transmission Methods and Devices

6.1 INTRODUCTION

Power transmission is a process to transmit motion from one shaft to another shaft by using some connection between them, Like belt rope chain and gears.

To connect the shafts there may be two types of connector—one is flexible and the other is rigid.

Flexible Connector
In this, there is relative velocity between the shaft and connectors due to slip and strain produced in the connectors. These are generally used where distance between the shafts is large. Example: Belt, rope and chains.

Rigid Connectors
In this, there is no relative velocity between the connector and shaft. These are used when distance between the shafts is very small.

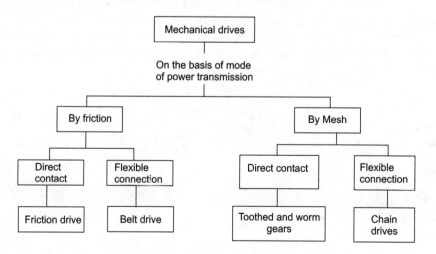

Fig. 6.1.

Example, gears in belt and rope drive the velocity of two shafts can be varied by variation of diameter of pulley on which belt or ropes are mounted. In chain and gear drive, the velocity of two shaft is varied by variation in no. of teeth on sprocket and gear respectively.

6.2 BELT DRIVES

Belts are used to transmit power between two parallel shafts. The shafts must be separated by a certain minimum distance.

The figure shows an open belt drive where pulley A is driver an pulley B is follower or driven pulley (because it is driven by pulley A). When the driver rotates, it carries the belt because of friction that exists between the pulley and the belt.

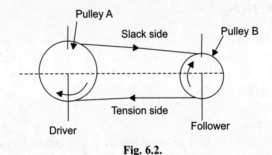

Fig. 6.2.

Types of Belt
Different types of belts that are commercially used are —

(1) Flat belt: They are having a narrow rectangular cross-section and are efficient at high speed. The load carrying capacity depends upon its width. In this pulley is slightly crowned which helps to keep the belt running centrally on the pulley rim as shown in figure. Rubber belts are used in damp conditions.

Leather belts are generally used for flat belt drive because they have best pulling capacity and can be used both in dry and wet places at ordinary temperature but are costly.

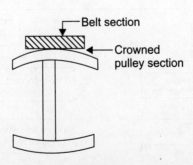

Fig. 6.3.

(2) V Belts: In this grooves are made on the rim of the pulley for wedging action. The belt does not touch the bottom of the groove as shown in figure, V belt need little adjustment and transmit more power without slip as compared to at belts. In multiple V belt system using more than one belt on the pulleys can be used to increase the power transmission capacity.

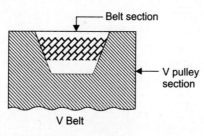

V Belt

Fig. 6.4

(3) Round belt: These are used when low power is to be transmitted such as in instruments, household appliances, tabletop machine tools and machinery of the clothing industry.

Fig. 6.5: Round belt

Types of Flat Belt Drives

(1) Open belt drives: It is used for same direction of rotation as the direction of rotation of driving shafts. In this tension in the lower side is more than the tension on in upper side.

(2) Cross-belt drive: It is used for opposite direction of rotation of driver and driven shaft. At the point where the belt crosses it rubs against itself and wears.

Other than this, we can have
  (a) Quarter cross-belt drive
  (b) Quarter cross-belt drive with guide pulley.
  (c) Quarter cross-belt drive with idler pulley.
  (d) Belt drive with many (more than two pulles) pulleys.
  (e) Cone or stepped pulley drive.

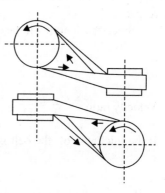

Quarter-twist belt drive

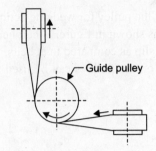

Quarter-twist belt drive with guide pulley

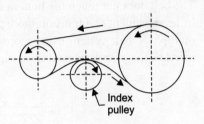

Belt drive with an idler pulley

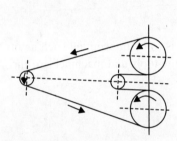

Belt drive with many pulleys

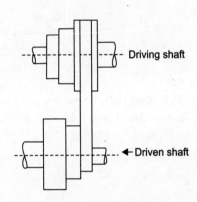

Stepped or cone pulley drive

Fig. 6.6

Velocity Ratio (VR)

Velocity ratio is the ratio of speed of the driven pulley to that of the driving pulley. Let D_1 and D_2 be the diameters of the driver pulley and driven pulley and N_1 and N_2 be the speed in driven pulley and driven pulley respectively. In one minute, the length of belt that passes over. The driver and follower will be $\pi D_1 N_1$ and $\pi D_2 N_2$ respectively.

Since this length will be equal.

$$\pi D_1 N_1 = \pi D_2 N_2$$

Velocity ratio = $\dfrac{N_2}{N_1} = \dfrac{D_1}{D_2}$

If the thickness of the belt is taken into account, then

$$\frac{N_2}{N_1} = \frac{D_1 + t}{D_2 + t}$$

Power Transmission Methods and Devices

Effect of Slip

The effect of slip is to decrease the speed of belt on driving shaft and then driven shaft.

The linear speed of rim of pulley and that of the belt on it are different. The difference between these two is known as slip of the belt drive. Slip will take place because of the motion of driver without carrying belt over it or motion of belt without taking follower with it.

Let w_1 = angular velocity of the driving pulley.
w_2 = angular velocity of the driven pulley.
S_1 = percentage slip between the driving pulley and the belt.
S_2 = percentage slip between the driven pulley and the belt.
S = total percentage slip.

Peripheral speed of driving pulley = $\dfrac{W_1 D_1}{2}$

Speed of belt on driving pulley = $\dfrac{W_1 D_1}{2}\left(\dfrac{100 - S_1}{100}\right)$

This is also the speed of the belt on the driven pulley.

Peripheral speed of driven pulley = $\dfrac{W_1 D_1}{2}\left(\dfrac{100 - S_1}{100}\right)\left(\dfrac{100 - S_2}{100}\right)$

If S is total percentage

Peripheral speed of driven pulley = $\dfrac{W_1 D_1}{2}\left(\dfrac{100 - S}{100}\right)$

or $\left[\dfrac{W_1 D_1}{2}\right]\left[\dfrac{100 - S_1}{100}\right]\left[\dfrac{100 - S_2}{100}\right] = \left[\dfrac{W_1 D_1}{2}\right]\left[\dfrac{100 - S}{100}\right]$

$\dfrac{(100 - S_1)(100 - S_2)}{100 \times 100} = \dfrac{(100 - S)}{100}$ or $S = S_1 + S_2 - 0.01 S_1 S_2$

$$VR = \dfrac{D_1}{D_2}\left(\dfrac{100 - S}{100}\right) = \dfrac{N_2}{N_1}$$

Length of Belt

For open belt
Length of the belt consists of three parts.

(i) Length of belt in contact with driver.
(ii) Length of belt in contact with follower.
(iii) Length of belt when belt which is not in contact with either of the pulleys.

Let L = length of belt for open drive.
  r = radius of smaller pulley.
  R = radius of larger pulley.
  c = centre distance between pulleys
  β = angle subtended by each common tangent (CD or EF)
  AB = line of centres of pulley.

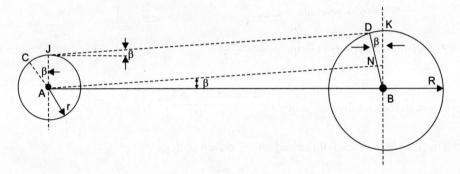

Fig. 6.7

Draw AN parallel to CD so that $\angle BAN = \beta$ and $BN = R - r$

As CD is tangent to two circles

AC and BD both are perpendicular to CD or AN.

Now $AB \perp BK$ and $AN \perp BD$

Similarly, as $BA \perp AJ, NA \perp AC$

$$\angle CAJ = \angle NAB = \beta$$

$$L = 2\left[\text{Arc } GC + CD + \text{Arc } DH\right]$$

$$0 = 2\left[\left(\frac{\pi}{2} - \beta\right)r + AN + \left(\frac{\pi}{2} + \beta\right)R\right]$$

$$= \pi(R + r) + 2\beta(R - r) + 2c \cos \beta$$

for small angle β

$$\beta = \sin \beta = \frac{R - r}{c} \text{ and } \cos \beta = \sqrt{1 - \sin^2 \beta} \simeq \left(1 - \frac{1}{2}\sin^2 \beta\right)$$

$$\simeq 1 - \frac{1}{2}\left(\frac{R-r}{c}\right)^2$$

Putting the values of β and $\cos \beta$

$$L = \pi(R+r) + 2\frac{(R-r)^2}{c} - \frac{(R-r)^2}{c} + 2c$$

$$L = \pi(R+r) + 2c + \frac{(R-r)^2}{c}$$

Cross Belt

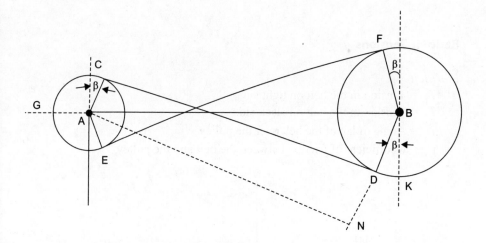

Fig. 6.8

Let A and B be the pulley centres and CD and EF, the common tangents (crossed) to the two pulley circles.

Draw AN parallel to CD meeting BD produced at N so that

$$\angle BAN = \beta = \angle CAJ = \angle DBK$$

$$L = 2(\text{Arc } GC + CD + \text{Arc } DH)$$

$$= 2\left[\left(\frac{\pi}{2} + \beta\right)r + c\cos\beta + \left(\frac{\pi}{2} + \beta\right)R\right]$$

$$= (\pi + 2\beta)(R+r) + 2c\cos\beta$$

164 *Mechanical Engineering: Fundamentals*

$$\beta = \sin \beta = \left(\frac{R+r}{c}\right) \text{ as } \beta \text{ is very small}$$

$$\cos \beta = \left(1 - \frac{1}{2}\beta^2\right) = 1 - \frac{1}{2}\left(\frac{R+r}{c}\right)^2$$

$$L = \left[\pi + 2\left(\frac{R+r}{c}\right)\right](R+r) + 2c\left[1 - \frac{1}{2}\left(\frac{R+r}{c}\right)^2\right]$$

$$L = \pi(R+r) + 2c + \frac{(R+r)^2}{c}$$

Ratio of Tensions

Flat Belt

T_1 = Tension in the belt on tight side
T_2 = Tension in the belt on slack side
θ = Angle of lap of the belt over the pulley.
μ = Coefficient of friction between the belt and the pulley.

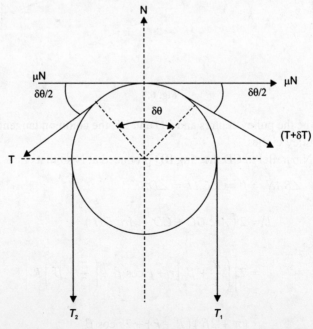

Fig. 6.9

Consider a short length of belt subtending an angle $\delta\theta$ at the centre of the pulley.
Let N = Normal reaction between the element length of a belt and the pulley.
  T = Tension on slack side of the element.
  δT = Increase in tension on tight side than that on slack side.
  $T+\delta T$ = Tension on tight side of the element. Resolving the force in tangential direction.

$$\mu N + T \cos \frac{\delta\theta}{2} - (T + \delta T) \cos \frac{\delta\theta}{2} = 0$$

As $\delta\theta$ is small

$$\cos \frac{\delta\theta}{2} \simeq 1$$

$$\mu N + T - T - \delta T = 0 \text{ or } \delta T = \mu N \qquad \text{(i)}$$

Resolving the forces in the radial direction

$$N - T \sin \frac{\delta\theta}{2} - (T + \delta T) \sin \frac{\delta\theta}{2} = 0$$

As $\delta\theta$ is small $\sin \dfrac{\delta\theta}{2} = \dfrac{\delta\theta}{2}$

$$N - T \frac{\delta\theta}{2} - \frac{T\delta\theta}{2} - \frac{\delta T \delta\theta}{2} = 0$$

Neglecting product of two small quantities

$$N = T\delta\theta \qquad \text{(ii)}$$

From (i) and (ii) $\delta T = \mu T \delta\theta$ or $\dfrac{\delta T}{T} = \mu\delta\theta$

Integrating between proper limits

$$\int_{T_2}^{T_1} \frac{dT}{T} = \int_0^\theta \mu d\theta$$

$$\log_e \left(\frac{T_1}{T_2}\right) = \mu\theta \text{ or } \frac{T_1}{T_2} = e^{\mu\theta}$$

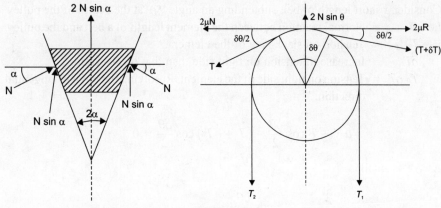

Fig. 6.10

V-Belt and Rope Drive

In case of V-belt or rope, there are two normal reactions so that the radial reaction is equal to $2 N \sin \alpha$.

Thus, total frictional force = $2 \mu N$

Resolving the forces tangentially

$$2\mu N + T \cos \frac{\delta\theta}{2} - (T + \delta T) \cos \frac{\delta\theta}{2} = 0$$

For small angle of $\delta\theta$

$$\cos \frac{\delta\theta}{2} \approx 1$$

$$\delta T = 2\mu N. \qquad \text{(i)}$$

Resolving the forces radially

$$2N \sin \alpha - T \sin \frac{\delta\theta}{2} - (T + \delta T) \sin \frac{\delta\theta}{2} = 0$$

As $\delta\theta$ is small $\sin \frac{\delta\theta}{2} \simeq \frac{\delta\theta}{2}$

$$2N \sin \alpha - T \frac{\delta\theta}{2} - T \frac{\delta\theta}{2} = 0$$

$$N = \frac{T\delta\theta}{2 \sin \alpha} \qquad \text{(ii)}$$

From equations (i) and (ii) μ we get $\delta T = 2\mu \dfrac{T\delta\theta}{2 \sin \alpha}$

$$\frac{\delta T}{T} = \frac{\mu \delta \theta}{\sin \alpha}$$

Integrating between proper limits.

$$\int_{T_2}^{T_1} \frac{dT}{T} = \int_0^\theta \frac{\mu d\theta}{\sin \theta} \text{ or } \log e \frac{T_1}{T_2} = \frac{\mu \theta}{\sin \alpha}$$

$$\frac{T_1}{T_2} = e^{\mu\theta / \sin \alpha}$$

The expression is similar to that for a flat belt drive except μ is replaced by $\mu / \sin \alpha$, i.e. the coefficient of friction is increased by $1 / \sin \theta$.

Rope Drives

Rope drives are classified as:
 (i) Fibre ropes
 (ii) Wire ropes.

Fibre ropes are made of manila or cotton. Wire ropes are made of steel wire. Number of wires make a strand and strands make a rope. When the distance between the two shafts is large then such type of drives are used. The groove angle varies from 40° to 60° but it is generally 45°.

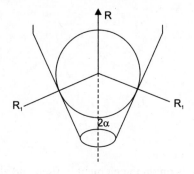

Fig. 6.11

The ratio of driving tensions can be represented by $\dfrac{T_1}{T_2} = e^{\mu\theta \, \text{cosec} \, \alpha}$.

Power Transmitted

Work done by the belt per second = $\begin{bmatrix} \text{Effective tension} \\ \text{or} \\ \text{force} \end{bmatrix} \times \begin{bmatrix} \text{Distance} \\ \overline{\text{time}} \end{bmatrix}$

$$= (T_1 - T_2) \times v. \text{ Nm/sec.}$$

or power transmitted = $\dfrac{(T_1 - T_2) \times v}{1000}$ kW.

T_1 = Tension on tight side in Newton.
T_2 = Tension on slack side in Newton.
θ = Velocity of belt in m/sec.
T_c = Centrifugal tension acting tangentially.

Centrifugal Tension

When velocity of the belt is more than 10 m/sec, the centrifugal force due to self weight of the belt is predominant. Below 10 m/sec the centrifugal tension can be neglected.

The effect of the centrifugal force is to increase the tension at both sides of the belt. The tension caused by centrifugal force is called centrifugal tension.

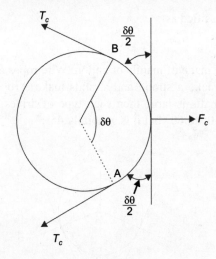

Fig. 6.12

m = mass of belt per unit
T_c = centrifugal tension on the element in tight and slack side
r = radius of pulley
F_c = centrifugal force on the element.
V = velocity of the belt
$\delta\theta$ = angle of lap

Mass of elemental length $AB = m\,(r\,\delta\theta)$

$$F_c = (mr\delta\theta)\dfrac{v^2}{r} = mv^2\delta\theta \tag{1}$$

Element is under equilibrium under the forces.
Resolving the forces vertically,

$$F_c = T_c \sin \frac{\delta\theta}{2} + T_c \sin \frac{\delta\theta}{2}$$

$$F_c = T_c \frac{\delta\theta}{2} + T_c \frac{\delta\theta}{2} = T_c \delta\theta \qquad (2)$$

From (1) and (2)

$$T_c \delta\theta = mv^2 \delta\theta$$

$$T_c = mv^2$$

$$P = T_1\left(1 - \frac{T_2}{T_1}\right) \times v \quad \text{watt}$$

$$= T_1\left(1 - \frac{1}{e^{\mu\theta}}\right) \times v$$

$$= T_1\left(\frac{e^{\mu\theta} - 1}{e^{\mu\theta}}\right) \times v$$

Now maximum tension in the belt

$$T = T_1 + T_c$$

$$T_1 = T_1 - T_c$$

Thus,
$$P = (T - T_c)\left(\frac{e^{\mu\theta} - 1}{e^{\mu\theta}}\right) \times v$$

$$\left(\frac{e^{\mu\theta} - 1}{e^{\mu\theta}}\right) = A$$

$$P = (T - T_c) A \times v$$

$$P = \left(T - mv^2\right) A \times v$$

for maximum power

$$\frac{dP}{d\theta} = AT - Am3v^2 = 0$$

$$T = 3mv^2 = 3T_c$$

$$T_c = \frac{1}{3}T$$

Maximum permissible velocity $V_{max} = \sqrt{\dfrac{T}{3m}}$

Maximum power $= \dfrac{2}{3}T\left(\dfrac{e^{\mu\theta}-1}{e^{\mu\theta}}\right) \times \sqrt{\dfrac{T}{3m}}$ watt

⇒ Total tension on tight side = friction tension + Centrifugal tension

$$T = T_1 + T_c$$

Total tension on slack side $= T_2 + T_c$

Initial Tension

When belt is stationary on the pulley the tension in the belt is known as static tension or initial tension T_0. During motion tension in tight side increases to $T_0 + \delta T$ and in slack side decreases to $T_0 - \delta T$

Hence, $\quad T_1 = T_0 + \delta T$

$$T_2 = T_0 + \delta T$$

$$T_0 = \frac{T_1 + T_2}{2}$$

Mean tension of tight and slack side during motion.

6.3 CHAIN DRIVES

Chain drives are used because they provide constant velocity ratio, no slippage or creep, long life and the ability to drive a number of shafts from a single source of power. The wheel over which chains are run, corresponding to the pulley in a belt drive is known as sprocket. It has projected teeth that fit into the recesses in the chain. They find practical application in cycles, motorcycles, agricultural machinery, rickshaw, etc.

Definition used in Chain Drives
(1) **Pitch:** Distance between roller centres of two adjacent links is known as pitch (P) of the chain.

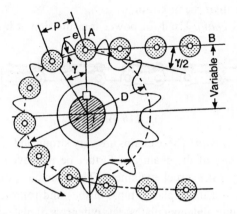

Fig. 6.13

(2) **Width:** It is the space between the inner link plates.

(3) **Pitch circle:** A circle drawn through roller centres of a wrapped chain round a sprocket is called the pitch circle and its diameter as pitch circle diameter.

(4) **Angle of articulation:** The angle $\dfrac{\alpha}{2}$ through which the link swings as it enters into contact, is called angle of articulation.

Types of Chain

(1) **Hoisting chains:** These are used for lower speed and it consists of oval links and stud links.

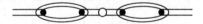

Fig. 6.14

(2) **Conveyer chain:** Conveyer chains may be hook joint type or detachable type or of closed end pintle type. The sprocket teeth are so shaped and spaced that the chain could run onto and off the sprockets smoothly and without interference. Such chains are used for low speed agriculture machinery.

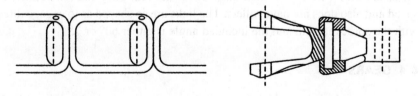

(a) Hook joint type   (b) Closed-end pintle type

Fig. 6.15

(3) Power transmission chain

(a) Block chain: Used for power transmission at low speeds.

Fig. 6.16: Block chain

(b) Roller chain: These chains are very strong and are quiet in operation. The capacity of the chain can further be improved by increasing pitch and using larger sprocket diameter. The main components of these chains are outer and inner plates, pins, bushes and rollers, etc. There are two rows of outer and inner plates. The outer row of plates is called as coupling link or pin link and inner row of plates is called as roller link.

A pin passes through the bush which is secured in the holes of the roller between the two sides of the chain. There is only sliding motion between the pin and the bushing. The roller is made of a hardened material and is free to turn on the bushing. A good roller chain is quite and wearless as compared to a block chain.

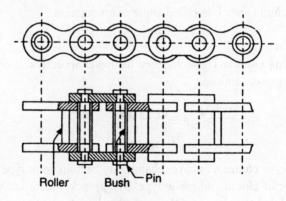

Fig. 6.17: Bush-roller chain

Silent Chain (Inverted Tooth Chain)

These are used where maximum quietness is desired. They can run at fairly high speed and also does not have rollers. The links are so shaped as to engage directly with the sprocket teeth and the included angle is either 60° or 75°.

6.4 GEARS

Gears are toothed disc which transmit power from one shaft to other shaft by meshing with teeth of other gear. Gear drive is widely used means of transmitting

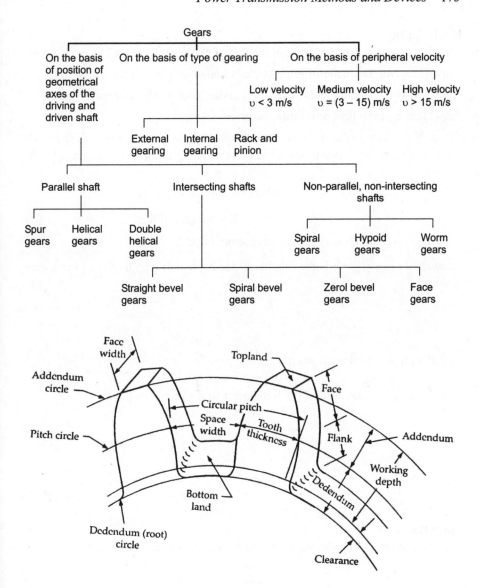

Fig. 6.18

power when the change in speed or change in direction between an input and output shaft is desired. Special feature of this drive is that a positive drive and constant speed ratio without any slippage is obtained by this drive. Practically they are used in gear boxes of vehicles, metal cutting tools, rolling mills, twisting machinery lathes etc.

Pitch Point
The point of contact of two pitch circles is known as pitch point.

Pitch Circle
The circle passing through point of contact of two gears is known as pitch circle.

Pitch Diameter: Diameter of pitch circle is known as pitch diameter.

Pinion: It is the smaller and usually the driving gear of a pair of mated gears.

Rack: It is a gear wheel of infinite diameter.

Circular pitch: It is the distance measured along the circumference of the pitch circle from a point of one tooth of the corresponding point on the adjacent tooth.

$$P_c = \frac{\pi d}{T} \qquad \text{where } d = \text{pitch diameter}$$
$$T = \text{number of teeth}$$

The two mating gears should have same pitch.

Diametral pitch (P_d): It is the number of teeth per unit length of the pitch circle diameter

$$P_d = \frac{T}{d}$$

Circular pitch × Diametral pitch $= \dfrac{\pi d}{T} \times \dfrac{T}{d} = \pi$

Module (m): It is the ratio of pitch diameter to the number of teeth

$$m = \frac{d}{T}$$

Also $m = \dfrac{d}{T} = \dfrac{1}{P_d} = \dfrac{P_c}{\pi}$ or $P_c = \pi m.$

Gear Ratio
It is the ratio of number of teeth on gear to number of teeth on pinion

$$G = \frac{T}{t}$$

Velocity ratio (VR): It is the ratio of the angular velocity of the following or driven gear to driving gear.

$$VR = \frac{w_2}{w_1} = \frac{N_2}{N_1} = \frac{d_1}{d_2} = \frac{T_1}{T_2}$$

Subscription 1 for driving gear
               2 for driven gear

Addendum circle: It is a circle passing through the tips of teeth.

Addendum: It is the radial height of a tooth above the pitch circle. Its standard value is one module.

Dedendum circle (or Root circle): It is a circle passing through the roots of the teeth.

Dedendum: It is the radial depth of a tooth below the pitch circle. Its standard value is 1.157 m.

Clearance: It is radial difference between the addendum and the dedendum of the tooth.

$$\text{Addendum circle diameter} = d + 2 \text{ m.}$$
$$\text{Dedendum circle diameter} = d - 2 \times 1 - 1577 \text{ m}$$
$$\text{Clearance} = 1.157 \text{ module.}$$

Space width: It is the width of the tooth space along the pitch circle.

Types of Gears
The commonly used forms of toothed gears are:

(1) Spur gears
(1) Most commonly used gear drive for transmitting motion.
(2) The gear has its teeth parallel to the axis of the shaft and is similar in profile throughout.
(3) The spur gears are used for transmitting motion between two shafts whose axis are parallel and coplaner.
(4) The spur gear may be internal or external. The external gear rotate in opposite directions while internal gears rotate in same direction.
(5) They have 96–98% efficiency of power transmission and are free from any axial thrust during tooth engagement.

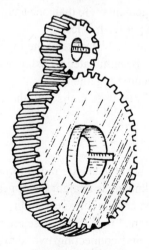

Fig. 6.19: Spur gear

176 *Mechanical Engineering: Fundamentals*

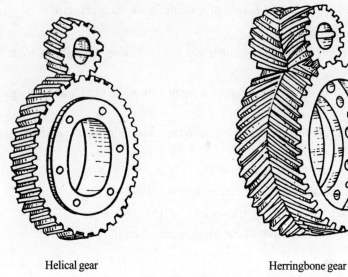

Helical gear   Herringbone gear

Fig. 6.20

(6) The disadvantages of these gears are they are noisy in operation, wear out readily and develop backlash.

(2) Helical gear
  (1) In helical gear tooth traces are straight helices, teeth are inclined at an angle to the gear axis.
  (2) The line of contact between two mating teeth is not parallel to the teeth but inclined to ensure gradual engagement of the teeth.
  (3) The gradual engagement makes the helical gearing very smooth and noise free at high speeds. The angle made by teeth with the axis of the wheel is known as helix angle.
  (4) In this gearing arrangement more than one teeth are in contact at a time so the load distribution per teeth is less.
  (5) A lateral thrust is produced because of the inclination of teeth.
  (6) The lateral thrust can be eliminated by using herringbone gears which are two helical gears secured together.
  (7) The effect of axial thrust is neutralised in these gears.

Intersecting Shafts

(i) Straight bevel gears
  (1) If the teeth of these bevel gears are radial to the point of intersection of shaft axis, they are known as straight bevel gear.
  (2) The teeth are tapered both in thickness and tooth height.
  (3) These gears are designed in pairs so these are not interchangeable.
  (4) Gears of same size and connecting two shafts at right angles to each other are known as mitre gears.

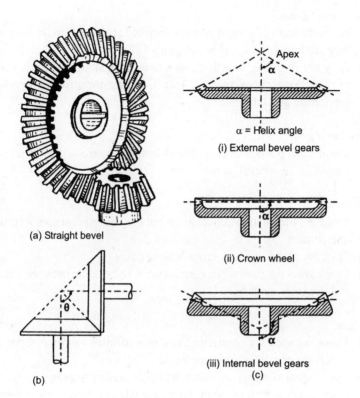

(a) Straight bevel

(b)

(i) External bevel gears
α = Helix angle

(ii) Crown wheel

(iii) Internal bevel gears
(c)

Fig. 6.21: Straight bevel gears

Spiral bevel

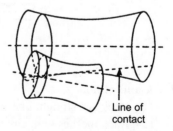

Spiral gears

Fig. 6.22

(ii) Spiral bevel gear
(1) When the teeth of a bevel gear are inclined at an angle to the face of bevel, they are known as spiral bevel gear.
(2) They are smoother in action and quieter than straight tooth bevel.
(3) In this, there exists an axial thrust calling for stronger bearings and supporting assemblies.

(iii) Zero bevel gear
(1) Spiral bevel gears with curved teeth but with a zero degree spiral angle are known as zero bevel gear.

Spiral gears
(1) These are identical to helical gears but they have point contact rather than line contact.
(2) They are also known as cross helical gears.
(3) These gears are used when connection is to be made between intersecting and coplanar shaft.

Worm gear
(1) These are used for connecting two non-parallel, non-intersecting shafts, which are usually at right angles.
(2) The system consists of worm which is basically part of a screw. The worm meshes with the teeth on a gear wheel called worm wheel.
(3) Their use is recommended when high-speed reduction (more than 10:1) is required.

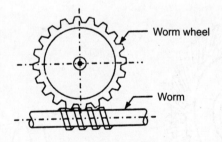

Fig. 6.23: Worm gear

Rack and pinion
(1) Rack is a straight line spur gear of infinite diameter.
(2) Rack meshes with circular wheel called pinion.
(3) It is used where rotary motion to be converted into linear motion (as in steering of four wheelers) and vice versa.

6.5 GEAR TRAINS

Any combination of gears made to transmit motion at a desired velocity from one shaft to another is called a gear train.

Various types of gear train are:
   (1) Simple gear train
   (2) Compound gear train
   (3) Reverted gear train
   (4) Epicyclic gear train

Simple Gear Train

If all gears are mounted on separate gear shaft and gear axes remain fixed, then such type of train is known as simple gear train.

Important features
   (1) All paired gears moves in opposite direction.
   (2) It is very useful when the distance between driver shaft and driven shaft is great.

Consider the figure
   Gear 1 Driver gear or input gear.
   Gears 2 and 3 are idler gears.
   Gear 4 Driven gear or output gear.

Let N_1, N_2, N_3 and N_4 be the speeds of gears 1, 2, 3 and 4. T_1, T_2, T_3 and T_4 be the number of teeth on gears 1, 2, 3 and 4. The velocity ratio of gear 1 and gear 2

$$= \frac{N_1}{N_2} = \frac{T_2}{T_1} \tag{1}$$

Likewise the intermediate gears 2 and 3 are in mesh and accordingly

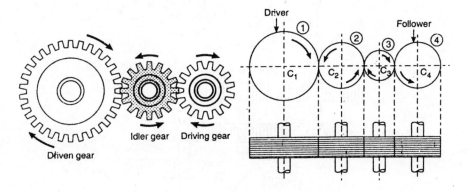

Fig. 6.24: Simple gear train

$$\frac{N_2}{N_3} = \frac{T_3}{T_2} \qquad (2)$$

Further velocity ratio for gears 3 and 4

$$\frac{N_3}{N_4} = \frac{T_4}{T_3} \qquad (3)$$

From equations (1), (2) and (3), we get

$$\frac{N_1}{N_2} \times \frac{N_2}{N_3} \times \frac{N_3}{N_4} = \frac{T_2}{T_1} \times \frac{T_3}{T_2} \times \frac{T_4}{T_3}$$

$$\frac{N_1}{N_4} = \frac{T_4}{T_1}$$

Velocity ratio = $\dfrac{\text{Speed of the driving wheel}}{\text{Speed of the driven wheel}} = \dfrac{\text{Number of teeth on the driven wheel}}{\text{Number of teeth on the driving wheel}}$

Compound Gear Train

A compound gear is the gear which carries two wheels mounted on the same shaft. In this gear train between the driving gear 1 and driven (follower) gears 6 two intermediate compound gears have been provided. Gears 2 and 3 are mounted on the same shaft so are the gears 4 and 5.

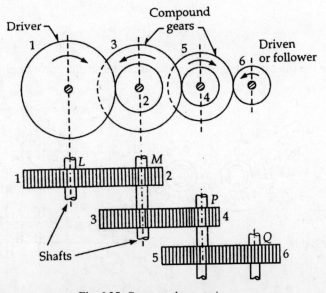

Fig. 6.25: Compound gear train

VR of gear 1 and gear 2 = $\dfrac{N_1}{N_2} = \dfrac{T_2}{T_1}$

Similarly, for gear 3 and gear 4 = $\dfrac{N_3}{N_4} = \dfrac{T_4}{T_3}$

for gears 5 and 6 = $\dfrac{N_5}{N_6} = \dfrac{T_6}{T_5}$

From equations (1, (2) and (3)

$$\dfrac{N_1}{N_2} \times \dfrac{N_3}{N_4} \times \dfrac{N_5}{N_6} = \dfrac{T_2}{T_1} \times \dfrac{T_4}{T_3} \times \dfrac{T_6}{T_5}$$

Now $N_2 = N_3$ and $N_4 = N_5$.
Since both the two wheels are mounted on same shaft.

$$\dfrac{N_1}{N_2} \times \dfrac{N_2}{N_4} \times \dfrac{N_4}{N_6} = \dfrac{T_2}{T_1} \times \dfrac{T_4}{T_3} \times \dfrac{T_6}{T_5}$$

$$\dfrac{N_1}{N_6} = \dfrac{T_2}{T_1} \times \dfrac{T_4}{T_3} \times \dfrac{T_6}{T_5}$$

$$\dfrac{\text{Speed of driver}}{\text{Speed of driven (last follower)}} = \dfrac{\text{Product of teeth on the follower}}{\text{Product of teeth on the drivers}}$$

Reverted Gear Train
If the axis of first and last wheels of a compound gear coincides, it is called reverted gear train.

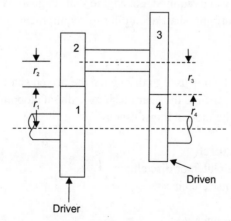

Fig. 6.26

In this

$$\frac{N_4}{N_1} = \frac{T_1 \times T_3}{T_2 \times T_4} \text{ (as in the compound gear train)}$$

$$r_1 + r_2 = r_3 + r_4$$

Epicyclic or planetary gear train
(1) If axis of at least one gear in gear train also moves relative to fixed axis or frame, such type of gear train is known as planetary or epicyclic gear train.
(2) In this gear train one gear rotates over other gear's pitch circle.
(3) In the figure if the arm a is fixed the wheels P and Q constitute a simple gear train.
(4) If the wheel Q is fixed so that the arm can rotate about the axis of Q the wheel P would also move around Q. This is an epicyclic gear train.

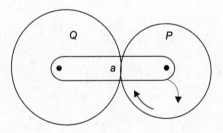

Fig. 6.27

6.6 CLUTCHES

Clutch is a mechanism which is used to transmit the power from one shaft to another shaft as and when required, i.e. engine shaft to gearbox shaft. It is capable of disengaging the driving shaft according to requirement without stopping the driving shaft.

or

Clutch is a device which is used to disengage the driven shaft from driving shaft during the motion to change the gear meshing without stopping the driving shaft and again continue the motion transmission.

Classification of Clutch
(1) Single plate clutch or disc clutch
(2) Multiplate disc clutch
(3) Cone clutch.
(4) Centrifugal clutch

(1) Single plate clutch

Three important parts of single plate clutch are:
1. Flywheel or (driving plate).
2. Clutch plate or friction plate or driven plate.
3. Pressure plate.

The friction plate is held between the flywheel and the pressure plate.

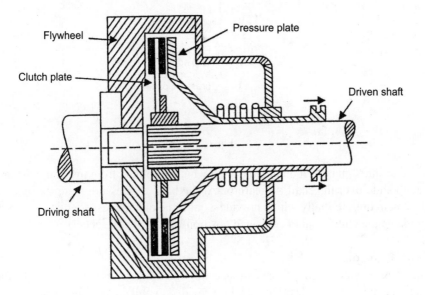

Fig. 6.28: Single plate clutch disengaged

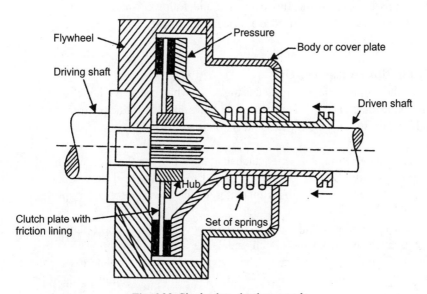

Fig. 6.29: Single plate clutch engaged

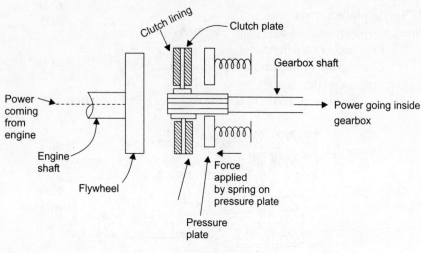

Fig. 6.30

The clutch plate is mounted on a hub which is splined from inside and is thus free to slide over the shaft of gearbox. At the back of pressure plate springs are placed circumferentially which provide axial force on the pressure plate which in turn keep the clutch plate in engaged position with flywheel wheel.

Basic Principle

The transmission of power takes place through friction facing attached to the friction plate on both sides. The pedal is provided to pull the pressure plate against the spring force whenever it is required to be disengaged. At this instant the clutch plate slips back from the flywheel face and follower shaft becomes stationary. When the pressure is released from pedal the thrust bearing moves back due to action of springs.

(2) Multiplate disc clutch

When the torque to be transmitted is higher or when space available is limited then multiplate clutch is used. This type of clutch is suitable for heavy transport vehicle and racing cars. In motorcycles and scooters, it is used because space available is limited.

Here the friction plates are attached at the top to the flywheel and are also free to move axially. Since they are connected to the flywheel they rotate with the flywheel. There are also discs or plates which are supported on splines of the driving shaft.

They can slide axially and are placed between the friction plates. When the foot is taken off from the clutch pedal, the set of springs press the discs into contact with the friction plates. The flywheel, the friction plate discs and whole assembly rotate as one unit because of the presence of tight gripping.

Hence, the power will be transmitted.

Power Transmission Methods and Devices   185

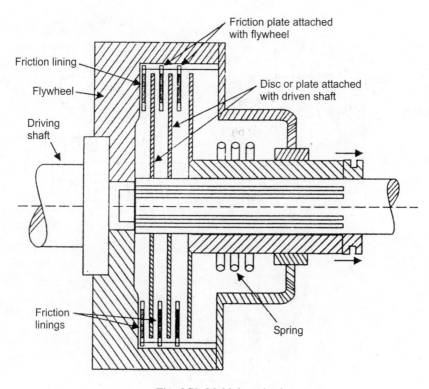

Fig. 6.31: Multiplate clutch

(3) Cone clutch

(1) Here the contact surfaces are in the form of cone.
(2) Because of the spring action the male part fits into the female part and the clutch gets engaged.

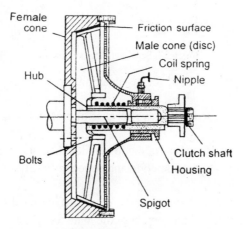

Fig. 6.32: Cone clutch

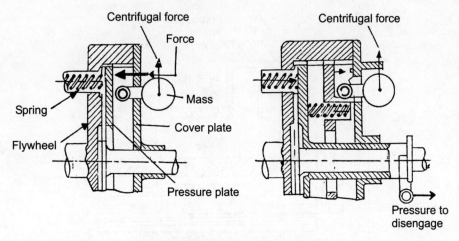

Fig. 6.33: Principle of centrifugal clutch

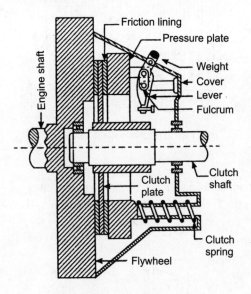

Fig. 6.34: Semi-centrifugal clutch

(3) 4 leaver system is used for disengaging the clutch is operated by clutch pedal.

(4) These are not used nowadays because of more advantages of single plate clutch.

(5) If the angle of the cone is made smaller than about 20° the male cone tends to bind the female cone and it becomes difficult to disengage the clutch.

(6) A small amount of wear on the cone surface results in a considerable amount of the axial movement of the male for which it will be difficult to allow.

6.7 BRAKES

Introduction

The main function of the broke is to bring the body into rest while it is in motion or to hold a body in a state of rest by applying resisting torque.

The brake in its operation, absorbs kinetic energy of moving member and converts it into heat energy. A good brake can apply large braking torques to the brake drums and can dissipate large quantities of heat without large temperature rise.

Types of Brakes

Depending upon the means employed for converting the energy by the braking elements the brakes can be classified as follows:
  (1) Mechanical brakes
  (2) Hydraulic brakes
  (3) Electric brakes

(1) Mechanical brakes

(i) Simple block or shoe brake: It consists of a block or shoe which is pressed against the rim of a revolving brake wheel drum. The block is pressed against the wheel by a force applied to one end of a lever to which the block is rigidly fixed as shown in figure. The other end of the lever is pivoted on a fixed fulcrum. Due to the frictional force between the block and the wheel, a tangential force comes into existence which acts on the wheel and retard the rotation of the wheel. Such types of brakes is commonly used in railway trains and tram cars.

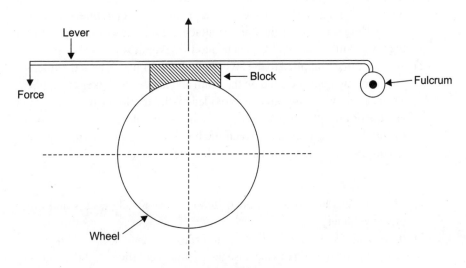

Fig. 6.35: Simple block or shoe brake

Double shoe brake: Due to pressure applied by single shoe, there is a side thrust on the shaft of the drum. To counterbalance the side thrust, two shoes may be used opposite to each other. The brake is set by a spring which pulls the upper ends of the brake arms together. When a force is applied to the bell crank lever, the spring is compressed and the brake is released. This type of system is often used on electric cranks.

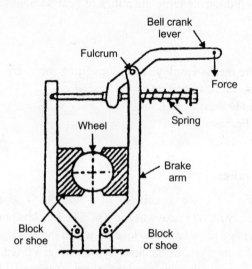

Fig. 6.36: Double block or shoe brake.

(iii) Simple band brake:
(1) It consists of a flexible band of leather, one or more ropes, or a steel lined with friction material, which embraces a part of the circumference of the drum. Here one end of the band is attached to a fixed pin or fulcrum of the lever while other end is attached to the lever at A.
(2) When a force is applied to the lever at B, the lever turns about the fulcrum pin O and tightens the band on the drum and hence the brakes are applied. The necessary braking force is provided by the friction between the band and the drum. When the band brake is lined with blocks of wood or other material and the friction between the blocks and drum provides braking action, the brake is termed as band and block brake.

(iv) Internal expanding brakes:
(1) These brakes are mostly used in scooters, motorbikes, cars and other motor vehicles.
(2) It consists of two shoes. The outer surfaces of the shoes are lined with some friction material, to increase the frictional coefficients and to prevent wearing away of the metal.

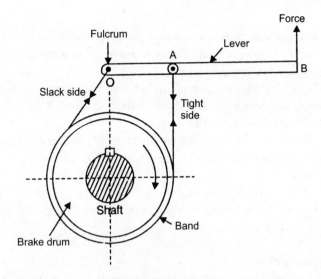

Fig. 6.37: Simple band brake

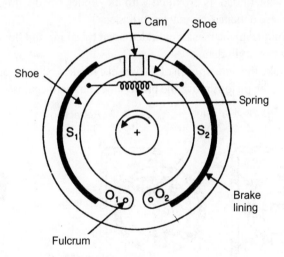

Fig. 6.38: Internal expanding brake

(3) Each shoe is pivoted at one end about a fixed fulcrum and made to contact a cam at the other end.
(4) When the cam rotates, the shoes are pushed outwards against the rim of the drum. The friction between the shoes and drum produces the braking torque and consequently speed of the drum reduces.
(5) The entire system is enclosed in the drum and thus prevents the system from dust and moisture.

(v) Disc brakes:
(1) In a simple disc brake, a cast iron disc is bolted to the wheel hub and a stationary housing called calliper is connected to some stationary part of the vehicle like the axle casing.
(2) Calliper contains two pistons. In between the piston and disc there is a friction pad held in position by retaining pins so that they are prevented from rotating with the disc.
(3) When the brakes are applied the hydraulically actuated pistons move the friction pad into contact with the disc applying equal and opposite forces on the disc.
(4) When the brake is released the rubber sealing ring between the cylinder and piston act as return spring and retract the pistons and friction pads away from the discs.

(2) Hydraulic brakes
(1) In a hydraulic brake system force is applied to a piston in a master cylinder. The brake pedal operates the piston by a linkage.
(2) Each wheel brake is provided with a cylinder fitted with opposed piston which are actuated by master cylinder.
(3) The fluid from the master cylinder is sed by taking and flexible hose to the four wheel cylinder.
(4) The brake fluid enters each of the wheel cylinders between opposed pistons, making the pistons move the brake shoes outward against the brake drum.

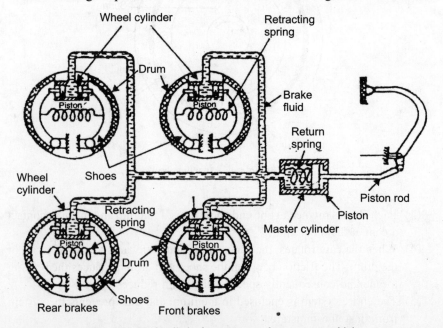

Fig. 6.39: Hydraulic brake system used on a motor vehicle.

(5) As pressure on pedal is increased greater hydraulic pressure is built up within the wheel cylinders, which ensures effective braking of the moving wheel.

6.8 DYNAMOMETER

The function of the dynamometer is to measure the torque and hence the power of the engine by applying the brake. Thus dynamometer measures the resisting force applied by the brake on the moving body.

Types of Dynamometers

(1) Absorption Dynamometers
In these dynamometers, the entire energy or power produced by the engine is absorbed by the friction resistance of the brake and is transformed into heat, during the process of measurement. Under this heading the two main dynamometers are:
  (a) Prony brake dynamometer.
  (b) Rope brake dynamometer.

(a) Prony brake dynamometrs: It consists of two wooden block, one of which is attached with lever. The pulley rotates between the block. The revolving pulley is known as brake drum. Brake drum is fixed to a shaft of the engine. The two blocks can be drawn together by means of bolts, cushioned by springs, so as to increase the pressure on pulley. The lever carries an adjustable load W at its other end. The load is adjusted to maintain the lever in horizontal position for the required speed of the engine, as the friction between blocks and pulley tends to rotate the lever in the direction of shaft rotation. To limit the motion of the lever, two stops T_1 and T_2 are applied.

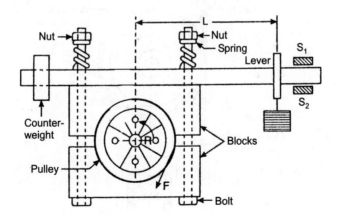

Fig. 6.40: Prony brake dynamometer

Calculation of Torque and Power

Let W = Weight at the outer end of lever.
    l = Horizontal distance of W from centre of pulley.
    R = Radius of pulley
    F = Frictional resistance between the blocks and pulley.
    N = Rotational speed of shaft in r.p.m.
Braking torque on shaft = $\tau = F.R. = Wl$.

$$\text{Power} = T \times W = Wl \times \frac{2\pi N}{60} \text{ watt}$$

Rope Brake Dynamometer: It consists of one or more ropes wrapped around the flywheel of an engine whose power is to be measured. Two or three ropes wound around the rim of the drum which is keyed to the shaft of the engine. The ropes are prevented from slipping off by means of three or four U-shaped wooden blocks at different points on the rim of the wheel. One end of the rope is connected to the spring balance while the other is connected to the dead weight. The rope tightens because of friction between rope and flywheel which in turn induces a force in spring balance. In order to keep the flywheel cool, their rims are provided channel sections so as to receive water which is whirled round the wheel.

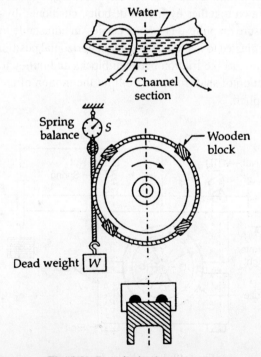

Fig. 6.41: Rope brake dynamometer

Let W be the dead, S be the spring balance reading D be the brake drum diameter and d be the rope diameter. Then effective radius of brake drum $R_{eff} = \dfrac{D+d}{2}$

$$\text{Brake load or net load} = (W - S) \text{ N}$$

$$\text{Braking torque } T = (W - S) R_{eff} \text{ Nm}$$

$$\text{Brake power} = \dfrac{2\pi N}{60} \times T. \text{ Nm/sec.}$$

$$\text{Brake power} = \dfrac{2\pi N}{60 \times 1000} \times (w - s) R_{eff} \text{ in kW.}$$

(2) Transmission Dynamometers

In these dynamometers energy is not absorbed or dissipated but is used for doing useful work. In these dynamometers the power of the engine is measured and then transmitted through them.

Important dynamometers in this category are.
 (a) Epicyclic—train dynamometer
 (b) Belt transmission dynamometer.
 (c) Bevis-Gibson or torsion dynamometer.

(a) Epicyclic—Train Dynamometer: It consists of an epicyclic train of spur gears. Three import parts of this dynamometer are a spur gear, an annular gear (gear having internal teeth) and a pinion.

Engine (driving) shaft is connected to spur gear which rotates in anticlockwise direction and the annular gear keyed to the driven shaft rotates in clockwise direction. The pinion meshes with both the annular and spur gear. The pinion revolves freely on a lever which is pivoted to the common axis of the driving and driven shafts. Two stops are provided to keep the lever in horizontal position. Also, a counterweight is provided to balance the brake when unloaded.

The force exerted by spur gear on the pinion (P)

= Tangential reaction by annular wheel on the pinion P

Since, both these force act in upward direction the net upward force on lever can be assumed to act through the axis of the pinion. The above force tends to rotate the lever about its fulcrum in clockwise direction. This motion is presented by dead weight (w).

When the lever is in equilibrium plot in continuation
About fulcrum 0 take moments.

$$2P \times a - W.L. = 0$$

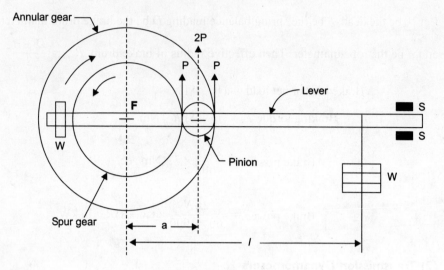

Fig. 6.42: Epicyclic gear train

$$F = \frac{WL}{2a}$$

$$W = mg$$

$$P = \frac{mgL}{2a}$$

$$\tau = p.r = \frac{mgrL}{2a}$$

Brake power $\quad B.P. = \tau w = \dfrac{mgrL}{2a} \times \dfrac{2\pi N}{60}$ watt

- m = mass of dead weight.
- g = acceleration due to gravity
- r = radius of spur wheel.
- L = distance between fulcrum and dead weight
- N = speed in rpm

(b) Belt Transmission Dynamometer
  (1) Also called Froude or Throneycraft transmission dynamometer.
  (2) Driving pulley is rigidly fixed to the shaft of an engine whose power is to be measured.
  (3) A driven pulley mounted on another shaft to which the power from pulley is transmitted.

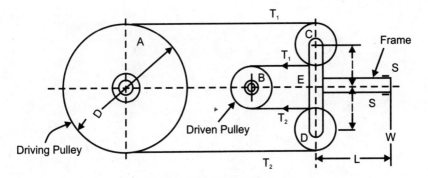

Fig. 6.43: Belt transmission dynamometer

(4) The pulleys (driving pulley and driven pulley) are connected by means of a continuous belt passing round two pulleys which are mounted on a T-shaped frame.
(5) The frame is pivoted at E and its movement is controlled by two stops.
(6) Since T_1 (tension on tight side) > T_2 (tension on slack side)

Force on pulley C = $2T_1$
Force on pulley D = $2T_2$

$$2T_1 > 2T_2$$

because of this frame caves movement about pivot E in anticlockwise direction. In order to balance it a weight W is applied at a distance L from E on the frame.

Taking moments about the pivot E

$$2T_1 \times l = 2T_2 \times l + W \times L$$

$$2l(T_1 - T_2) = W \times L$$

$$(T_1 - T_2) = \frac{WL}{2l}$$

D = diameter of driving pulley.
N = speed of engine shaft rpm.
Then work done/revolution = $(T_1 - T_2)\pi D$ N-m

Work done/minute = $(T_1 - T_2)\pi DN$ N-m

Brake power of the engine $B.P. = \dfrac{(T_1 - T_2)\pi DN}{60 \times 1000}$ kW

(c) Bevis Gibson Flash Light Torsion Dynamometer: The principle of which this dynamometer is based is that torque transmitted is directly proportional to the angle of twist.

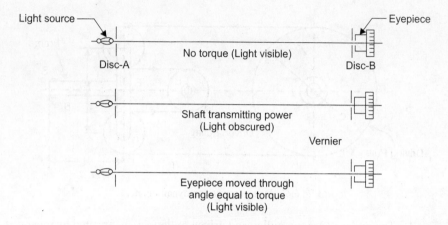

Fig. 6.44: Flash light torsion dynamometer

(1) It consists of two discs A and B which are fixed to the shaft at a certain distance.
(2) Two discs revolve with the shaft. Each disc has a narrow radial slot and the two slots are in line when there is no torque transmission along the shafts.
(3) A masked lamp is placed behind the disc A so as to throw a narrow beam of light parallel to the axis of the shaft and at the same distance from the axis as the radial slots in discs A and B.
(4) Behind the disc an eyepiece is supported on fixed bracket, but by the vernier adjustment it can be moved along an arc concentric with axis of the shaft.
(5) When shaft rotates without transmission of any torque all four slots are in a straight line and light can be seen through the eyepiece after each rotation.
(6) When torque is transmitted by the shaft the disc B twisted with shaft and light cannot be seen through eyepiece since light does not passes through the four slots in this situation.
(7) However, if the eyepiece is moved circumferentially by an equal amount of twist, the flat light will again be visible once in each rotation of shaft.
(8) From the angular twist and the shaft constant I_p, l and C torque can be determined.

$$T = I_p \times \frac{C\theta}{l}$$

$I_p = \frac{\pi}{32} d^4$ = polar moment of inertia of a shaft of diameter d.

θ = twist in radians over length l of the shaft
C = modulus of rigidity of shaft material.

EXERCISE

1. Derive an expression for the length of flat belt between two pulleys.
   (1) for open belt drive
   (2) for crossed belt drive.
2. If T_1 is the tension on the light side and T_2 is the tension on slack side and μ is the limited coefficient of friction, between belt and pulley and θ is the angle of contact of belt our pulley prove that
$$\frac{T_1}{T_2} = e^{\mu\theta}$$
3. Write short notes on.
   (a) Gear Drive
   (b) Chain Drive
   (c) Function of clutches
   (d) Dynamometers
4. Explain how gear trains are classified.
5. Classify the types of brakes. Explain each type of brake with the help of neat sketches.
6. Classify the types of transmission dynameters. Explain each type of transmission dynamometers with the help of neat sketches.
7. Differentiate between hoisting chain and conveyer chain
8. Define the following with respect to gear.
   (a) Pitch circle
   (b) Addendum circle
   (c) Module
   (d) Herringbone gear
   (e) Diametral pitch (f) bevel gears.

OBJECTIVE TYPE QUESTIONS

1. Crowning on pulleys helps
   (a) in increasing velocity ratio
   (b) in decreasing the slip of the belt
   (c) for automatic adjustment of belt position so that belt runs centrally
   (d) increase belt and pulley life
2. Idler pulley is used
   (a) for changing the direction of motion of the belt
   (b) for applying tension
   (c) for increasing velocity ratio
   (d) all of the above
3. In multi V-belt transmission, if one of the belt is broken we have to change the

(a) broken belt
(b) broken belt and its adjacent belts.
(c) all the belts
(d) all the weak belts
4. Creep in belt drive is due to
(a) material of the pulley
(b) material of the belt
(c) large size of the driver pulley
(d) uneven extensions and contractions due to varying tensions
5. The horse power transmitted by a belt is dependent upon
(a) tension on tight side of belt
(b) tension on slack side of belt
(c) radius of pulley
(d) all of the above
6. To transmit power from one rotating shaft to another whose axis are neither parallel nor intersecting use.
(a) spur gear
(b) spiral gear
(c) bevel gear
(d) worm gear
7. In a gear drive, module is equal to

(a) $\dfrac{1}{\text{diametral pitch}}$

(b) $\dfrac{1}{\text{circular pitch}}$

(c) $\dfrac{\text{circular pitch}}{\pi}$

(d) $\dfrac{\text{diametral pitch}}{\pi}$

8. Addendum is given by

(a) $\dfrac{\pi}{\text{circular pitch}}$

(b) diametral pitch
(c) one module
(d) 1.25 module
9. To obviate axial thrust, following gear drive is used
(a) double helical gears having opposite teeth
(b) double helical gears having identical teeth
(c) single helical gear in which one of the teeth of helix angle α is more
(d) mutter gears

10. Rope brake dynamometer uses
    (a) oil as lubricant
    (b) water as lubricant
    (c) grease as lubricant
    (d) no lubricant
11. The most commonly used dynomometer for tests in the laboratory is
    (a) rope brake dynomometer
    (b) prony brake dynamometer
    (c) froude water vortex dynamometer
    (d) amsler dynamometer
12. The following dynamometer is widely used for absorption of wide range of power at wide range of speeds.
    (a) hydraulic
    (b) belt transmission
    (c) rope brake
    (d) electric generator
13. The brake commonly used on train bogies is
    (a) internal expanding
    (b) band brake
    (c) band and block brake
    (d) shoe brake
14. Length of cross belt, in addition to centre length depends
    (a) only on the sum of the radii of pulleys
    (b) on the sum and difference of the radii of the pulleys
    (c) square of difference of radii of the pulleys
    (d) square of sum of radii of pulleys
15. The belting can transmit maximum power when maximum total tension in belt equals
    (a) twice the centrifugal tension
    (b) thrice the centrifugal tension
    (c) four times the centrifugal tension
    (d) centrifugal tension
16. If T_1 and T_2 be the tensions in kg on tight and slack sides of a belt and v be its velocity in m/sec, then h.p. transmitted is equal to
    (a) $\dfrac{(T_1 - T_2)V}{4500}$
    (b) $\dfrac{T_1 V}{75}$
    (c) $\dfrac{(T_1 - T_2)V}{75}$
    (c) $\dfrac{(T_1 - T_2)V}{5500}$

17. The circle passing through the bottom of the teeth of gear is known as
    (a) inner circle
    (b) prime circle
    (c) base circle
    (d) addendum circle
18. Pitch circle diameter of an involute gear is
    (a) independent of any other element
    (b) dependent of pressure angle
    (c) constant for a set of meshing gear
    (d) proportional to base diameter
19. Intermediate gear are used for
    (a) obtaining rotation in desired direction
    (b) reducing the size of the individual gear
    (c) bridging the gap between the first and last wheels of the train
    (d) any one of the above
20. If D and T be the pitch circle diameter and no of teeth of a gear then its circular pitch $P =$
    (a) $\dfrac{D}{T}$
    (b) $\dfrac{T}{D}$
    (c) $\pi\dfrac{T}{D}$
    (d) $\pi\dfrac{D}{T}$
21. The difference between dedendum of addendum is known as
    (a) backlesh
    (b) clearance
    (c) flank
    (d) tooth space
22. The product of circular pitch and diametral pitch is equal to
    (a) module
    (b) unity
    (c) π
    (d) $\dfrac{1}{\pi}$
23. Bevel gears have their teeth
    (a) straight over the wheel rim
    (b) inclined to wheel rim
    (c) Curved over the wheel rim
    (d) cut on the surfaces of the frusts of cones

24. Helical gears have their teeth
    (a) straight over the wheel rim
    (b) inclined to wheel rim
    (c) curved over the wheel rim
    (d) none of the above
25. The gear train in which the first and last gear are on the same axis is known as
    (a) epicyclic gear train
    (b) simple gear train
    (c) compound gear train
    (d) compound gear train
    (d) reverted gear train

ANSWERS

1. c	2. c	3. c	4. d	5. d
6. d	7. a	8. c	9. a	10. d
11. b	12. b	13. d	14. a	15. c
16. c	17. b	18. b	19. d	20. d
21. b	22. c	23. d	24. b	25. d

UNIT 7

Stresses and Strain

7.1 INTRODUCTION

Strength of materials is the science which deals with the relations between externally applied loads. There are certain behaviours of all materials under the influence of external force. Stress and strain are one of the measures to show these behaviours. If the forces which produce the deformation of the body are gradually removed, the body returns or try to return to its original shape. During this return, the stored potential energy can be recovered in the form of external work. Strength, stiffness and stability of various load carrying members are the main concerns of this subject.

Basic Concept About Stresses

When an external force is applied on a body, internal resistance is developed within the body to balance the effect of externally applied forces. This resistive force per unit area is called stress.

These stresses are inclined with respect to sectional plane. But in engineering practice, it is customary to resolve the force perpendicular and parallel to the section investigated. If the stress developed in the material is perpendicular to cross-section, it is known as direct stress or normal stress and if it is tangential or parallel to cross-section it is known as shear stress.

In Fig. 7.1 the bar will remain in equilibrium if the resistance R equals the external load P. This resistance R against deformation is called as stress.

Assuming resistance to be uniform across the section than resistance per unit area at the section is called the unit stress or intensity of stress.

$$(\text{Stress})\tau = \frac{R}{A} = \frac{P}{A} \text{ since } R = P$$

Unit of stress is N/m^2 or Pascal.

$$1 \text{ Pascal (Pa)} = 1 \text{ N/m}^2$$

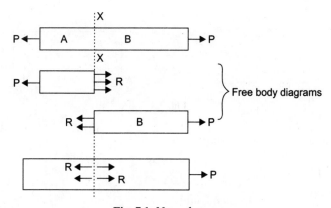

Fig. 7.1: Normal stress.

Normal stress or Direct stress
In Fig.7.1 the resistance R is acting perpendicular to the cross-section and that is why it is called as normal stress or direct stress.

$$\text{Direct stress} = \tau = \frac{P}{A} \ (\text{N/m}^2)$$

If the normal stress tries to create an extension in length of rod, it is called tensile stress and in case it tries to reduce the length of rod it is termed compressive stress.

Shear Stress
If the stresses developed in the material is tangential or parallel to cross-section, it is known as shear stress.

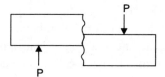

Fig. 7.2: Shear stress.

In Fig. 7.2 external forces are acting parallel to the cross-section and because of these forces there will be resistance just opposite to external forces which is parallel to cross-section and thus known as shear force.

When a body is subjected to two equal and opposite parallel forces but not in same line, it tends to shear off across the resisting section.

The stress induced in the body is called shear stress. Most commonly this pure shear occurs in cotter and riveted joints.

Consider a rectangular block ABCD fixed at one face as shown in Fig. 7.3 and is subjected to force P after application of force it distorts through an angle ϕ and occupies new position $AD_1 \ C_1 \ B$.

 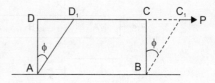

Fig. 7.3

Then

$$\text{Shear strain} = \frac{\text{Deformation}}{\text{Original length}}$$

$$= \frac{CC_1}{l} = \phi \qquad (\text{Area} = AB \times 1)$$

$$\text{Shear stress} = \frac{P}{AB}$$

7.2 STRAIN

Strain is a dimensionless quantity. It gives us a measure of deformation in the shape of body on application of force.

Broadly strains can be classified into two types:
(1) Normal or longitudinal strain
(2) Shear strain

Normal Strain
When a body is acted upon by tensile or compressive loading, its dimensions will increase or decrease along the line of action of load applied. This deformation or change in length per unit original length is called longitudinal strain.

$$\text{Longitudinal strain} = \frac{\text{Change in length}}{\text{Original length}}$$

$$e = \frac{\delta l}{l}$$

It is a dimensionless quantity. This strain can be further classified into two types depending upon the type of force.

(a) Tensile strain: When equal and opposite axial tensile forces are applied on the member then the forces create an elongation. The ratio of elongation in the length to the original length of the member is known as tensile strain.

(b) Compressive strain: When equal and opposite axial compressive forces are applied on the member, there is reduction in length. The ratio of this reduced length to the original length of the member termed as compressive strain.

Shear Strain

When a body is subjected to two equal and opposite parallel forces not in same line, it tends to shear off across the resisting section.

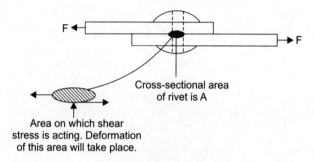

Fig. 7.4: Shear strain

In the Fig. 7.4 shear strain is mentioned which is given as

$$\text{Shear strain} = \frac{\text{Deformation}}{\text{Original length}}$$

$$= \frac{CC_1}{L} = \phi$$

7.3 POISSON RATIO

When a bar is subjected to normal tensile stress, it produces tensile strain in the direction of normal stress and the body elongates in the direction of stress. There is a contraction in cross-sectional area or decrease in the diameter of the bar. The contraction produces a lateral strain defined as

$$\text{Lateral strain} = \frac{D - D'}{D}$$

$$\text{Longitudinal strain} = \frac{L - L'}{L}$$

within the proportionality limit

$$(\text{Poisson Ratio}) = \frac{\text{Lateral strain}}{\text{Longitudinal strain}}$$

Its value ranges from 0.25 to 0.35.

7.4 STRESS AND STRAIN IN SIMPLE AND COMPOUND BAR UNDER AXIAL LOADING

Consider a bar made up of different lengths and having different cross-sections as shown in Fig. 7.5. For such type of section each section is subjected to the same external pull or push and total change in length is equal to the sum of changes of individual lengths.

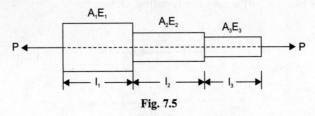

Fig. 7.5

That is $P_1 = P_2 = P_3 = P$ and

$$\delta l = \delta l_1 + \delta l_2 + \delta l_3$$

$$= \frac{\sigma_1 l_1}{E_1} + \frac{\sigma_2 l_2}{E_2} + \frac{\sigma_3 l_3}{E_3}$$

$$= \frac{P_1 l_1}{A_1 E_1} + \frac{P_2 l_2}{A_2 E_2} + \frac{P_3 l_3}{A_3 E_3}$$

If the bar segments are made of same material, then

$$E_1 = E_2 = E_3 = E.$$

In that case

$$\delta l = \frac{P}{E}\left[\frac{l_1}{A_1} + \frac{l_2}{A_2} + \frac{l_3}{A_3}\right]$$

7.5 THE STRESSES IN BARS OF UNIFORMLY TAPERING CIRCULAR CROSS-SECTION

Consider a circular bar of uniformly tappering section.

Let
  P = Force applied on bar (tensile) (N)
  L = Length of bar (m)
  d_1 = Diameter of the bigger end of the bar (m)
  d_2 = Diameter of the smaller end of the bar (m)

Now focus the attention on an elementary strip of length dx at distance x from the bigger end.

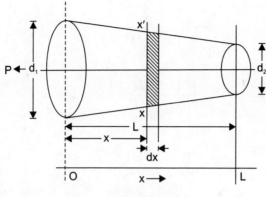

Fig. 7.6

Diameter of the elementary strip

$$dx = d_1 - (d_1 - d_2)\frac{x}{L}$$

$$k = \frac{d_1 - d_2}{L}$$

$$d = d_1 - kx$$

Thus, cross-sectional area of the bar at this section.

$$A = \frac{\pi}{4}(d_1 - kx)^2 \tag{1}$$

$$\text{Stress, } \sigma = \frac{P}{\frac{\pi}{4}(d_1 - kx)^2} = \frac{4P}{\pi(d_1 - kx)^2} \tag{2}$$

strain $$\varepsilon = \frac{\sigma}{E} = \frac{4P}{\pi E(d_1 - kx)^2} \tag{3}$$

∴ elongation of the elementary length = $\varepsilon.dx = \dfrac{4Pdx}{\pi E(d_1 - kx)^2}$ (4)

Therefore, total extension of the bar may be found out by integrating the equation (4) between the limits $x = 0$ and $x = L$.

$$\int_0^x \varepsilon.dx = \delta L \int_0^L \frac{4Pdx}{\pi E(d_1 - kx)^2}$$

$$= \frac{4P}{\pi E} \int_0^L \frac{dx}{(d_1 - kx)^2}$$

$$= \frac{4P}{\pi E} \left[\frac{(d_1 - kx)^{-1}}{-1(-k)} \right]_0^L$$

$$= \frac{4P}{\pi EK} \left[\frac{1}{d_1 - kx} \right]_0^L$$

Substituting the value of k

$$\delta L = \frac{4P}{\pi E \left(\frac{d_1 - d_2}{L} \right)} \left[\frac{1}{d_1 \left(\frac{d_1 - d_2}{L} \right) L} \cdot \frac{1}{d_1} \right] = \frac{4PL}{\pi E (d_1 - d_2)} \left[\frac{1}{d_2} - \frac{1}{d_1} \right]$$

$$= \frac{4PL}{\pi E (d_1 - d_2)} \left[\frac{1}{d_2} - \frac{1}{d_1} \right] = \frac{4PL}{\pi E (d_1 - d_2)} \left[\frac{d_1 d_2}{d_1 d_2} \right]$$

$$\delta L = \frac{4PL}{\pi E d_1 . d_2}$$

The stress in a bar of uniformly tapering rectangular cross section.

Consider a bar of constant thickness and uniformly tapering in width from end to the other.

Let P = Axial load on the bar (tensile)
L = Length of bar (m)
a = Width at bigger end (m)
b = Width at smaller end (m)
E = Young's modulus.
t = Thickness of bar.

Width of the bar at a distance x from origin O.

at
$$x - x' = a - \frac{(a-b)}{L} \cdot x$$

$$= a - kx$$

$$k = \left(\frac{a-b}{L} \right)$$

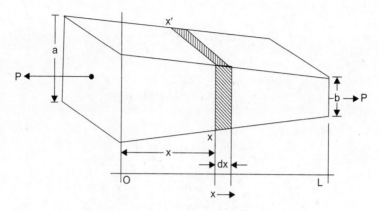

Fig. 7.7

where

thickness of bar at section $x - x' = t$

∴ Area of section $\quad x - x' = (a - kx) \times t$

Stress on the section $\quad x - x' = \dfrac{P}{(a - kx).t}$

and strain at section $\quad x - x' = \dfrac{\sigma}{E}$

⇒ Extension of the small elemental length dx

$$= \text{strain} \times dx$$

$$= \dfrac{\left(\dfrac{P}{(a-kx).t}\right)}{E} dx$$

$$= \dfrac{P.dx}{E(a-kx).t}$$

∴ Total extension of the bar is obtained by integrating the above equation between the limits $x = 0$ and $x = L$

$$= \int_0^L \dfrac{P}{E(a-kx)t} .dx$$

$$= \dfrac{P}{Et} \int_0^L \dfrac{dx}{(a-kx)}$$

$$= \frac{P}{E.t} \ln [(a-kx)]_0^L \left(-\frac{1}{k}\right)$$

$$= -\frac{P}{Et.k}[\ln(a-kL) - \ln a]$$

$$= \frac{P}{Etk}[\ln a - \ln(a-kL)]$$

$$= \frac{P}{Etk}\left[\ln\left(\frac{a}{a-kL}\right)\right]$$

Putting the value of k.

$$\delta L = \frac{P}{Et\left(\frac{a-b}{L}\right)}\left[\ln\left(\frac{a}{a-\left(\frac{a-b}{L}\right).L}\right)\right]$$

$$\delta L = \frac{P.L}{E.t(a-b)}\ln\left(\frac{a}{b}\right)$$

7.6 EXTENSION OF BAR DUE TO SELF WEIGHT

For the bar shown in the Fig. 7.8
    A = cross-sectional area
    l = length of bar which is hanging freely under its own weight
    dy = small element at a distance y from lower end.
    w = specific weight (weight per unit volume)
Total tension at section BB'

$$P = w \text{ (weight per unit volume)} \times \text{(volume)}$$

$$= w \times A \times y.$$

Because of this load the elemental length dy elongates by a small amount Δx

$$\Delta x = \frac{Pdy}{AE} = \frac{wAy}{AE}dy = \frac{w}{E}y\,dy$$

By integrating the above expression between the limits $y = 0$ and $y = 1$, we can calculate the total change in length of the bar due to self weight

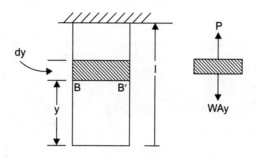

Fig. 7.8

$$\delta l = \int_0^l \frac{w}{E} y \, dy = \frac{w}{E}\left[\frac{y^2}{2}\right]_0^l$$

$$= \frac{w}{E}\frac{l^2}{2}$$

If W is the total weight of the bar ($W = wAl$), then $w = \dfrac{W}{Al}$. Putting this in the above equation,

$$\delta l = \left(\frac{W}{Al}\right)\frac{l^2}{2E} = \frac{Wl}{2AE}$$

Thus, total extension of the bar due to self weight is equal to the extension that would be produced if one half of the weight of the bar is applied at its end.

7.7 STRESS-STRAIN DIAGRAM

Stress-strain diagram is a graphical plot of stress versus strain. The graph is obtained on a machine called universal testing machine or tensile testing machine in labs. For plotting the curve a metallic bar of uniform cross-section is subjected to a gradually increasing tensile load till failure of the bar occurs. The testing specimen is shown below. The specimen has collars provided at both the ends for gripping it firmly in the fixtures of the machine. The central portion of the test specimen is somewhat smaller than the end regions and this central section constitutes the gauge length over which elongations are measured. A dial gauge is used to measure very small changes in length. After that vernier scale on the machine is used to measure extension. Load and extension are simultaneously recorded till the specimen breaks.

Following calculations are made:

$$\text{Stress} = \frac{\text{Load}}{\text{Original cross-sectional area of the test specimen}}$$

$$\text{Strain} = \frac{\text{Extension of a given length}}{\text{Original unstrained length}}$$

$$\text{Strain } \varepsilon = \frac{\delta l}{l} \text{ stress } \sigma = \frac{P}{A}$$

l = gauge length
A = original cross-sectional area

Stress-strain curve is then plotted by having obtained numerous pairs of values of stress and strain—stress plots as ordinate and strain as abscissa on a graph taking suitable scale.

Proportionality limit: (Point A) This is the point on stress-strain curve till which Hook's law is followed or stress $\propto$ strain. It is also called limit of proportionality.

Elastic limit: Beyond point A stress and strain depart from straight line relationship. The material however remains elastic upto state point B. Stress at B is called the elastic limit stress. This represents the maximum unit stress to which a material can be subjected and is still able to return to its original form upon removal of load.

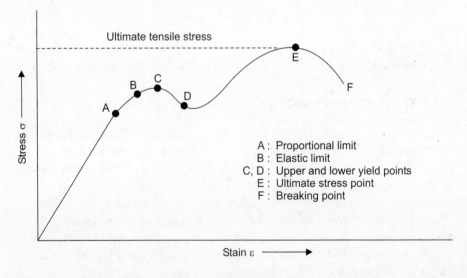

Fig. 7.9

Yield point: Beyond yield point the behaviour of the material is inelastic, i.e., there is no tendency of the material to return to its original shape. This onset of plastic deformation is called yielding of the material. Yielding pertains to region C-D and there is drop in load at the point D. The point C is called the upper yield point and point D is the lower yield point.

Ultimate strength: After yielding the material becomes strain hardened (strength of the specimen increases) and an increase in load is required to make the material to its maximum stress at point E. Point E represents the maximum ordinate of the curve and the stress at this point is known as ultimate stress.

Breaking strength: In the portion EF, there is falling off the load (stress) from the maximum until fracture takes place at F. The point F is referred to as the fracture or breaking point and the corresponding stress is called the breaking stress.

The above-mentioned curve is valid for most of the ductile material.

Stress-strain Curve for Brittle Materials

Materials which show very small elongation before they fracture are called brittle materials such as C.I. tool steel etc.

7.8 HOOKE'S LAW AND ELASTIC CONSTANTS

Hooke's law states that upto the proportionality limit the stress is proportional to strain or the experimental values of the stress vs. strain lie on a straight line.

Stress α strain

$$\sigma \, \alpha \, \varepsilon$$

$$\sigma = E \, \varepsilon$$

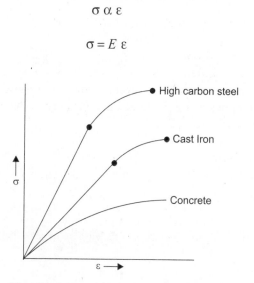

Fig. 7.10: Stress-strain curve for brittle materials.

E = modulus of elasticity or Young's modulus.

In case of shear stress and strains, the shear stress is proportional to shear strain and the Hooke's law can be defined as

$$\tau \alpha \gamma$$

$$\tau = C \gamma$$

C = modulus of rigidity or shearing modulus of elasticity

If a cuboid is subjected to three mutually perpendicular normal stresses σ_x, σ_y and σ_z of equal intensity on its faces so that it gets volumetric strain ε_v. Then the bulk modulus is defined as:

$$\text{Bulk modulus } K = \frac{\text{Volumetric stress}}{\text{Volumetric strain}}$$

$$\text{Volumetric strain } \varepsilon_v = \frac{\text{Change in volume of cuboid}}{\text{Original volume of cuboid}}$$

$$\sigma = \frac{\sigma_x + \sigma_y + \sigma_z}{3}$$

$$\text{(Bulk modulus) } K = \frac{\sigma_x + \sigma_y + \sigma_z}{3 C_v}$$

Relation between elastic constants E K and C.

(i) Relation between E, K and v

Consider a cubical element subjected to volumetric stress σ which acts simultaneously along the mutually perpendicular x, y and z-direction.

The resultant strains along the three directions can be worked out by taking the effect of individual stresses.

Strain in the x-direction

ε_x = strain in x-direction due to σ_x–strain in x–direction due to σ_y–strain in x-direction due to σ_z

$$= \frac{\sigma_x}{E} - v\frac{\sigma_y}{E} - v\frac{\sigma_z}{E}$$

But $\sigma_x = \sigma_y = \sigma_z$

$$\varepsilon_x = \frac{\sigma}{E} - v\frac{\sigma}{E} - v\frac{\sigma}{E} = \frac{\sigma}{E}(1-2v)$$

Similarly,

$$\varepsilon_y = \frac{\sigma}{E}(1-2v) \text{ and } \varepsilon_z = \frac{\sigma}{E}(1-2v)$$

Volumetric strain $\varepsilon_v = \varepsilon_x + \varepsilon_y + \varepsilon_z = \frac{3\sigma}{E}(1-2v)$

Now bulk modulus $K = \dfrac{\text{Volumetric stress}}{\text{Volumetric strain}}$

$$= \frac{\sigma}{\frac{3\sigma}{E}(1-2v)} = \frac{E}{3(1-2v)}$$

$$E = 3K(1-2v)$$

Relation between *E*, *C* and v

Consider a square block ABCD of side L and of thickness unity perpendicular to the plane of drawing as shown in figure. Block ABCD is fixed at the bottom face and subjected to shearing force at the top face.

Due to these stresses the cube is subjected to some distortion; such that the diagonal *AC* will get elongated and diagonal *BD* will be shortened.

Longitudinal strain in diagonal *AC*

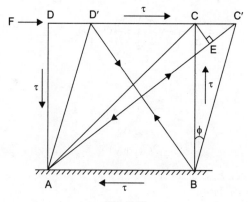

Fig. 7.11

$$= \frac{AC' - AC}{AC} = \frac{AC' - AE}{AC} = \frac{EC'}{AC}$$

where CE is perpendicular from C into AC'

Since extension CC' is small, $\angle AC'B$ can be assumed to be equal to $\angle ACB$ which is 45°. Therefore,

$$EC' = CC'\cos 45° = \frac{CC'}{\sqrt{2}}$$

$$\text{Longitudinal Strain} = \frac{CC'}{\sqrt{2}AC} = \frac{CC'}{\sqrt{2}\times\sqrt{2}BC} = \frac{CC'}{2BC}$$

from triangle BCC' : $\dfrac{CC'}{BC} = \tan\phi$

$$\therefore \text{longitudinal strain} = \frac{\tan\phi}{2} = \frac{\phi}{2} \tag{1}$$

where $\phi = \dfrac{CC'}{BC}$ represents the shear strain.

In terms of shear stress τ and modulus of rigidity C, shear strain = $\dfrac{\tau}{C}$

$\therefore$ Longitudinal strain of diagonal $AC = \dfrac{\tau}{2C}$.

The strain in diagonal AC is also given by

= Strain due to tensile stress in AC − strain due to compressive stress in BD.

$$= \frac{\tau}{E} - \left(-\nu\frac{\tau}{E}\right) = \frac{\tau}{E}(1+\nu) \tag{2}$$

From (1) and (2)

$$\frac{\tau}{2C} = \frac{\tau}{E}(1+\nu)$$

$$E = 2C(1+\nu)$$

(iii) Relation between E, C and K.
From the above relation.

$$E = 2C(1+v) = 3K(1-2v)$$

To eliminate v from these two expressions for E, we have

$$v = \frac{E}{2C} - 1 \text{ and } E = 3K\left[1 - 2\left(\frac{E}{2C} - 1\right)\right]$$

$$E = 3K\left[1 - \left(\frac{E}{C} - 2\right)\right] = 3K\left[3 - \frac{E}{C}\right] = 9K - \frac{3KE}{C}$$

$$E + \frac{3KE}{C} = 9K; \quad E\left(\frac{C + 3K}{C}\right) = 9K$$

$$E = \frac{9KC}{C + 3K}$$

SOLVED PROBLEMS

Problem 1

A Specimen for tensile has a circular cross section and enlarged ends as shown in figure. The total elongation is 0.1 mm. The length of the enlarged ends are equal to 50 mm and diameter 25 mm for each end. The tensile load applied at the ends are 90 kN and stress produced in the middle portion $E = 2 \times 10^5$ MPa.

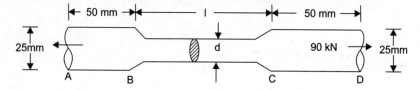

Solution

Cross sectional area of the middle portion $= \dfrac{\pi d^2}{4}$

Stress produced in the middle portion

$$= 45 \times 10^6 \ N/m^2 = \frac{90 \times 10^3}{\dfrac{\pi d^2}{4}} = \frac{4 \times 90 \times 10^3}{\pi d^2}$$

or

$$d^2 = \frac{4 \times 90 \times 10^3}{\pi \times 450 \times 10^6} = 2.546 \times 10^{-4}$$

$$d = 0.015 \ m = 15 \ mm \ \text{Answer}$$

elongation produced in end parts = $2 \times \left[\dfrac{\sigma_{AB}}{E} \times 0.05 \right]$

$$\delta l_{(AB+CD)} = 2 \times \left\{ \dfrac{90 \times 10^3}{2 \times 10^5 \times 10^6} \times 0.05 \right\}$$

$$\delta l_{BC} = \dfrac{450 \times 10^6}{2 \times 10^5 \times 10^6} \times l = 2.25 \times 10^{-3} * \text{lm}$$

Total elongation = $0.1 \times 10^{-3} = 4.5 \times 10^{-8} + 2.25 \times 10^{-3}$ l

l = 0.0444 m = 44.42 mm

Problem 2
A brass rod in static equilibrium is subjected to axial load as shown in figure. Find load P and change in length of the bar if its diameter is 100 mm. Take E = 80 GN/m²

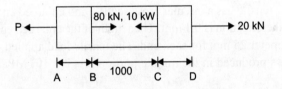

Solution
Drawing free body diagram for each part

$P - 80 = 10$

$P = 90$ kN.

Change in length = $\dfrac{1}{EA}[P_{AB} \times l_{AB} + P_{BC} \times l_{BC} + P_{CD} \times l_{CD}]$

$$= \dfrac{1}{A}\left[\dfrac{90 \times 10^3}{80 \times 10^9} \times 0.5 + \dfrac{10 \times 10^3}{80 \times 10^2} \times 1.0 \times \dfrac{20 \times 10^3}{80 \times 10^9} \times 0.5 \right]$$

$$\delta l = \frac{4 \times 10^4}{80 \times 10^9 + \pi \times (0.1)} \{9 \times 0.5 + 1 + 2 \times 0.5\}$$

$$\delta l = 9.549 \times 10^{-3} \text{ m (Ans)}$$

Problem 3
A rod tapers uniformly from 50 mm to 20 mm diameter in a length of 500 mm. If the rod is subjected to an axial load of 10 kN find the extension of the rod. Assume $E = 2 \times 10^5$ Mpa.

Solution

$d_1 = 0.05$ mm

$d_2 = 0.02$ m

$l = 0.5$ m

$P = 10 \times 10^3$ N

$$\delta l = \frac{4Pl}{\pi E d_1 d_2}$$

(since equivalent area of cross section = $\frac{\pi d_1 d_2}{4}$)

$$\delta l = \frac{4 \times 10 \times 10^3 \times 0.5}{\pi \times 2 \times 10^{11} \times 0.05 \times 0.02} = 3.1830 \times 10^{-5} \text{ m.}$$

=0.0318 mm

Problem 4
For a given material, Young's modulus is 110 GN/m² and shear modulus is 42 GN/m². Find the bulk modulus and lateral contraction of a round bar of 37.5 mm diameter and 2.4 m long when stretched 2.5 mm.

Solution

$E = 110$ GN/m² $C = 42$ GN/m²

$E = 2c(1+v)$

$$v = \frac{E}{2c} - = \frac{110}{2 \times 42} - 1 = 0.30$$

$E = 3K(1-2v)$

$$K = \frac{E}{3(1-2\nu)} = \frac{110}{3(1-2\times 0.3)} = 91.66 \text{ GN/m}^2$$

$$K = 91.66 \text{ GN/m}^2$$

$$\nu = \frac{\text{Lateral strain}}{\text{Longitudinal strain}} = \frac{\delta d/D}{\delta l/L} = \frac{\delta d \times L}{\delta l \times D}$$

$\delta d = 0.0117$ mm.

EXERCISE

1. Define stress and strain? (M.D.U. Jan 2010)
2. What is Hook's law? Derive an expression for relationship between elastic constants. (M.D.U. Jan 2010)
3. Explain stress and strain. (M.D.U. Jan 2010)
4. Etablish a relation between E (Young's modulus of elasticity), G (modulus of Rigidity) and bulk modulus. (DRIU Dec 2009_
5. Define Poission's ratio. (MRIU Dec. 2009)
6. Draw stress diagram for a ductilematerial (MRIU Dec 2009)
7. Two copper rods one steel rod lie in a vertical plane and togather support a load of 40 kN as shown in figure. Eachrod is 25 mm in diameter, length of steel rod is 3 m and length of eachcopper rod is 2 m. If modulus of elaticity of steel is twice that of copper make calculations for the stress induced in each rod. It may be presumed that each rod deforms by the same amount.

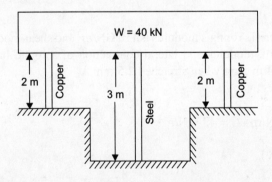

8. A bar 2 cm × 4 cm in cross section and 40 cm long is subjected to radia tensile load of 70 kN. It is found that the length increases by0.176 mm and lateral dimension of 4 cm decreases by0.0044 mm Find

(a) Young's modulus
(b) Poisson ratio
(c) change in volume of the bar
(d) Bulk Modulus.

OBJECTIVE TYPE QUESTIONS

1. Strain is defined as the ratio of
   (a) change in volume to original volume
   (b) change in length to original length
   (c) change in cross sectional area to original cross sectional area
   (d) any one of the above
2. Hooke's law holds good upto
   (a) yield point
   (b) limit of proportionality
   (c) breaking point
   (d) elastic limit
3. Young's modulus is defined as the ratio of
   (a) Volumetric stress and volumetric strain
   (b) lateral stress and lateral strain
   (c) longitudinal stress and longitudinal strain
   (d) shear stress to shear strain
4. The unit of Young's modulus is
   (a) mm/mm
   (b) kg/cm
   (c) kg
   (d) kg/cm^2
5. Deformation per unit length in the direction of force is known as
   (a) strain
   (b) lateral strain
   (c) kg
   (d) kg/cm^2
6. Deformation per unit length in the direction of force is known as
   (a) strain
   (b) lateral strain
   (c) linear strain
   (d) linear stress
7. If equal and opposite forces applied to a body tend to elongate it, the stress so produced is called
   (a) internal resistance
   (b) tensile stress
   (c) transverse stress
   (d) compressive stress

8. Modulus of rigidity is defined as the ratio of
    (a) longitudinal stress and longitudinal strain
    (b) volumetric stress and volumetric strain
    (c) lateral stress and lateral strain
    (d) shear stress and shear strain
9. The intensity of stress which causes unit strain is called
    (a) unit stress
    (b) bulk modulus
    (c) modulus of rigidity
    (d) modulus of elasticity
10. Which of the following has no unit
    (a) kinematic viscosity
    (b) surface tension
    (c) bulk modulus
    (d) strain
11. The value of Poisson's ratio for steel is between
    (a) 0.01 to 0.1
    (b) 0.23 to 0.27
    (c) 0.25 to 0.33
    (d) 0.4 to 06
12. The property of a material by virtue of which a body returns to its original shape after removal of the load is called
    (a) plasticity
    (b) elasticity
    (c) ductility
    (d) malleability
13. Poisson's ratio is defined as the ratio of
    (a) longitudinal stress and longitudinal strain
    (b) longitudinal stress and lateral stress
    (c) lateral stress and longitudinal stress
    (d) lateral stress and lateral strain
14. The change in the unit volume of a material under tension with increase in its poission's ratio will
    (a) increase
    (b) decrease
    (c) remain same
    (d) increase initially and then decrease
15. If a material expands freely due to heating it will develop
    (a) thermal stresses
    (b) tensile stress
    (c) bending
    (d) no stress

16. In the tensile test, the phenomenon of slow extension of the material i.e. stress increases with the time at a constant load is called
    (a) creeping
    (b) yielding
    (c) breaking
    (d) plasticity
17. The stress developed in a material at breaking point in extension is called
    (a) breaking stress
    (b) fracture stress
    (c) yield point stress
    (d) Ultimate tensile stress
18. The elasticity of various materials is controlled by its
    (a) Ultimate tensile stress
    (b) proof stress
    (c) stress at yield point
    (d) stress at elastic limit
19. Five specimens of MS have their lengths and diameters as l, d; $2l, 2d$; $3l, 3d$; $4l, 4d$ and $5l, 5d$. Which of these will have largest extension when the same tension is applied to all of them
    (a) first
    (b) second
    (c) third
    (d) fourth
20. The energy absorbed in a body, when it is strained within the elastic limits, is known as
    (a) strain energy
    (b) re silence
    (c) proof resilience
    (d) modulus of resilience
21. Resilience of a material is considered when it is subjected to
    (a) frequent heat treatment
    (b) fatigue
    (c) creep
    (d) modulus of resilience
22. The relationship between modulus of elasticity E, modulus of rigidity G is
    (a) $E = G(1 + \mu)$
    (b) $G = E(2 - \mu)$
    (c) $G = \dfrac{E}{2(1+\mu)}$
    (d) $G = \dfrac{E}{1+2\mu}$

23. Volumetric strain of a rectangular body subjected to an axial force in terms of linear strain e and Poisson's ratio μ is equal to
    (a) $e(1 - 2\mu)$
    (b) $e(1 - \mu)$
    (c) $e(1 - 3\mu)$
    (d) $e(1 + \mu)$

24. A cube is subjected to three mutually perpendicular tensile stresses s, the volumetric strain will be
    (a) $\dfrac{3s}{E}(1 - 2\mu)$
    (b) $\dfrac{E}{2s}(1 - 2\mu)$
    (c) $\dfrac{E}{2s}(2\mu - 1)$
    (d) $\dfrac{3s}{E}(2\mu - 1)$

25. Bulk modulus K is in terms of modulus of elasticity E and Poisson's ratio μ is given as equal to
    (a) $\dfrac{E}{3(1 - 2\mu)}$
    (b) $E(1 - 2\mu)$
    (c) $3E(1 - 2\mu)$
    (d) $\dfrac{E}{3}(1 + 2\mu)$

ANSWERS

1. d	2. b	3. c	4. d	5. c
6. b	7. d	8. d	9. d	10. b
11. b	12. c	13. b	14. d	15. a
16. a	17. d	18. a	19. a	20. a
21. c	22. c	23. a	24. a	25. a

UNIT 8

Manufacturing System

8.1 INTRODUCTION

A manufacturing system can be defined as a collection of integrated equipment and human resources whose function is to perform one or more processing and/or assembly operations on a starting raw material, part or set of parts.

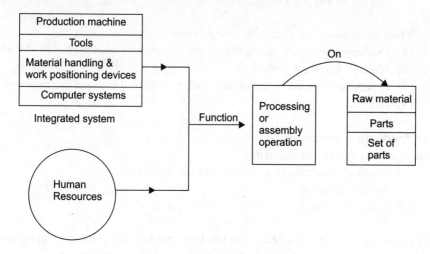

Fig. 8.1: Manufacturing system.

The manufacturing system is where the value-added work is accomplished on the part or product.

The examples of manufacturing system are:
(1) A team of workers producing raw tyre in a production line.
(2) A group of automated machines working on automated cycles to produce a family of similar parts.
(3) One worker working on one machine, which operates on semi-automatic cycle.
(4) Worker working in tractor assembly plant and assembling the engine.

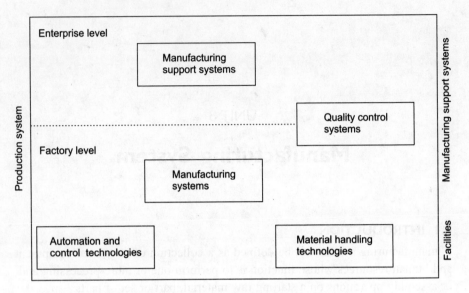

Fig. 8.2: Components of a manufacturing system.

Main components of a manufacturing system are:
(1) Production machines along with tools, fixtures and other related hardware.
(2) Material handling system.
(3) Computer systems to coordinate and/or control the above components.
(4) Human resources.

(1) Production machines
The machines can be classified as:
 (1) Manually operated (e.g., lathes, milling machines).
 (2) Semi-automated.
 (3) Fully automated.

(1) Manually operated machine: The machine provides the power for operation and the worker provides the control, e.g., lathes, milling machines etc.

(2) Semi-automated machine: It performs a portion of the work cycle under some form of program control, and a human worker tends to the machine for the remainder of the cycle, by loading and unloading it or performing some other task during each cycle.
Example: CNC lathe

(3) Fully automated machine: It has a capacity to operate for extended periods of time with no human attention. A worker is not required to be present during each cycle.

Example—fully automatic injection moulding machine. In this periodically collection bin full of moulded parts at the machine must be taken away and replaced by an empty bin.

(2) Material handing system
For this following units have to be provided:
   (a) Loading and unloading work units.
   (b) Positioning the work units at each station.
   (c) Transporting work units.
   (d) A temporary storage which ensures that work is always present for the stations.

Loading involves moving the work units into the production machine. Positioning provides for the part to be in a known location and orientation relative to the workhead or tooling that performs the operation. Unloading when the production operation has been completed, the work unit must be unloaded. Transportat means work transport between stations

(3) Computer controlled system
The function includes:
   (a) Communicate instructions to workers
   (b) Download part programs to computer-controlled machines (e.g., CNC machines)
   (c) Material handling system control
   (d) Schedule production
   (e) Failure diagnosis
   (f) Safety monitoring
   (g) Quality control
   (h) Operations management

(4) Human resources
In manufacturing systems, human performs some or all of the value-added work that is accomplished on the parts or products.

8.2 CLASSIFICATION OF MANUFACTURING SYSTEM

The various ways of classification of manufacturing system are:
(1) Type of operations performed
   (a) Processing operation
   (b) Assembly operation
(2) Number of workstations and system layout
   (a) One station
   (b) Many stations

(3) Level of automation
   (a) Manual
   (b) Semi-automated
   (c) Fully automated
(4) Part or product variety
   (a) Identical units
   (b) Variations in work units

8.3 FLEXIBILITY IN MANUFACTURING SYSTEM (FMS)

Flexibility is the term used for the attribute that allows a mixed model manufacturing system to cope with a certain level of variation in part or product style without interruptions in production for changeovers between models. Flexibility is generally a desirable feature of a manufacturing system. Systems that possess it are called flexible manufacturing systems. They can produce different part styles or can readily adopt to new part styles when the previous ones become obsolete.

FMS possesses following features:
   (a) Identification of different work units
   (b) Quick changeover of operating instructions
   (c) Quick changeover of physical set up.

8.4 FUNDAMENTALS OF NUMERICAL CONTROL (NC)

Introduction

The simplest definition of numerical control (NC) given by Electronic Industries Association (EIA) is "A system in which the actions are controlled by direct insertion of numerical data at some point. The system must automatically interpret at least some portion of this data."

NC machines are controlled by prepared programs which consist of blocks or series of numbers, alphabet or alphanumerics. These numbers define the required position of each machine slide, feed, cutting speed and depth of cut. For some functions such as coolant on/off, tool change, etc. certain codes are used. The data for preparing the coded instructions, called part program, is taken from the finished component drawing. In cases of the manufacture of complicated parts, the system calculates the additional data points automatically.

NC Machines

In the initial stages, the NC machine tools had NC systems added to the machine but only to control the position of the workpiece relative to the cutting tool. Further with passage of time the capabilities of machine tools improved and in addition to maintaining cutter/workpiece relationship, the material removal was also controlled by the numerical control system.

The mechanical design of the machine tool was also improved with the development of recirculating ball screw and better slideways.

The instructions to the NC machines are fed through an external medium, i.e., paper tape or magnetic tape. The information read from the tape is stored into the memory of the control system called buffer storage and is processed by the machine step by step. So when the machine is working on one instruction block, the next block is read from the tape and stored in the memory of the machine control system. Since the part cannot be produced without a tape being run through the machine, the tape has to be run repeatedly depending on the number of components to be produced. For even a minor change in the design component, the tape has to be discarded and new tape with changed program has to be produced.

Constructional details of N.C. Machines

The basic units of NC machines are:

(a) Software
(b) Machine Control Unit (MCU)
(c) Machine Tool (MT)

(a) Software: The programs or set of instructions, languages, punched cards, magnetic tape, punched paper tape and other such information processing items are referred to as software.

The software controls the sequence of movement of an N.C. That is why, these NCs are sometimes called software controlled machines and the skill required in producing a part by NC lies entirely in the programming. Programming of a numerical operation is executed mostly manually. The programmer plans the operations are their sequence from seeing the drawing and writes instructions in tabulated blocks of information, known as part programme on a programme manuscript. Then these instructions are punched on the control tape. Tape reader reads the code and sends it to MCU which conversely converts them into the machine movements of machine tool.

The part program is written in coded form, which is punched on tape in binary format. No complete standardization of codes has yet been achieved, but more or less all codes use binary numbers.

Special Purpose Manufacturing Machines (SPMM). To make the product, within control tolerances and with least wastage and requires special purpose or automatic machines. It is also called fixed automation and is feasible only for long mass productions, such as automobile, home appliances, newspapers etc. It may use machines like automatic lathes having fixed sequence of operations.

(b) Machine Control Unit (MCU): Every NC machine tool has a main unit which is known as MCU, it consists of some electronic circuit called hardware that reads the NC programme, interprets it and conversely translates it for mechanical actions of the machine tool.

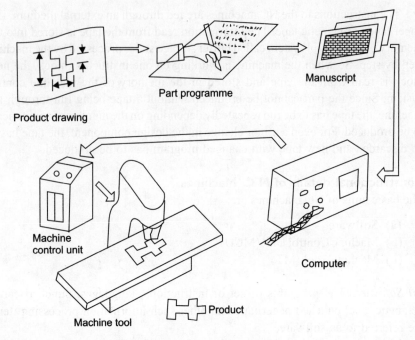

Fig. 8.3: Process layout.

The MCU may be of three types:
   (i) Housed MCU
   (ii) Swing around MCU
   (iii) Stand alone MCU

(i) Housed MCU: This MCU, itself may be mounted on the machine tool or may be built in the casing of the machine.

(ii) Swing around MCU: This MCU is directly mounted on the machine which can swing around it and can be adjusted as per requirement of the operator's position. This arrangement provides large working space around the machine.

(iii) Stand-alone MCU: This MCU is enclosed in a separate cabinet which is installed at some remote or some place near to the machine.

Different parts of MCU: A typical MCU may consist of the following units:
   (1) Input/reader unit
   (2) Memory
   (3) Processor
   (4) Output channels and actuators
   (5) Control panel
   (6) Feedback channels and transducers.

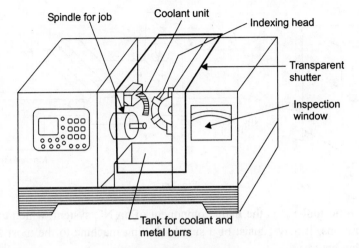

Fig. 8.4: House MCU.

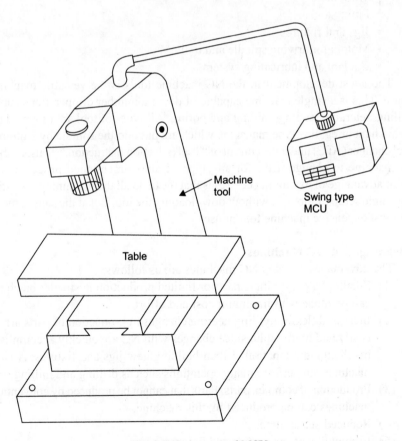

Fig. 8.5: Swing around MCU.

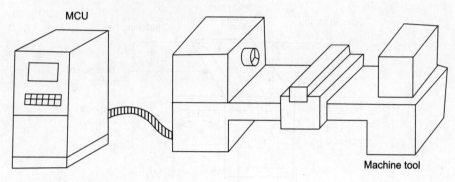

Fig. 8.6

(c) Machine tools: It is the main component of an NC system, which executes the operations. It may consist of a simple drilling machine to the most flexible machining centres.

It includes different part/sub-assemblies:
- Work table
- Fitting tools
- Jigs and fixtures
- Motors for driving spindle and tool
- Coolant and lubricating system.

The latest development in the NC machine tools is the versatile machining centre. This is a single machine capable of doing a number of operations such as drilling, reaming, tapping, milling and boring. All types of tool are mounted on a drum/chain or eggbox type magazine which are put into the spindle by Automatic Tool changer (ATC) under the control of "Tool Selection Instruction." Thus, enabling the machine to change itself for any type of above-mentioned processes. The major advantage of machining centres is that it can do all the machining operations in a single setup of job, i.e., without downloading the job. But at the same time it is the most expensive machine tool in use.

Advantages of NC Machines

The advantages of using NC machines are as follows:
(1) Smaller batches: Single parts or limited production in smaller batch runs can be made with minimum cost and effort.
(2) Increased flexibility: Engineering changes in production of parts are less costly and more rapid, since changes with NC are quickly accomplished by changing a tape rather than building new jigs and fixtures. A single machine can perform many operations such as drilling and milling on it.
(3) Production of complex parts: Parts that cannot be produced by conventional machines can be produced by this machine.
(4) Reduced set-up time.
(5) Elimination of special jigs and fixtures.

(6) Machine utilisation: The NC machine is utilised for the 80% of time as it requires a little of the operator's attention. Hence, the operator in the mean time is free to do the non-productive activities while the machining is going on such as presetting of tools and setting of other jobs.
(7) Machining accuracy: The consistent products lead to better assembly and reduced fitting costs.
(8) Lesser scrap: The reliability of the system eliminates most of the human error associated with manual operation, e.g., in correct hole location errors are entirely eliminated. This reduces the cost of scrap by more than 50%.
(9) Reduced inspection: Inspection costs are greatly reduced because of the reliability of the NC system and do not needful inspection. Hence, the cost to transport parts to inspection site is also reduced.
(10) Longer tool life: Speeds and feeds of the tool can be selected accurately according to the material of the workpiece. This reduces the wear and tear of the tool giving it a longer cutting life.
(11) Tool storage cost. As the tools are more readily located in a computer controlled tool index system and hence there is no need of separate storage system of the tools causing reduction in tool storage cost.
(12) Power labour cost: One operator can look after several machines at a time or multipallet machines resulting in reduction of labour cost.
(13) Cost of modification: The production is not stopped while a modification is introduced because a new tape can be quickly prepared with very less cost.
(14) Reduced floor space: Although the floor space occupied by on NC machine is usually greater than a conventional machine but it will do different machining operations in a single set-up, i.e., the output of one NC machine is equivalent to the output from several conventional machines. Hence, it requires an overall less floor space.
(15) Less human error: The possibility of human error is greatly reduced as the management instructions are converted on a paper tape or card and this paper tape controls the machine. Hence, there is no chance for the human emotions or fatigue to come in affecting the quality of product.
(16) Easy and effective production planning: The production planning is easier and more effective with NC equipment because manufacturing capacity is more constant, predictable and efficient. Cost estimates are improved because of the reliability and efficiency of NC machines.
(17) Reduction in transportation cost: The machining centre reduces the number of times a component has to be moved from machine to machine, thus reducing the transport cost during the production. The time spent in awaiting the availability of the machine is also eliminated.

Advances in NC Systems

The computers were used in place of control units of NC increasing the capacity and capability of it and reducing the size of machine control hardware. The use of computers emphasised on controlling the NC with software instead of hardware. There are two types of computer control for NC machines:
(1) Computer Numerical Control CNC
(2) Direct Numerical Control DNC.

Computer Numerical Control (CNC) Machines

Definition

It is self-contained NC system for a single machine tool which uses a dedicated minicomputer, controlled by the instructions stored in its memory, to perform all the basic numerical control functions.

As given by EIA

The numerical control system where a dedicated stored program computer is used to perform some or all of the basic numerical control functions in accordance with control programs stored in read/write memory (RAM) of the computer.

Nowadays CNC units have more memory, processing speed and more intelligence built into the computer. A typical CNC may need only the drawing specifications of a part to be manufactured and the computer automatically generates the part program for the loaded part.

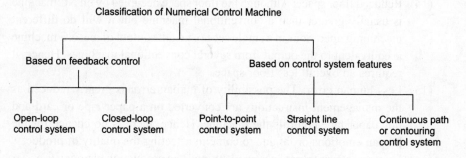

Fig. 8.7: Classification of CNC.

Classification based on Feedback Control

Open-loop control system

Machine tool in which there is no provision to compare the actual position of the cutting tool or workpiece with the input command value are called open-loop system.

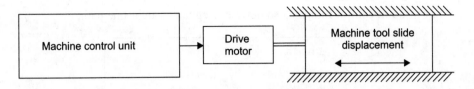

Fig. 8.8: Block diagram of an open-loop system

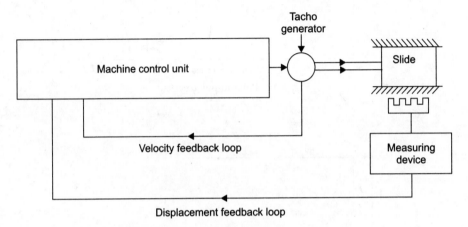

Fig. 8.9: Block diagram of a closed-loop system.

Closed-loop control system
In a closed-loop control system, the actual output from the system, i.e., actual displacement of the machine slide is compared with the input signal.

Classification based on Control System Features

(i) Point-to-point control system
Point-to-point control is one where accurate positional control is required only to place the machine slides in fixed position and the machine tool slide is required to reach a particular fixed coordinate point in the shortest possible time.

(ii) Straight line control system
In this there is a provision of machining along a straight line as in the case of milling and turning operations. This is obtained by providing movement at controlled feed rate along the axis in the line of motion.

Continuous path or contouring system

The contouring system generates a continuously controlled motion of the tool and workpiece along different coordinate axis. This control system enables the machining of profiles, contours and curved surface.

Point-to-point control follows a somewhat irregular straight line path.

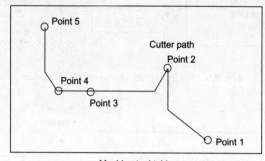

Fig. 8.10: Point-to-point system

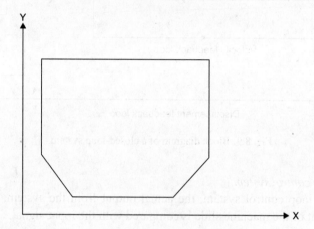

Fig. 8.11: Straight line system

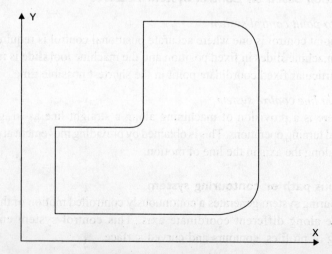

Fig. 8.12: Contouring system.

Comparision of CNC and NC

		CNC	NC
(1)	Flexibility	More	Less
(2)	Date reading error	Less	More
(3)	Editing of program	Easily edited	Difficult
(4)	Diagnostic	By diagnosing the program it can detect the machine malfunctioning even before the part is produced.	Cannot diagnose.
(5)	Conversion of units	Inches can be converted into SI units in memory only	Have to be changed
(6)	Cost	High	Low
(7)	Maintenance cost	More	Less
(8)	Operator required	Skilled	Less skilled
(9)	Long-run application	Not suitable	Suitable

EXERCISE

1. Explain NC machines in details? (MDU Jan 2010)
2. Write down difference between NC & CNC machines (MDU January 2010)
3. Explain manufacturing system (MDU Jan 2010)
4. What is MCU in NC machines (MDU Jan 2010)
5. What are the different waysof classification of manufacturing system?
6. What is flexibility in manufacturing system
7. What are the advantages of NC machines
8. How can CNC be classified.

OBJECTIVE TYPE QUESTIONS

1. For a functional N.C. system doing all the operations following basic elements are required.
   (a) Software
   (b) Machine Control Unit (MCU)
   (c) Machine Tool
   (d) all of the above.
2. E I A stands for
   (a) Electronic Industries Association.

(b) Electronic Intel Association.
(c) Essar Integrated Association.
(d) Elements of Industries Association.
3. NC machines used following type of codes.
   (a) E I A codes.
   (b) Iso/Asc IF code.
   (c) Both the above two.
   (d) None of the above.
4. Optical reader uses _____ for light sensing
   (a) photo cells.
   (b) papers.
   (c) plate.
   (d) electrodes.
5. M C U stands for
   (a) Matter control unit.
   (b) Motor control unit
   (c) Machine control unit
   (d) Mood control unit
6. Floppy discs are coated with
   (a) Magnetic oxide.
   (b) Calcium oxide.
   (c) Ammonium oxide.
   (d) Hydrogen oxide.
7. CNC stands for
   (a) Computer Numerical control.
   (b) Computer Nortan control.
   (c) Case Numerical control.
   (d) Computer Not control.
8. Following is the advantage of CNC
   (a) Reduced lead time.
   (b) lower labour cost.
   (c) longer tool life.
   (d) All of the above.
9. Following are the disadvantages of CNC machines.
   (a) Higher Investment cost.
   (b) Higher Maintenance cost.
   (c) Costlier CNC personnel.
   (d) All of the above.
10. The following factors should be considered while selecting components for machining on CNC machine tools.
    (a) Operations are very complex.
    (b) Size of batches is medium.

(c) Labour cost for the component is high.
(d) All of the above.
11. The basic components of an operational numerical control system are
    (a) Programme of instructions
    (b) Controller unit
    (c) Machine tool.
    (d) All of the above.
12. NC/CNC control systems are classified as.
    (a) point to point control system.
    (b) Straight line control system.
    (c) Continuous path or contouring control system.
    (d) All of the above.

ANSWERS

1. d	2. a	3. c	4. a	5. c
6. a	7. a	8. d	9. d	10. d
11. d	12. d			

UNIT 9

Shear Force and Bending Moment Diagram

9.1 BEAM

Any member of a machine or structure whose one dimension is very large as compared to the other two dimensions and which can take lateral forces in the axial plane is called a beam. The lateral forces acting on the beam produce bending in the beam.

Special feature in the beam is that internal forces called shear forces and the internal moments called bending moments are developed in it, to resist the external load.

9.2 SHEAR FORCE

The shear force at any section of a beam shows the tendency to slide one layer of the beam over other layer laterally. It is a tangential force.

Shear Force Diagram (SFD)
It is a graphical method to show the variation of shear force along the length of beam.

Sign convention of (SFD)
Shear force will be positive when the resultant of the forces to the left is upwards or to the right is downward and negative if resultant of the forces to the left is downward and to the right is upward, with reference to a section.

Bending Moment
An internal resisting moment or a couple developed within the cross-sectional area of the cut, to counteract the moment caused by external forces, is termed as bending moment.

Shear Force and Bending Moment Diagram

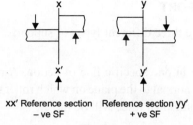

XX' Reference section   Reference section YY'
     – ve SF   + ve SF

Fig. 9.1

Bending moment at any section of the beam shows the tendency to bend the beam and it is equal to algebraic sum of all the moments due to all forces acting either side of the section.

Bending moment diagram (BMD)
Bending moment diagram (BMD) is a graphical representation of variation of bending moment along the length of the beam.

Sign Convention
Positive B.M.:
  (a) Concavity at the top of the beam
  (b) Moment on the left portion is clockwise and on the right portion is anti-clockwise.
  (c) Also called sagging bending moment.

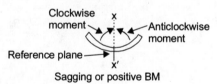

Fig. 9.2

Negative B.M.
  (a) Convexity at the top of the beam.
  (b) Moment on the left-hand side is anticlockwise and on the right-hand clockwise.
  (c) Also known as hogging bending moment.

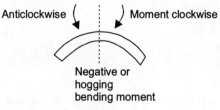

Fig. 9.3

9.3 TYPES OF SUPPORT

The beams usually have three different types of support.

(1) Roller support

It can resist force only in one specific line of action. A roller can resist only a vertical force or a force normal to the plane on which roller moves. On this support only a single reaction unknown force will be there.

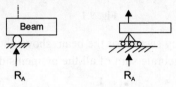

Fig. 9.4

(2) Hinged or pin support

In hinged support the reaction has two components. One in horizontal and another in vertical. Two equations of statics are used to determine these two components. A hinged support no translation movement is possible but beam is free to rotate. If there is no horizontal component of external force then horizontal component of reaction at hinge support will be zero. The hinged and roller supports are also termed as simple supports.

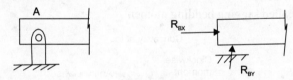

Fig. 9.5

(3) Fixed support

It is capable of resisting a force as well as couple or moment in any direction, e.g. beam fixed in wall. A system of three forces can exist at such a support. (i.e., two components of force and a moment)

9.4 TYPES OF BEAMS

Based on support condition.
(1) Cantilever beam
(2) Simply supported beam
(3) Overhanging beams
(4) Propped beams
(5) Fixed beams
(6) Continuous beams

Cantilever beam: A cantilever beam is supported at one end only with a suitable restraint to prevent rotation at that end.

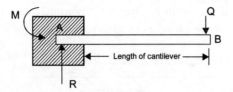

Fig. 9.6

End A is rigidly fixed in support and other end B is fixed.

Simply supported beam: A beam having both the ends freely resting on supports, is called a simply supported beam. The reactions act at point A and B which are the ends of effective span of the beam as shown in figure.

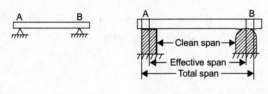

Fig. 9.7

Overhanging beams: If the beam extends beyond the supports than the beam will be called overhanging beam.

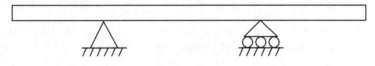

Fig. 9.8

Propped beam: One end is restrained and other end is provided support. Such beams are also called as restrained beams as an end is restrained from rotation.

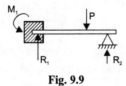

Fig. 9.9

Fixed beam: In this case, both the ends of the beam are fixed rigidly.

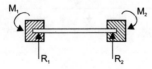

Fig. 9.10

Continuous beam: If the beam is supported by more than two simple supports. Then such a beam will be called as continuous beam.

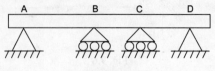

Fig. 9.11

The supports at the end are called end supports, while all the other supports are called intermediate supports.

9.5 TYPES OF LOADING

Variety of ways are available for the distribution of load on the beams. Following are the different ways for the purpose of analysis.
(1) Point or concentrated load
(2) Uniformly distributed load
(3) Uniformly varying load
   (a) Uniformly increasing load ⎫
   (b) Uniformly decreasing load ⎬ Triangular loading
   (c) Parabolic loading
   (d) Cubic loading
(4) Couple
  (i) Point or concentrated load: If the load is concentrated at a point on the beam it is known as point load or concentrated load.

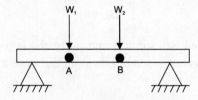

Fig. 9.12

  (ii) Uniformly distributed load: It is generally abbreviated as U.D.L. and its units is kN/m. In this load variation is constant for the particular length of the beam.

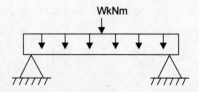

Fig. 9.13

(iii) **Uniformly varying load:** If the intensity of loading increases or decreases at a constant rate or follows a particular equation along the length then it is called uniformly varying load.

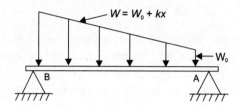

Fig. 9.14

(a) Uniformly increasing load: formulae followed

$$W = W_0 + kx$$

W_0 = loading at reference point

(b) Uniformly decreasing load

$$W = W_0 - kx$$

W_0 = loading at reference point

Parabolic loading

Formulae used

$$W = W_0 + k_1 x + k_2 x^2$$

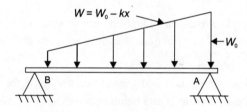

Fig. 9.15

Cubic loading
formulae used

$$W = W_0 + k_1 x + k_2 x^2 + k_3 x^3$$

Couple

A beam can also be subjected to a couple M at any point.

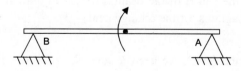

Fig. 9.16

9.6 METHOD TO DRAW THE SFD AND BMD FOR SIMPLY SUPPORTED BEAM

Step 1: Apply the condition of equilibrium

i.e., $\quad \Sigma F_x = 0, \ \Sigma F_y = 0, \ \Sigma M_A = 0$

Take moments due to all forces about any support point and equate to zero.

Two equations will be formed {one because of equilibrium of force and second because of moments}.

Solve two unknown (i.e. R_A and R_B) and find R_A and R_B.

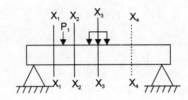

Fig. 9.17

Step 2: Take a section between one support say A and the nearest load. Draw free body diagram for the section. Find the shear force for the section and bending moment about that section $X_1 X_1$.

Step 3: Similarly, draw section between each force and find the shear force and bending moment for the section.

Shear force and bending moment equations are in term of x. Put the different values of x and find the shear force and bending moment at different point.

Step 4: Plot graph between
   (1) length of beam on x axis and shear force on y axis for SFD.
   (2) length of beam on x axis and bending moment on y axis for BMD.

The above-mentioned method can be easily understood with the help of an example described below.

Problem
A horizontal beam AB of span 10 m carries a uniformly distributed load of intensity 160 N/m and a point load of 400 N at the left end A. The beam is supported at a point C which is 1 m from A and at D which is on the right half of the beam. If the point of contraflexure is at the mid-point of the beam, determine the distance of support at D from the end B of the beam. Proceed to draw the shear force and bending moment diagrams for the arrangement.

Solution
To draw the SFD and BMD, we need R_C and R_D. from the condition of static equilibrium.

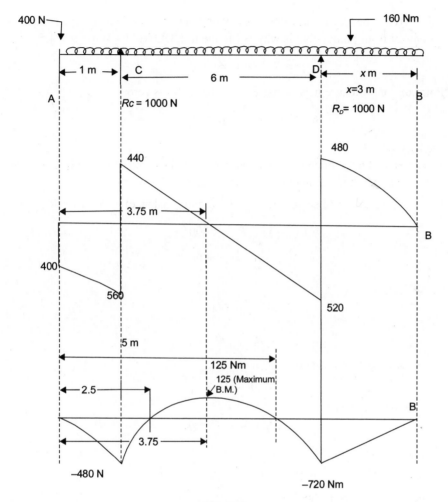

Fig. 9.18

$$R_C + R_D = 160 \times 10 + 400$$
$$R_C + R_D = 2000 \tag{1}$$

By taking moment of all the forces about the point C, we get.

Let us consider the distance from point B to support be x.

$$400 \times 1 + 160 \times 1 \times 1/2 = 9 \times 160 \times 4.5 - R_D(9 - x).$$

for U.D.L. first we take the total mass on the span i.e., (160×1) and then multiply it with half the distance from the reference point because we consider that load is acting at the centre of the disturbed load on the span. AC distance is 1 metre. UDL will be acting at 1/2 m from point C in the form of a concentrated load.

Similarly, the total load on the right side of point C because of UDL will be 160 × 9 [160 N/m × 9 (distance over which it is distributed)]. This load could be considered to be like a concentrate load acting in midway of 9 metre span that is 4-5 m away from point c that is why B.M. due to UDL will be 160 × 9 × 4.5

$$\Rightarrow R_D(9-x) = 6000 \tag{2}$$

Point of contraflexure means that B.M at that point is zero.

$$160 \times 5 \times 2.5 - (5-x) \times R_D = 0$$

Explanation

It is given that point of contraflexure is midway of beam so BM at midway is zero. Taking bending moment towards right side

160 (UDL) × 5 (distance right) × 2.5 (centre distance) × (5 – x) × R_D = 0

$$(5 - x) \times R_D = 2000$$

$$R_D \times (9 - x) = 6000$$

Solving

$$x = 3$$

$$\Rightarrow R_D = 1000$$

$$\Rightarrow R_C = 1000$$

SF equation between A and C

$$-400 - 160\,x$$

SF just toward left of C

$$-400 - 160 \times 1 = 50$$

SF just toward right

$$-400 - 160 \times 1 + 1000 = 440$$

SF equation between C and D

$$-400 - 160 \times x + 1000$$

for locating the point where S.F. is zero. We put the equation equal to zero.

$$-400 - 160 \times x + 1000 = 0$$

$$x = \frac{600}{160} = 3.75$$

Point x is at 3.75 distance away from point A

at point just left of D, shear force = – 520

at point just right of D, shear force = 480

See figure for diagram

BM diagram

BM between A and C hogging –ve

$$-400 \times x - \frac{160x^2}{2}$$

Between C and D

$$-400x - \frac{160x^2}{2} + 1000(x-1) \qquad (3)$$

for checking where the BM is zero putting the value equation to zero

$$-400x - \frac{160x^2}{2} + 1000(x-1) = 0$$

$$-80x^2 + 600x - 1000 = 0 \qquad (4)$$

Solution of quartic equation is given by

$$x = \frac{7.5 \pm \sqrt{(7.5)^2 - 4 \times 12.5}}{2 \times 1}$$

$$x = 2.5$$
$$x = 5$$

That means B.M. line will cut the axis at two points i.e. at $x = 2.5$ and at $x = 5$ from the point A. or BM will be zero at $x = 2.5$ and $x = 5$ m from the point A.

For knowing at what point maximum BM occur's differentiating equation and then putting that value equal to zero.

$$-80 \times 2x + 600 = 0$$

$$x = \frac{600}{160}$$

$$= 3.75 \text{ m}$$

for calculating the value of maximum BM put the value of x in equation 3

$$-400 \times 3.75 - 80 \times (3.75)^2 + 2750 = 125 \text{ Nm}$$

value of BM at $x = 7$ put the value in equation 3

$$-400 \times 7 - 80 \times 49 + 1000 \times 6$$

$$-2800 - 3920 + 6000$$

$$-720 \text{ Nm.}$$

9.7 POINT OF CONTRAFLEXURE (OR POINT OF INFLECTION)

It is basically a point where bending moment changes its sign and the beam is not called upon to resist any moment i.e. the bending moment is zero. It is a point of transition on the elastic curve where the curve moves into reverse curvature. We can locate the inflection point by setting up an algebraic expression for the moment in the beam for a segment, and solving this relation equated to zero.

9.8 METHOD TO DRAW SFD AND BMD FOR CANTILEVER BEAM

Always start solving the problem from free end and then follow the steps mentioned for simply supported beam from (ii) to (iv).

Let us try to find the SF and BM for cantilever beam with the help of example mentioned below.

Example

A cantilever beam is subjected to various loads as shown in figure. Draw the SFD and BMD for the beam.

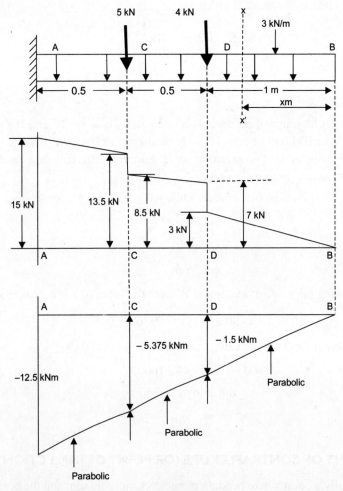

Fig. 9.19

Solution
First draw two horizontal lines equivalent to the length of beam i.e. of 2 m one for SFD other for BMD just below the given diagram.

Total load on the beam.
$$5 + 4 \;+\; 3 \;+\; 2 = 15 \text{ kN.}$$
$$\uparrow \;\uparrow \qquad \uparrow \qquad \uparrow$$

for point load   UDL per metre length   total length of beam   [This means that $R_A = 15$ kN]

Consider a section $(x - x^1)$ at a distance x from B.

SF between B and D

$$\text{Shear force } F_x = + Wx$$

[see sign convention for shear force with respect to reference xx^1 in the figure load W per unit length is acting in downward direction toward the right side of xx^1 so + ve sign]

Now put the value of x.

$$\text{for } x = 0 \text{ m. } F_B = 0$$

$$\text{for } x = 1 \text{ m } F_D \text{(just right)} = 3 \times 1 = 3 \text{ kN}$$

Shear force between D and C

$$F_x = + Wx + 4$$

[+4 is added because + 4 kN is point load acting at point D toward right side in downward direction]

$$\text{at } x = 1 \text{ m } F_D \text{ (just left)} = 3 \times 1 + 4 = 7 \text{ kN}$$

$$\text{at } x = 1.5 \text{ m } F_C \text{ (just right)} = 3 \times 1.5 + 4 = 8.5 \text{ kN.}$$

Shear force between C and A.

$$F_x = + Wx + 4 + 5$$

[+5 is added because +5 kN is point load acting at point c towards right side in downward direction]

$$\text{at } x = 1.5 \text{ m } F_c \text{ (just left)} = 3 \times 1.5 + 4 + 5$$

$$= 13.5 \text{ kN}$$

$$\text{at } x = 2 \text{ m } F_A = 3 \times 2 + 4 + 5 = 15 \text{ kN.}$$

Now draw the SF diagram just below the given diagram.

[Our answer is checked because total load on the beam is 15 kN and reaches $R_A = 15$ kN]

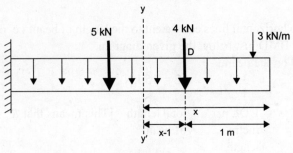

Fig. 9.20

For drawing bending moment diagram.
BM at xx^1 between BD

$$M_x = -(Wx) \cdot \frac{x}{2}$$

–ve sign because hogging ⌒ beam is bending like this. total load on the distance x = W × x. Now this load is acting midway of the distance x that mean x/2 so. B.M. is

$$\underset{\underset{\text{load}}{\uparrow}}{(Wx)} \quad \underset{\underset{\text{midway}}{\text{Acting}}}{\overset{\uparrow}{\times x/2}}$$

$$M_x = \frac{Wx^2}{2}$$

at x = 0

$$M_B = 0$$

at x = 1

$$M_D = \frac{-3 \times (1)^2}{2} = -1.5 \text{ kNm}$$

Bending moment between D and C

$$M_x = -\frac{Wx^2}{2} - 4(x-1)$$

Explanation

We are considering BM toward right side of yy′ total load = 3 × x (distance) = 3 x.
This load is acting midway that means bending moment

$$= \text{load} \times \text{distance}$$

$$\Rightarrow -3x \times \frac{x}{2} = -\frac{3x^2}{2}$$

[–ve because of sign convection hogging BM]
Distance between yy' and D = (x – 1)
from yy' load of 4 kN is acting at a distance of (x – 1) so BM due to 4 kN load will be

$$= -4 \times (x - 1)$$

so the total BM at yy'

$$= -\frac{3x^2}{2} - 4(x - 1)$$

[– ve sign because hogging BM ⌒ according to the sign convention of BM]

SOLVED PROBLEMS

Problem 1
Draw the shear force and bending moment diagram for the beam as shown below and mark the position of the maximum bending moment and determine its value.

(MRIU Dec 2009)

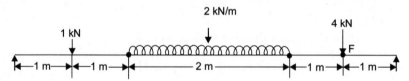

Solution
Applying condition for equilibrium

$\Sigma F_y = 0$ $R_A + R_B = 1 + 2 \times 2 + 4 = 9$ kN

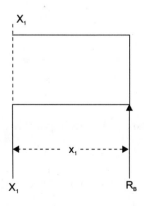

$\Sigma M_A = 0 \qquad R_B \times 6 - 4 \times 5 - 4 \times 3 - 1 \times 1 = 0$

$$R_B = \frac{33}{6} = 5.5 \text{ kN}$$

$$R_A = 9 - 5.5 = 3.5 \text{ kN}$$

For part B.F.
Take section $x_1 - x_1$ at a distance of x_1 from point B.

$$F_{x_1 - x_1} = -R_B = -5.5 \text{ kN}$$

$$F_{F\, x_1 = 1} = -R_B = -5.5 \text{ kN}$$

$$BM_{x_1 x_1} = R_B \times x_1$$

$$BM_{\text{at point B}} = R_B \times 0 = 0$$

$$BM_{F\, x_1 = 1} = RB \times 1 = 5.5 \text{ kNm}$$

For part FE
Take section $x_2 - x_2$ at a distance x_2 from point B.

$$F_{x_2 - x_2} = 4 - R_B = 4 - 5.5 = -1.5 \text{ kN}$$

$$F_E = -1.5 \text{ kN}$$

$$BM_{x_2 - x_2} = R_B \times x - 4x(x_2 - 1)$$

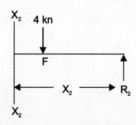

For part ED
Take section $x_3 \times x_3$ at a distance of x_3 from point B

$$Fx_3 - x_3 = 2*(x_3 - 2) + 4 - R_B$$

$$F_D x_3 = 4 = 4 + 4 - 5.5 = 2.5 \text{ kNm}$$

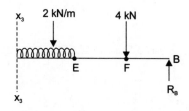

$$BM_{x_3-x_3} = -2 \times \frac{(x_3-2)}{2} - 4(x_3-1)R_B \times x_3 \quad \text{(parabolic)}$$

$$BM_{x_3-x_4} = -4 - 12 + 5.5 \times 4 = 6 \text{ kNm}$$

Part DC

Take section $x_4 - x_4$ at a distance of x_4 from B

$$F_{x_4-x_4} = 4 + 4 - R_B$$

$$F_C = 8 - 5.5 = 2.5 \text{ kN}$$

$$BM_{x_4-x_4} = R_B \times x_4 - 4(x_4 - 1)$$

$$-(2 \times 2) \times (x_4 - 3)$$

$$= 5.5 \times 6 - 4 \times 4 - 4 \times 2$$

$$BM_C x_4 = 27.5 - 16 - 8 = 3.5 \text{ kNm}$$

For part CA

Take a section $x_4 - x_5$ at a distance of x_5 from point B

$$F_{x_5-x_5} = 4 + 4 + 1 - R_B$$

$$F_A = 9 - 5.5 = 3.5 \text{ kN}$$

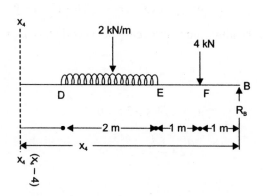

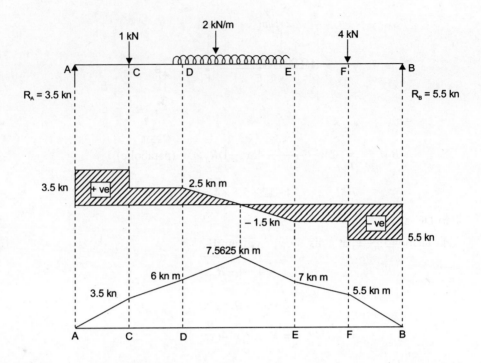

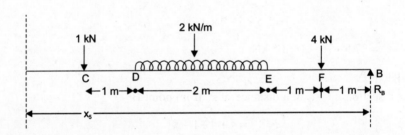

$$BM_{x_5-x_5} = R \times x_5 - 4(x_5 - 1) - 4x(x_5 - 3) - 1 \times (x_5 - 5)$$

$$BM_A = 5.5 \times 6 - 4 \times 5 - 4 \times 3 - 1 = 33 - 20 - 12 - 1 = 0$$

BMD: Maximum bending moment occurs between E and D where shear force changes its sign

i.e., $$F_{x_3-x_3} = 0 = 2(x-2) + 4 - R_B$$

or $$2(x-2) + 4 - 5.5 = 0$$

$$x = 2 + \frac{1.5}{2} = 2.75 \text{ m}$$

and maximum BM $= 2 \times \dfrac{(2.75-2)^2}{2} - 4(2.75-1) + 5.5 \times 2.75$

$= -0.5624 - 7 + 15.125 = 7.5625$ kNm

Problem 2

Draw the shear force and bending moment diagram for a cantilever beam of length 6 m carries two point loads of 2 kN and 3 kN at a distance of 1 m and 6 m from the fixed end respectively. In addition to this the beam also carries a uniformly distributed load of 1 kN/m over a length of 2 m of at a distance of 3 m from the fixed end clearly mark the position of the maximum bending moment and determine it value.

Solution

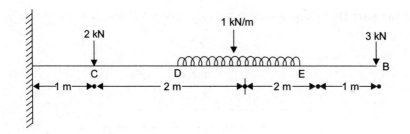

For part BE: Draw section x_1 x_1 between B and E at a distance x_1 from point B

$F_{x_1 x_1} = 3$ kN

$F_E = F_B = 3$ kN

$M_{x_1 - x_1} = 3x_1$

$M_{B=0}$ $M_{x_1 = 1 \, m} = -3$ kNm

For part ED: Draw $x_2 - x_2$ between E and D at a distance of x_2 from B.

$F_{x_2 - x_2} = 3 + 1(x_2 - 1)$

$F_{D_{x_2=3}} = 3 + 1 \times 2 = 5$

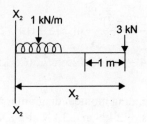

$$BM_{x_2 x_2} = 3x_2 - \frac{1(x_2-1)^2}{2}$$

$$BM_{D_{x_2=3}} = -3 \times 3 - \frac{1 \times 4}{2} = -9 - 2 = -11 \text{ kNm}$$

For the part DC: Draw a section $x_3 - x_3$, between D and C at a distance of x_3 from point B

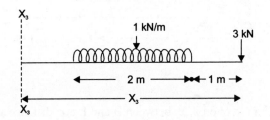

$$F_{x_3 x_3} = 3 + 1 \times 2 = 5 \text{ kN}$$

$$F_c = 5 \text{ kN}$$

$$BM_{x_3 - x_3} = -3 \, x_3 - 1 \times 2 \times (x_3 - 2)$$

$$BM_{C \, x_3=5} = -3 \times 5 - 1 \times 2 \times 3 = -15 - 6 = -21 \text{ kN/kNm}$$

For the part CA: Draw a section $x_4 x_4$ between C and A at a distance x_4 from point B

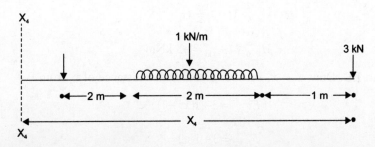

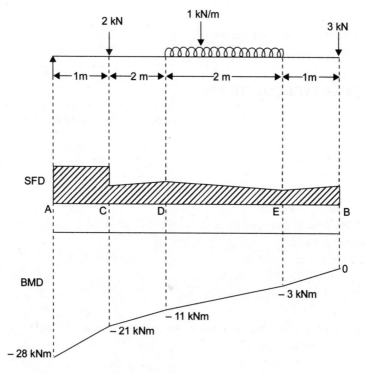

Maximum BM is produced at fixed end and its value is −28 kNm

$$F_{x_4 x_4} = 3 + 1 \times 2 + 2 = 7 \text{ kN}$$

$$F_A = 7 \text{ kN}$$

$$BM_{x_4 x_4} = -3x_4 - 2(x_4 - 2) - 2(x_4 - 5)$$

$$BM_{A_{x_4=6}} = -3 \times 5 - 2 \times 4 - 2 \times 1 = -18 - 8 - 2 = -2.8 \text{ kNm}$$

EXERCISE

1. What do you understand by shearforce and bending moment diagram
2. What are the various types of beam? Distinguish clearly the difference betwee
   (a) overhanging and continuous beam
   (b) simply supported and rigidly fixed beam
3. What is the difference between uniformly distributed load and uniformly varying load.
4. Explain the sign convention for shearforce and bending moment
5. What is the significance of point of contra flexure? Explain with reasons if it exists in.

(i) Contilever
(ii) simply supported beam
(iii) overhanging beam

OBJECTIVE TYPE QUESTIONS

1. The moment diagram for a cantilever beam subjected to bending moment at end of beam will be
   (a) rectangle
   (b) triangle
   (c) parabola
   (d) cubic parabola.
2. If the load at free end on a cantilever is increased so as to cause rupture, same will occur
   (a) below the load
   (b) at fixed end
   (d) between fixed end and centre.
   (d) at centre.
3. Shear force at any point on the beam is the algebraic sum of
   (a) all vertical forces
   (b) all horizontal forces
   (c) forces on either side of the point.
4. Bending moment at any point is equal to the algebraic sum of
   (a) all vertical forces.
   (b) all horizontal forces.
   (c) forces on either side of the point.
   (d) moments of forces on either side of the point.
5. The rate of change of shearing force at any section is equal to the rate of
   (a) loading at that section.
   (b) change of deflection at that section.
   (c) change of bending moment at that section.
   (d) integration of bending moment at that section.
6. The rate of change of BM at any section is equal to the
   (a) shearing force at that section.
   (b) rate of change of shearing force at that section.
   (c) deflection at that section.
   (d) rate of change of deflection at that section.
7. The moment diagram for a cantilever beam carrying uniformly distributed load will be
   (a) rectangle
   (b) triangle
   (c) parabola
   (d) cubic.

8. The reaction in the case of hinged support.
   (a) perpendicular to surface of hinge.
   (b) along the surface of hinge.
   (c) opposite to the direction of load.
   (d) In any direction depending upon the load.
9. If the shear force is zero along a section, the BM at that section will be.
   (a) minimum
   (b) maximum
   (c) zero
   (d) either minimum or maximum.
10. The point of contraflexure occurs only in.
    (a) Cantilever beams.
    (b) Overhanging beams.
    (c) Simply supported beams.
    (d) Continuous beams.
11. The point inflexion or contraflexure is the point where
    (a) bending moment diagram changes sign.
    (b) stress is minimum.
    (c) deflection changes sign.
    (d) bending moment is maximum.
12. The bending moment on a section is maximum where shearing force.
    (a) is maximum
    (b) is minimum
    (c) is equal
    (d) changes sign.
13. In the case of cantilever, irrespective of the type of loading the maximum bending moment and maximum shear force occurs at
    (a) free end
    (b) under the load
    (c) fixed end
    (d) middle.
14. In a continuous curve of bending moment the point of zero bending moment, where it changes sign is called
    (a) the point of inflexion.
    (b) the point of contraflexure.
    (c) the point of virtual hinge.
    (d) all of the above.
15. The shear force diagram for a cantilever beam carrying a uniformly distributed load over its length is a
    (a) triangle
    (b) rectangle
    (c) hyperbola
    (d) parabola.

16. The rate of change of shear force at any section is equal to.
    (a) bending moment.
    (b) loading
    (c) deflection
    (d) intensity of loading.
17. The bending moment diagram will be a cubic parabola in the case of a cantilever loaded as follows.
    (a) bending moment applied at free end.
    (b) concentrated load at the end.
    (c) uniformly distributed load.
    (d) varying load, zero at free end.
18. The moment diagram for a cantilever beam carrying linearly varying load from zero at free end to maximum at supported end will be
    (a) rectangular
    (b) triangle
    (c) parabola
    (d) cubic parabola.
19. The reactions of each support of beam can be determined from following condition of equilibrium.
    (a) algebraic sum of all vertical forces is zero.
    (b) algebraic sum of all horizontal forces is zero.
    (c) algebraic sum of moments about any point is zero.
    (d) all of the above.
20. When the external forces and moments that support an object can be found by the equations of statics alone, the object is said to be
    (a) free body
    (b) statistically determinate
    (c) statistically indeterminate
    (d) homogeneous

ANSWERS

1. a	2. b	3. c	4. d	5. a
6. a	7. c	8. d	9. d	10. b
11. a	12. a	13. c	14. d	15. b
16. d	17. d	18. d	19. d	20. b

UNIT 10

Fabrication Process

10.1 INTRODUCTION

Ferrous metals and alloys are commonly used because that can produce different alloys and grades which provide different range of properties that are not found in any other family of materials. Further the alloys produced are cheaper and economical per unit of mass produced as compared to their competitors.

The Latin word of iron is ferrum. So metals which contain iron are all classified as ferrous metals.

Following is the general classification of ferrous metals.
(1) Pig iron
(2) Cast iron
(3) Wrought iron
(4) Carbon steel
(5) Alloy steel

Pig Iron

All iron and steel products are derived originally from pig iron. This is the raw material obtained from the chemical reduction of iron ore in a blast furnace. The process of reduction of iron ore to pig iron is known as smelting.

Cast Iron

(a) Cast iron is pig iron remelted and thereby refined together with definite amount of lime stone, steel scrap and spoiled castings in a cupola or other form of remelting furnace and poured into suitable moulds of required shape.

(b) Cast iron contains 2 to 4 per cent of carbon, a small per cent of silicon, sulphur, phosphorus and manganese and certain amount of alloying elements, e.g., nickel, chromium, molybdenum, copper and vanadium.

Carbon in cast iron exists in two forms.

(1) Compound cementite (in chemical combination) also called white cast iron.
(2) Free carbon (mechanical mixture) also called grey cast iron.

An intermediate stage of the two varies is called mottled iron.
Quality of cast iron depends upon.
(1) Amount of carbon
(2) In what form carbon exists

Different varieties of cast iron are:
(1) Grey cast iron
(2) White cast iron
(3) Malleable cast iron
(4) Nodular cast iron
(5) Chilled cast iron
(6) Alloy cast iron
(7) Mechanite cast iron

Chemical compositions (approximate) of iron and steel products

Materials	Carbon (per cent)	Silicon (per cent)	Manganese (per cent)	Sulphur (per cent)	Phosphorus (per cent)
Pig iron	3.00–4.00	0.50–3.00	0.10–1.00	0.02–0.10	0.03–2.00
Grey cast iron	2.50–3.75	1.00–2.50	0.40–1.00	0.06-0.12	0.10–1.00
Malleable cast iron	2.20–3.60	0.40–1.10	0.10–0.40	0.03–0.30	0.10–0.20
White cast iron	1.75–2.30	0.85–1.20	0.10–0.40	0.12–0.35	0.05–0.20
Wrought iron	0.02–0.03	0.10–0.20	0.02–0.10	0.02–0.04	0.05–0.20
Carbon steel	0.05–2.00	0.05–0.30	0.30–1.00	0.02–0.20	002–0.15

10.2 EFFECT OF CHEMICAL ELEMENTS ON IRON

Carbon (1) Carbon when dissolved in liquid iron and allowed to solidify slowly than fairly tough iron is obtained. It can be easily machined. (2) If the same iron is cooled quickly, we get hard, difficult to machine iron. It contains iron carbide called cementite.

Silicon It makes iron soft and easily machinable; produces sound castings free from blowholes, small percentage of silicon makes wrought iron hard and brittle.

Sulphur It makes the cast iron hard and brittle.

Manganese Reduces harmful effects of sulphur.

Phosphorus It aids fusibility and fluidity.

Nickel It resists heat and corrosion and have low expansivity.

Chromium It acts as a carbide stabilizer in cast iron.

Molybdenum It improves tensile strength, hardness and shock resistance of castings.

Copper It promotes formation of graphite.

	Obtained	Composition	Properties	Advantages
Grey cast iron	Obtained by allowing the moltent metal to cool and solidify slowly	Large quantities of carbon and relatively small quantities of the other elements	Brittle, free graphite in its structure seems to act as lubricant.	Cheap, low melting point (1150–1200°C) easily machined
White cast iron	Rapid cooling	Carbon in cementite form, large quantity of manganese, small quantity of silicon	Hard, brittle unmachinibality	Do not rust, used in the manufacture of wrought iron
Malleable cast iron	Cast iron with carbon in combined state packed in steel boxes and surrounded with hametite ore and heated slowly to temp of 950°C to 1000°C maailavel for 5 days and very slowly cooled	3.6–2.2 (C%) 1.1–0.4 (Si%) 0.1–0.4 (Mg%)	Have properties of toughness, less brittle	Hinges, door keys, spanners, crankshafts

10.3 STEELS

It is an alloy of iron and carbon. In steel, carbon content varies from 0 to 1.5 per cent. The carbon is distributed throughout the mass of the metal, not as elemental or free carbon but as a compound (chemical combination) with iron.

When the percentage of carbon increases beyond 1.5, a stage will come when no more carbon can be contained in the combined stage and any excess will have to be present as free carbon (graphite). It is at this stage when we start calling the composition as cast iron.

Important: Therefore, for a material to be classed as steel there must be no free carbon in its composition. As soon as the carbon starts appearing in the form of free graphite the steel is categorised into cast iron.

Commercial production of steel can be done by
 (1) Bessemer process
 (2) L-D process
 (3) Open hearth process
 (4) Crucible process
 (5) Electric process
 (6) Duplex process

Mechanical properties and applications of carbon steel

Type of steel	% Carbon	BHN number	Tensile strength MPa	Yield strength MPa	% Elongation	% Reduction in area	Uses
Dead mild	0.05–0.15	100–110	390	260	40	60	Chains, stampings, rivets, nails, sea-mwelded pipes, tin plate; automobile body steel, and materials subject to drawing and pressing.
Mild steel	0.10–0.20	120–130	420	355	36	66	Structural steels, universal beams, screw, drop forgings, case hardening steel.
	0.20–0.30	130–150	555	480	21	55	Machine and structural work, gears, free cutting steel, shafting and forgings.
Medium steel	0.30–0.40	150–160	700	550	18	51	Connecting rods, shafting, axels, crankhooks forging.
	0.40–0.50	350	770	580	20	53	Crankshafts, axels, gears, shafts, die-block rotors, tyres, skip wheels.
High carbon	0.50–0.60	350–400	1200	750	10	35	Loco tyres, rais, wire ropes.
	0.60–0.70	400–450	1235	780	12	40	Drop hammer dies, saws, screw-drivers.
	0.70–0.80	450–500	1420	1170	12	35	Band saws, anvil faces, hammers, wrenches, laminated springs, cable wire, large dies for cold presses.
	0.80–0.90	500–600	665	645	12	33	Cold chisels, shear blades, punches, rocks drills.
Tool steels	0.90–1.10	550–600	580	415	13	26	Axes, kinves, drill, taps, screw ring dies, picks.
(High carbon)	1.10–1.50	600–750	500	375	13	20	Ball bearings, files, broaches, razors, boring and finishing tools machine parts where resistance to wear is essential.

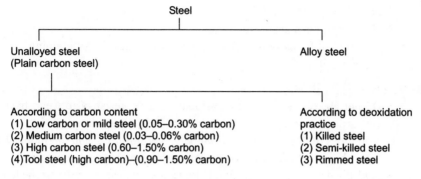

Fig. 10.1

10.4 INTRODUCTION TO FOUNDRY

The study of foundry deals with the process of making casting in moulds prepared by patterns. The total process of producing castings can be classified into five stages.
   (a) Pattern making
   (b) Moulding and core making
   (c) Melting and casting
   (d) Fettling
   (e) Testing and inspection

10.5 PATTERN

Except pattern making all other stages to produce castings are done in foundry shops.
   - Pattern may be defined as a model of any object, so constructed that it may be used for forming an impression called mould in damp sand or other suitable material. When this mould is filled with molten metal, and the metal is allowed to solidify, it forms a reproduction of the pattern and is known as casting.
   - The process of making a pattern is known as pattern making.

The different materials used for making pattern are:

(1) Wood
   - Wood can be cut and fabricated into numerous form by gluing, bending and curving.
   - Disadvantage is that it changes its shape when moisture dries out.
   - They are used when smaller number of castings are to be produced.

(2) Metal
   - Metal patterns are used when large number of castings are required or when conditions are too severe.

- Do not warp or shrink.
- Very helpful in machine moulding.
- Metals used for patterns include are cast iron, steel, brass, aluminium and white metal.

(3) Plastics
- Do not absorb moisture, dimensionally stable, resistant to wear, have a very smooth and glossy surface and are light in weight.
- Two types of plastics are used, namely, thermosetting and thermoplastic. In the thermosetting varieties epoxy resin has become very popular because of its good production qualities. In thermoplastic variety polystyrene foam is commonly used.

Rubbers: A special variety of rubber such as silicon rubber are favoured for forming a very intricate type of die for investment casting.

Plasters: Gypsum cement known as plaster of Paris is also used for making patterns and core boxes.

Waxes: Wax patterns are excellent for investment casting process. (A special type of casting.)

10.6 TYPES OF MOULDING SAND

The different types of sand used for preparing a mould are.
  (1) Green Sand
   - Silica – 70%
     Clay – 18 to 30%
     Water – 6 to 8%
   - The clay and water furnish the bond for green sand.
   - When the damp sand is squeezed in the hand, it retains the shape; the impression given to it under pressure.
  (2) Dry sand: Green sand that has been dried or baked after the mould is made is called dry sand.
  (3) Loam sand: In this percentage of clay is high and can reach upto 50%.
  (4) Facing sand:
   - It is made of fresh silica sand and clay.
   - It forms the face of the mould.
   - It is used directly next to the surface of the pattern and it comes into contact with the molten metal when molten metal is poured in the mould.
  (5) Backing sand:
   - Old, repeatedly used moulding sand is used to back up the facing sand and to fill the whole volume of flask. It is also called black sand.
  (6) Parting sand:
   - It is used to keep the green sand from sticking to the pattern.
   - It is clean clay free silica sand.

(7) Core sand:
- Sand used for making cores is called core sand.
- This is silica sand mixed with core oil (linseed oil, light mineral oil, resin etc.)

Sand Additives

These are the materials generally added to the sand mixture to develop special properties in the mould and consequently in castings.

Facing materials: (To provide a smooth surface on the casting):
  (i) Charcoal
  (ii) Gas carbon
  (iii) Coke dust
  (iv) Plumbago
  (v) Black lead
  (vi) Graphite
  (vii) Sea coal (finely powdered bituminous coal).

Properties of the Moulding Sand

Moulding sand must possess six properties. The properties are:
  (1) Porosity
  (2) Flowability
  (3) Collapsibility
  (4) Adhesiveness
  (5) Cohesiveness
  (6) Refractoriness

10.7 TYPES OF PATTERN

Selection of pattern for a particular casting will depend upon many conditions. Important factors are:
  (i) Difficulty of moulding operation
  (ii) Small or large number of castings are wanted.
  (iii) Type of moulding process.

Commonly used patterns are:
  (1) Single piece or solid pattern
  (2) Split pattern
  (3) Match plate pattern
  (4) Gated pattern
  (5) Cope and drag pattern
  (6) Loose piece pattern

Single piece or solid pattern
- A pattern that is made without joints, partings, or any loose pieces in its construction is called a single piece or solid pattern.

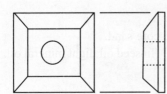
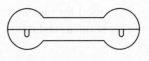

Fig. 10.2 Solid pattern.   **Fig. 10.3:** Two-piece split pattern.

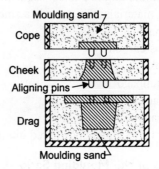

(a) Three-part pattern   (b) Three-part mould

Fig. 10.4: Three-piece split pattern.

- Also called loose pattern.
- Used for large castings of simple shapes.
- When using such pattern, the moulder has to cut his own runners and feeding gates and risers.
- Used only for limited production

Split pattern
- These patterns are usually made up of two parts.
- One part will produce the lower half of the mould and the other, the upper half.
- Two parts are held in their proper relative position by means of dowel pins fastened in one piece and fitted holes bored in the other.
- Pattern may be made of three or more parts for a complicated casting. This type of pattern is known as multipiece pattern.
- Examples of castings that are made of split patterns are, cylinders, steam values, water stopcocks, taps, bearings wheels, etc.

Match plate pattern
- When split patterns are mounted with one half on one side of plate and the other half directly opposite on the other side of the plate, the pattern is called a match plate pattern.
- A single pattern or a number of patterns may be mounted on a match plate.

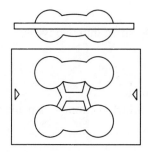

Fig. 10.5: Match plate pattern

- Aluminium is commonly used for metal match plate.
- The gates and runners are completed in one operation.

Cope and drag pattern
- In the production of large castings, the complete moulds are too heavy to be handled by a single operators so separate cope and drag patterns are built
- Cope holy is prepared separately so is the drag half.

Gated Pattern
- In mass production a number of castings are produced in a single multicavity mould by joining a group of patterns, and the gates or runners for the molten metal, are formed by the connecting parts between the individual patterns. So the patterns in which gates are attached are called gates patterns.

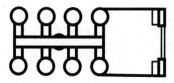

Fig. 10.6: Gated pattern.

Loose-piece pattern
- Patterns that are produced as assemblies of loose component pieces so that it can be removed easily are called loose-piece patterns.
- In this case, the main pattern is usually removed first and later the remaining parts.

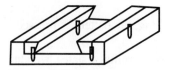

Fig. 10.7: Loose-piece pattern.

Moulding Processes

Broadly moulding can be classified into two types.
 (1) Hand moulding
 (2) Machine moulding

Other ways of classification are
 The type of material of which the mould is made up of
 (a) Green sand moulds
 (b) Dry sand moulds
 (c) Skin dried moulds
 (d) Loam moulds

According to the method used in making the mould
 (1) Bench moulding
 (2) Floor moulding
 (3) Pit moulding
 (4) Sweep moulding and plate moulding

10.8 MOULDING PROCESSES BASED ON SAND USE

Green Sand Mould

It is prepared with natural moulding sand or with mixtures of silica sand, bonding clay, and water. The sand must be properly tempered before it can be used. To check the sand for proper temper, a handful is grasped in the fist. The pressure is released, and the sand is broken in two sections. The sections of sand should retain their shape and the edges of the break should be sharp and firm.

Advantages of Green sand mould are:
 (1) Cheapest method of producing a mould.
 (2) Less distortion than in dry sand mould.
 (3) Flasks are ready for reuse.

Principal methods of green sand mould are:
 (1) Open sand method
 (2) Bedded-in method
 (3) Turn-over method.

Open sand method
 (1) Entire mould is made in the foundry floor.
 (2) No moulding box is necessary and the upper surface of the mould is open to air.
 (3) After proper levelling, the pattern is pressed in the sand bed for making mould.
 (4) Mainly used for grills, railings, gates, weights, etc.

Bedded-in method

(1) In this method pattern is pressed or hammered down into the sand of the foundry floor.
(2) Properly rammed mould cavity is obtained.
(3) Parting sand spread, a cope is placed over the pattern.
(4) Cope is rammed up, runners and risers cut and the cope box lifted, leaving the solid pattern in the floor.

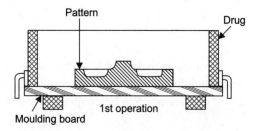

Fig. 10.8: Making a green sand mould (1st step).

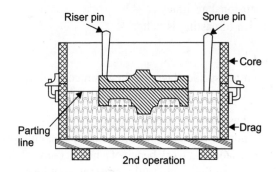

Fig. 10.9: Making a green sand mould (2nd step).

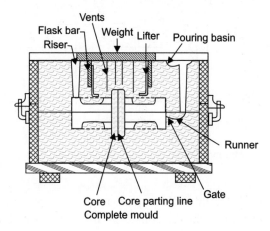

Fig. 10.10: Making a green sand mould (complete mould).

Turn-over method

(1) Mostly used method and most suitable for split pattern.
(2) One half of the pattern is placed with its flat side on a moulding board, and a drag is rammed and rolled over.
(3) It is now possible to place the other half of the pattern and a cope box in proper position.
(4) After ramming the cope is lifted off and the two halves of the pattern are rapped and drawn separately.
(5) The cope is next replaced on the drag to assemble the mould.

Dry Sand Mould

The moulding process involved in making dry sand moulds are similar to those employed in green sand moulding except that a different sand mixture is used and all parts of the mould are dried in an oven before being reassembled for casting.

Advantages

(1) They are stronger and thus are less susceptible to damage in handling.
(2) Better dimensional accuracy.
(3) Surface finish of casting is better.

Skin-dried Mould

In this moisture from the surface layer of the rammed sand is dried to a depth of about 23 mm or more by using gas torches or heaters.

Moulding processes based on the methods used.
(1) Bench and floor moulding: Bench moulding applies chiefly to moulds small enough to be made on a work bench of a height convenient to the moulder. In floor moulding floor acts as the drag for very heavy castings.

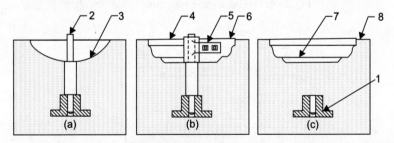

Fig. 10.11: Sweep moulding
1. Base, 2. Spindle, 3. Well-rammed excavation, 4. Foundry floor, 5. Sweep holder, 6. Sweep, 7. Mould, 8. Gate

(2) Pit moulding: Moulds of large jobs are generally prepared in a pit dug in the foundry floor which facilitates in lifting the pattern and casting the mould easily.

(3) **Sweep moulding:** Sweep mouldings are employed for moulding parts whose shape is that of a surface of revolution.

10.9 CASTING METHOD

Permanent Mould Casting

In the sand casting the moulds are destroyed after solidification of castings, the moulds are reused repeatedly in the permanent mould castings. Mould materials should have sufficient high melting point to withstand erosion by the liquid metal at pouring temperature, high strength so that mould may not deform by repeated use and a high thermal fatigue resistance to resist premature crazing (the formation of thermal fatigue cracks) that would leave objectionable marks on the finished castings.

Cast iron and alloy steels are used for making castings. Recently refractory metal alloys particularly molybdenum alloys, have found increasing application.

Zinc, copper, aluminium, lead, magnesium and tin alloys are most often cast by this process.

Permanent metal moulds can be advantageously used for small and medium-sized (upto 10 kg) non-ferrous castings, but would be impractical for large castings, and metals and alloys of very high melting temperature.

Different casting methods are:
  (1) Slush casting
  (2) Die casting
  (3) Centrifugal casting
   • True centrifugal casting
   • Semi-centrifugal casting
   • Centrifuged casting
  (4) Continuous casting

Slush Casting

(1) The principle involves pouring the molten metal into a permanent mould.
(2) After the skin has frozen, the mould is turned upside down to remove the metal still liquid.
(3) A thin walled casting results.
(4) Limited to tin zinc or lead based alloys.

Die Casting

(1) In this molten metal is forced under pressure into split metal dies which resembles a common type of permanent mould.
(2) The fluid alloy fills the entire die, including all minute details.
(3) Because of low temperature of die, the casting solidifies quickly, permitting the die halves to be separated and the casting ejected.
(4) Lead, magnesium, tin and zinc alloys are cast by this method.

Centrifugal Casting
  (1) In this molten metal is poured into the moulds while they are rotating.
  (2) The metal falling into the centre of the mould at the axis of rotation is thrown out by the centrifugal force under sufficient pressure towards the periphery.
  (3) Contaminants or impurities present being lighter in weight are also pushed towards the centre.
  (4) Hollow cylindrical bodies are cast by this method.

(a) *True centrifugal casting*
  - In this process, the casting is made in a hollow, cylindrical mould rotated about an axis common to both casting and mould.
  - The axis may be horizontal, vertical or inclined.
  - The casting have more or less a symmetrical configuration (round, square, hexagonal, etc.) on their outer contour and does not need any central core.
  - Casting cools and solidifies from outside towards the axis of rotation; so it results in good directional solidification. Hence, castings are free from shrinkage.

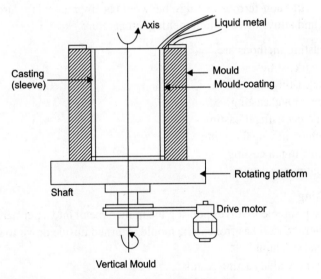

Fig. 10.12: True centrifugal casting.

Applications
  - Liners for IC engines
  - Pipes, rolls, cylinder sleews, piston ring stocks, bearings bushing, etc.

(b) *Semi-centrifugal casting*
  - In this type a sand core is used to form the central cavity (as in the hub of the wheel). So internal shapes are controlled which is not possible in true centrifugal casting.

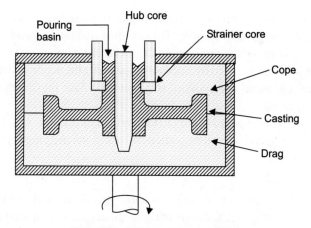

Fig. 10.13: Semi-centrifugal casting.

- Semi-centrifugal castings are normally made in vertical machines. The mould axis is concentric with the axis of rotation.
- Casing shapes more complicated than those possible for true centrifugal casting can be made.
- Parts produced are gears, flywheels and track wheels, etc.

(c) *Centrifuge Casting*
- In this casting, the axis of the mould and that of the rotation do not coincide with each other.

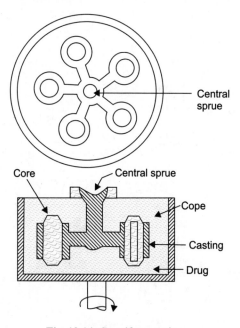

Fig. 10.14: Centrifuge casting.

- Parts are not symmetrical about any axis of rotation and cast in a group of moulds arranged in a circle. The set-up is revolved around the centre of the circle to induce pressure on the metal in the mould.
- Mould cavities are fed by a central sprue under the action of the centrifugal force. The metal is introduced at the centre and fed into the mould through radial ingates.
- Centrifuging is possible only in the vertical direction.
- Parts produced are valve bodies, valves, bonnets, plugs, yokes, pillow blocks, etc.

Investment Casting

This process uses wax pattern which is subsequently melted from the mould, leaving a cavity having all the details of the original pattern (required casting):

(1) Pattern wax is injected into metal die to form a dispensable pattern.

(2) Gating patterns are gated to a sprue to form a tree or cluster and base of the pattern material is attached to the tree.

(3) (a) Solid pattern: A metal flask is put over the cluster of pattern and sealed to a base plate to form a container. A hard setting moulding material is then poured into the flask, completing the pattern cluster completely.

(b) Shell pattern: The ceramic shell is formed by dipping the clustered patterns in a ceramic slurry and then sprinkling them with a refractory grain. This procedure is repeated until the required thickness of the shell is achieved.

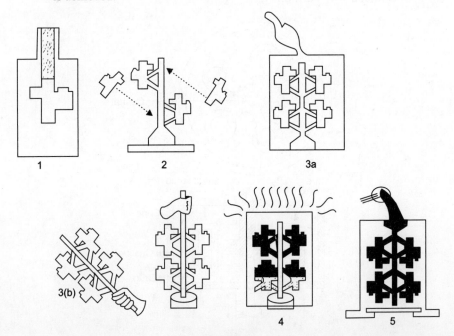

Fig. 10.15

(4) Pattern removal: The flask and shells are then placed in ovens to bake at a moderate temperature in order to slowly melt the embedded patterns. The cavity left in the mould will receive the poured molten metal.
(5) Casting: The flask or shell is inverted and the metal is poured into the hot mould.
  - Application of investment casting
  - Parts for sewing machines, locks, rifles, burner nozzles, milling cutters, and other type of tools, jewellery and art casting.
  - In dentistry and surgical implants
  - Parts of gas turbines.
  - Corrosion resistant and wear resistant alloy parts used in diesel engines, picture projectors and chemical industry equipment.

10.10 FORGING

Forging refers to the process of plastically deforming metals or alloys to a specific shape by a compressive force exerted by some external agency like hammer, press, rolls or by an upsetting machine of some kind. The portion of a work in which forging is done is termed the forge and the work is mainly performed by means of heavy hammers, forging machines and presses.

Different forging tools are:
  (1) The anvil
  (2) The swage block
  (3) Hand hammer
  (4) Tongs
  (5) Chisels
  (6) Swages
  (7) Fullers
  (8) Flatters
  (9) The set hammer
  (10) The punch and the drift

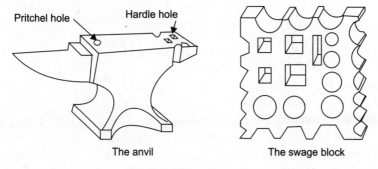

Fig. 10.16: Forging tools.

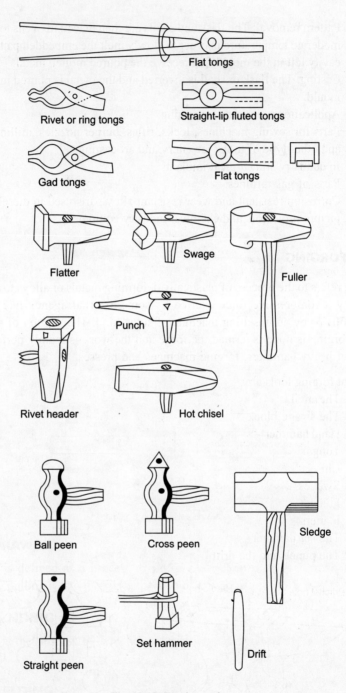

Fig. 10.17: Forging tools.

The typical forging operations are
  (1) Upsetting
  (2) Drawing down
  (3) Setting down
  (4) Bending
  (5) Welding
  (6) Cutting
  (7) Punching
  (8) Fullergeing

10.11 ROLLING

In rolling, a metal is formed through a pair of revolving rolls with plain or grooved barrels. The metal changes its shape gradually during the period in which it is in contact with the two rolls.

Two high reversing mill
It comprises two rollers rotating in opposite direction.

The various reductions in cross-sectional area are achieved by providing different sizes of grooves on both the upper and the lower rolls.

Three high mill
It consists of three rolls mounted one above the other.

Four high mill
It consists of four rolls, two smaller in size and two bigger.

10.12 EXTRUSION

Extrusion is defined as the process of pushing the heated billet or slug of metal through an orifice provided into a die thus forming an elongated part of uniform cross-section corresponding to the shape of the orifice. The pressure is applied either hydraulically or mechanically. Aluminium, nickel and their alloys are the metals used for extrusion directly at the elevated temperatures.

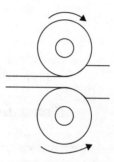

Two high reversing mill

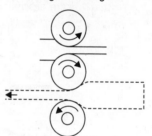

Three high mill

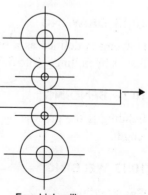

Four high mill

Fig. 10.18

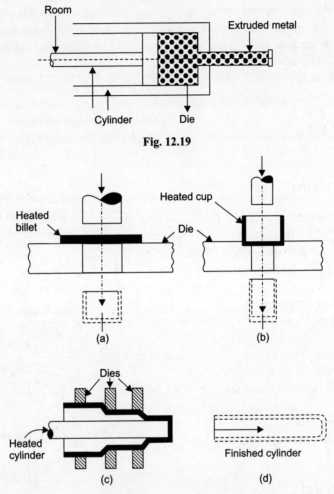

Fig. 12.19

Fig. 12.20: Hot drawing.

10.13 DRAWING

Drawing is defined as a process of making cup-shaped parts from sheet metal blanks by pulling it into dies with the help of a punch.

10.14 BENDING

Bending is the process by which a straight length is transformed into a curved length.

10.15 WELDING

Gas Welding

Gas welding is done by burning a combustible gas with air or oxygen in a concentrated flame of high temperature. The flame heats and melts the parent metal and filler rod at the joint. Gas welding can weld most common materials. The equipment used are inexpensive, versatile and serve adequately in many job and general repair shop.

The main types of gas welding are:
  (a) Oxyacetylene welding
  (b) Oxyhydrogen welding

Oxyacetylene Gas Welding

In oxyacetylene welding the two gases used for producing flame are oxygen and acetylene. Oxygen is used to support and intensify combustion. Acetylene can be easily produced by the chemical reaction between water and calcium carbide (CaC_2). It has to be produced and used at only low pressures as at high pressures; explosition might be there.

$$CaC_2 + 2H_2O \rightarrow C_2H_2 + Ca(OH)_2$$
Calcium carbide   water   Acetylene   Hydrated lime

Combustion

$$2C_2H_2 + 5O_2 \rightarrow 4CO_2 + 2H_2O + \text{Heat}$$
Acetylene   Carbon dioxide   Water vapour

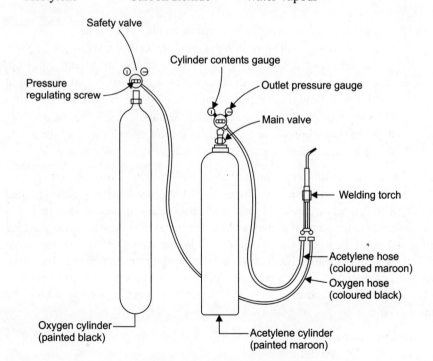

Fig. 10.21: Oxyacetylene welding set

The temperature developed in the flame as a result of these reactions can reach 3200°C to 3300°C. The neutral flame is desired for most welding operations but in certain cases a slightly oxidising flame, in which there is an excess of oxygen or slightly carburizing flame, in which there is an excess of acetylene is needed.

Welding Equipment
(1) Welding torch.
(2) Welding tip.
(3) Pressure regulator.
(4) Regulator pressure for gas welding.
(5) Hose and hose fittings.
(6) Goggles, gloves and spark lighter.
(7) Gas cylinders.

Oxyhydrogen Welding
(1) Not popularly used.
(2) The process is similar to oxygen—acetylene system with the only difference being a special regulator used in metering the hydrogen gas.

Arc Welding
(1) Arc welding is mostly used for joining metal parts. Here the source of heat is an electric arc.
(2) The arc column is generated between an anode, which is the positive pole of DC (direct current) power supply and the cathode, the negative pole. When these two conductors of an electric circuit are brought together and separated for a small distance (2 to 4 mm) such that the current continues to flow through a path of ionized particles (gaseous medium) called plasma an electric arc is formed. This ionized gas column acts as a high resistance conductor that enables more ions to flow from the anode to the cathode. Heat is generated as the ions strike the cathode.

At the centre of the arc temperature of about 6000–7000°C is produced.

The heat of the arc raises the temperature of the parent metal which is melted forming a pool of molten metal. The electrode metal is also melted and is transferred into the metal in the form of globules of molten metal. The deposited metal serves to fill and bond the joint or to fuse and built up the parent metal surface. About 66% of the heat is developed near the positive pole while the remaining is developed near the negative pole. That is why an electrode that is connected to the positive pole will burn away approximately 50% faster than when it is connected to the negative pole.

Arc welding equipment
(1) AC or DC Machine
(2) Electrode

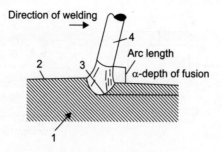

Fig. 10.22: A welding arc
1. Parent metal, 2. Deposited metal, 3. Crater, 4. Electrode.

(3) Electrode holder
(4) Cables, cable connectors
(5) Chipping hammer
(6) Earthing clamps
(7) Wire brush
(8) Helmet
(9) Safety goggles
(10) Hand gloves

Types of arc welding
(1) Carbon arc
(2) Metal arc
(3) Metal inert gas arc (MIG)
(4) Gas tungsten arc (T.I.G.)
(5) Atomic hydrogen arc
(6) Plasma arc
(7) Submerged arc
(8) Flux-cored arc
(9) Electro-slag welding

Resistance Welding

In resistance welding metal parts to be joined are heated to a plastic state over a limited area by their resistance to the flow of an electric current and mechanical pressure is used to complete the weld.

The different resistance welding methods are
(1) Butt welding
(2) Spot welding
(3) Seam welding
(4) Projection welding
(5) Percussion welding

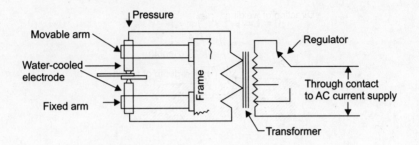

Fig. 10.23: The electrical circuit of a resistance welder.

10.16 SOLDERING

Soldering is a method of uniting two or more pieces of metal by means of a fusible alloy or metal called solder applied in the molten state.

For larger parts, the surfaces may be tinned first by cleaning, dipping in flux, and then by applying solder with a soldering iron or by dipping the parts in molten solder. The parts may then be assembled and heated together until the solder melt. The different compositions of solder for different purposes
 (1) Soft solder – Pb(lead), 37% (Sn), tin 63%
 (2) Medium solder – Pb 50% Sn 50%
 (3) Plumber's solder – Pb 70% Sn 30%
 (4) Electrician solder – Pb 58% Sn 42%

10.17 BRAZING

Brazing is similar to soldering but it gives a much stronger joint than soldering. In this harder filler material called spelter is used. This spelter fuses at temperature above red heat but below the melting temperature of the parts to be joined.

Brasses (copper and zinc) sometimes upto 20% tin are mostly used mainly for brazing the ferrous metals. Borax is widely used flux. Torch brazing in which heating is done by a blow torch is very common.

EXERCISE

1. Differentiate between cast Iron, pig Iron and steel.
2. Explain the effect of various chemical element on Iron
3. What are the variousmethods for commerical production of steel.
4. What are the different types of sand used for preparing a mould
5. Name the various patterns that are normally encountered in foundary practice. Mention the situations where each is advantageous used.
6. (a) Name the essential constitutents of moulding sand.
   (b) What properties a good moulding sand must possess?

7. Explain the difference between open sand and Bedde in method for preparing the mould
8. Explain the following
   (a) Slush casting
   (b) Centrfugal casting
   (c) Continuous casting
9. Name the various tools used inforging
10. Differentiate betweendrawing and bending.
11. Name the various welding equipments used in gas welding.
12. Name twotypes of Resistance welding methods.
13. Differentiate between soldering and brazing.
14. (a) Describe cast Iron with its composition, properties andaplications.
    (b) Explain the principle of resistance welding with an example.
    (c) Explain various steps involved in moulding process (MRIU Dec. 2009)
15. What is extrusion principle? (MRIU Dec. 2009)

OBJECTIVE TYPE QUESTIONS

1. Which of the following welding process uses non-consumable electrode.
   (a) LASER welding.
   (b) MIG welding.
   (c) TIG welding.
   (d) ion beam welding.
2. When welding is going on, arc voltage is of the order of.
   (a) 18-40 volts.
   (b) 40-95 volts.
   (c) 100-125 volts.
   (d) 130-170 volts.
3. Following gases are used in tungsten inert gas welding.
   (a) CO_2 and H_2
   (b) argon and neon.
   (c) helium and neon.
   (d) argon and helium.
4. Copper is
   (a) easily spot welded.
   (b) very difficult to be spot welded.
   (c) as good for spot welding as any other material.
   (d) preferred to be welded by spot welding.
5. In resistance welding, voltage used for heating is.
   (a) 1 v
   (b) 10 v
   (c) 100 v
   (d) 500 v.

6. In resistance welding, the pressure is released.
   (a) just at the time of passing the current.
   (b) after completion of current.
   (c) after the weld cods
   (d) during heating period.
7. Grey cast Iron is best welded by.
   (a) TIG
   (b) are
   (c) MIG
   (d) only-acetylene.
8. Seam welding is
   (a) multi-spot welding process.
   (b) continuous spot welding process.
   (c) used to form mesh.
   (d) during heating period.
9. The brazing metals and alloys commonly used are.
   (a) copper
   (b) copper alloys
   (c) silver alloys
   (d) all of the above.
10. Which of the following carbon steels is most weldable
    (a) 0.15% carbon steel
    (b) 0.30% carbon steel
    (c) 0.50% carbon steel
    (d) 0.70% carbon steel
11. In arc welding, temperature of the following order may he generated.
    (a) 100°C
    (b) 1500°C
    (c) 5500°C
    (d) 8000°C
12. Fluxes are used in welding in order to protect the molten metal and the surfaces to be jointed from
    (a) oxidation
    (b) carburizing
    (c) dirt
    (d) distortion and warping
13. Metal deposited on to the workpiece from the electrode
    (a) is forced across the arc
    (b) falls because of gravity
    (c) is attracted towards the workpiece due to the positive polarity of the workpiece
    (d) is attracted towards the workpiece due to negative

14. Weld spalter is
    (a) flux
    (b) electrode coating
    (c) welding defect
    (d) welding test
15. The following welding process uses consumable electrode
    (a) TIG
    (b) MIG
    (c) thermic
    (d) laser
16. Seam welding is
    (a) arc welding
    (b) multispot welding
    (c) continuous spot welding
    (d) used for forming sound bars
17. In resistance welding the electrode material is made of
    (a) carbon steel
    (b) stainless steel
    (c) copper
    (d) high speed steel
18. Metals like copper and brass can be welded by
    (a) oxidising flame
    (b) carburizing flame
    (c) neutral flame
    (d) any of the above flames
19. The commonly use flux for brazing is
    (a) resin
    (b) NH_4Cl
    (c) barox
    (d) soft iron
20. Draft on pattern for casting is
    (a) shrinkage allowance
    (b) identification number marked on it.
    (c) taper to facilitate its removal from mould
    (d) increase in size of mould cavity due to shaking of pattern
21. Casting process is preferred for parts having
    (a) few details
    (b) many details
    (c) no details
    (d) non-symmetrical shape

22. In order to facilitate the withdrawal of pattern
    (a) pattern is made smooth
    (b) water is applied on pattern surface
    (c) allowances are made on pattern
    (d) draft is provided on pattern
23. The mould is housed in a
    (a) flask
    (b) cope
    (c) drag
    (d) cheek
24. Cores are used to
    (a) make desired recess in castings
    (b) strengthen moulding sand
    (c) support loose pieces
    (d) remove pattern easily
25. Loose piece patterns are
    (a) a sort of split patterns
    (b) used when the pattern cannot be drawn from the mould
    (c) similar to core prints
    (d) never used in foundry work

ANSWERS

1. c	2. a	3. d	4. b	5. b
6. c	7. d	8. b	9. d	10. a
11. c	12. a	13. a	14. c	15. b
16. c	17. c	18. a	19. c	20. c
21. b	22. d	23. a	24. a	25. b

UNIT 11

Applied Mechanics

11.1 INTRODUCTION

Study of Mechanics enables an engineer to design members, elements or parts of a machine in such a way that these are manufactured with minimum resources and they perform strictly as predicted because an instantaneous failure may be disastrous as compared to the failure which is expected or known beforehand.

Mechanics

Mechanics is the branch of science which deals with the study of forces and motion of a physical system.

Application

Design of buildings, machinery, vehicles, etc.

Applied Mechanics

Applied mechanics is the branch of engineering which deals with the study of actual applications of principles of mechanics to solve engineering problems.

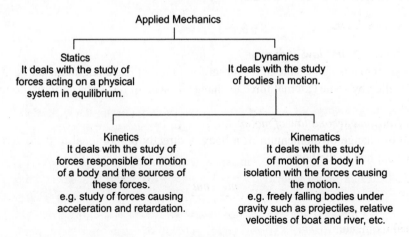

Fig. 11.1

Definitions

(1) **Rigid body:** A rigid body is one which does not undergo any deformation due to application of external forces.
(2) **Particle:** It is the body having zero dimensions or in other words, a body whose volume is negligibly small. It is assumed to be concentrated at a point.
(3) **Displacement:** It is defined as the change in position of a moving body. It is a vector quantity having magnitude as well as direction.
(4) **Speed:** It is defined as the rate at which a body describes its path with respect to time. It is a scalar quantity having SI units m/s.
(5) **Velocity:** Rate of displacement of a body in a specific direction. It is a vector quantity.
(6) **Acceleration:** It is defined as the rate of change of velocity of a moving body. It is a vector quantity having S.I. units m/s^2.
(7) **Force:** It is defined as the push or pull which changes or tends to change the state of rest or uniform motion of body. It is a vector quantity having S.I. units as Newton.
(8) **Weight:** It is defined as the force by which a body is attracted towards the centre of earth.

Scalar and Vector Quantities

A quantity is said to be scalar if it is completely defined by its magnitude alone. Examples are length, area and time.

A quantity is said to be vector if it is completely defined only when its magnitude and direction are specified. Hence force is a vector. The other examples of vectors are velocity and acceleration.

11.2 LAWS OF MECHANICS

Newton's Laws

(1) Newton's first law of motion
A body remains in the state of rest or uniform motion in a straight line unless it is compelled by some external force to change its state of rest or of uniform motion.

(2) Newton's second law of motion
Rate of change of momentum of a body is directly proportional to the external force applied to the body.

$$F \alpha \frac{mv - mu}{t}$$

Initial momentum = mu $\qquad\qquad$ m = mass

Final momentum = mv. v = final velocity
                                  u = initial velocity
                                  t = time

But $\left(\dfrac{v-u}{t}\right) = a = accleration$

$\Rightarrow$ $F \alpha\, m \times a$

                        $F = k.m \times a$

but $k = 1$

                        $F = m \times a$

(3) Newton's third law.
To each and every action there is an equal but opposite reaction.

Resultant Force

A number of forces F_1, F_2, F_3 ..., etc. acting simultaneously on a particle can be replaced by a single force R which, produces the same effect as is p produced by the given forces.

The single force R is called the resultant force and the forces F_1, F_2, F_3 ... etc. are called component forces.

If a single force equal to the resultant forces but opposite in direction is applied on the particle, it will cancel the effect of all forces acting on it such a force is called equilibrant as it brings the particle in equilibrium.

Parallelogram Law of Forces

It states that if two forces, acting simultaneously on a particle, be represented in magnitude and direction by the two adjacent sides of a parallelogram, their resultant may be represented in magnitude and direction by the diagonal of the parallelogram, which passes through their point of intersection.

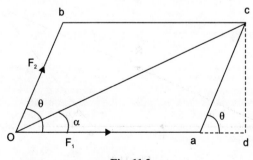

Fig. 11.2

Proof

From geometry $\angle cad = \theta$

$$ac = F_2$$

∴ $cd = F_2 \sin\theta$ and $ad = F_2 \cos\theta$

Now from right-angled $\Delta\ odc$, we get

$$oc = \sqrt{(od)^2 + (cd)^2} = \sqrt{(oa+ad)^2 + (cd)^2}$$

$$R = \sqrt{(F_1 + F_2 \cos\theta)^2 + (F_2 \sin\theta)^2}$$

$$= \sqrt{F_1^2 + F_1^2 \cos^2\theta + 2F_1 F_2 \cos\theta + F_2^2 \sin^2\theta}$$

$$= \sqrt{F_1^2 + F_2^2(\cos^2\theta + \sin^2\theta) + 2F_1 F_2 \cos\theta}$$

$$= \sqrt{F_1^2 + F_2^2 + 2F_1 F_2 \cos\theta} \qquad (\because \sin^2\theta + \cos^2\theta = 1)$$

$$\therefore R = \sqrt{F_1^2 + F_2^2 + 2F_1 F_2 \cos\theta}$$

Let the resultant make an angle α with F_1

then $\qquad \tan\alpha = \dfrac{cd}{od} = \dfrac{cd}{oa+ad} = \dfrac{F_2 \sin\theta}{F_1 + F_2 \cos\theta}$

Triangle Law of Forces

It states that if two forces acting simultaneously on a particle be represented in magnitude and direction by the two sides of a triangle taken in order, their resultant is represented, in magnitude and direction by the third side of the triangle taken in the opposite direction.

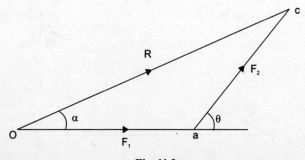

Fig. 11.3

Polygon Law of Forces

If any number of forces acting simultaneously on a particle be represented in magnitude and direction by the sides of a polygon taken in order, their resultant will be represented by the closing side of the polygon taken in opposite order.

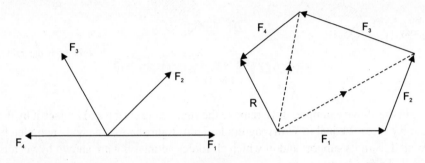

Fig. 11.4

Resolution of a Force into Components

A single force can be replaced by two forces acting in directions which will produce the same effect as the given force. This breaking up of a force into two parts is called the resolution of a force. The force which is broken into parts is called the resolved force and the parts are called component forces or the results.

Equilibrium Force

A force equal and opposite to the resultant of all the forces acting on a particle is called an equilibrant force. This is represented by Re.

Equilibrant of Coplanar Force System

Triangle law of equilibrium

It states that if three forces acting at a point in a body keep it in equilibrium, then they can be represented both in magnitude and direction by the three sides of a triangle taken in order.

Polygon law of equilibrium

It states that if a body be in a state of equilibrium under the action of a number of forces acting on it, then these forces can be vectorially represented by the sides of a closed polygon taken in order.

Free Body Diagrams

To study the equilibrium of a constrained body, we shall always imagine that we remove the supports and replace them by the reactions which they exert on the body.

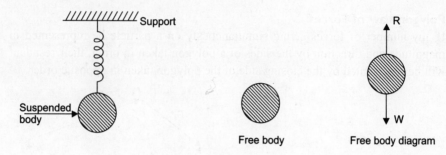

Fig. 11.5

In the above example, if we remove the supporting spring and replace it by the reactive force R equal to W in magnitude. Third figure shows the body completely isolated from its support and in which all forces acting on it are shown by vectors is called a free body diagram.

In the following figure, two cylinders are resting in a smooth trough. The free body diagrams of the two cylinders are also shown where the various forces will keep each cylinder in equilibrium.

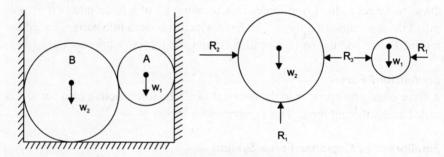

Fig. 11.6: Free body diagram.

Lami's Theorem

It states that if three forces acting on a point are in equilibrium, each force is proportional to the sine of the angle between the other two forces.

Let the three forces F_1, F_2 and F_3 acting at a point O be represented by OB, OA and OC respectively. Let the angle between F_1 and F_2 be α, between F_2 and F_3 be β and between F_3 and F_1 be γ. If these forces are in equilibrium, then according to Lami's theorem.

$$\frac{F_1}{\sin \beta} = \frac{F_2}{\sin \gamma} = \frac{F_3}{\sin \alpha}$$

Let us first consider the two forces F_1 and F_2 which are represented by the two sides *ob* and *oa* of the parallelogram *obda* as shown in figure. Then the resultant of these two forces will be given by *od* (the diagonal of the parallelogram) in

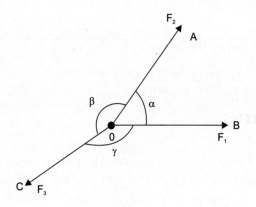

Fig. 11.7

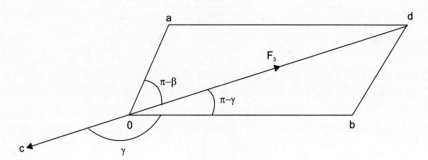

Fig. 11.8

magnitude and direction. This means od should be equal to F_3 in magnitude but in opposite direction to oc as F_1, F_2 and F_3 are in equilibrium.

Since $\quad\quad\quad ob \parallel oB, oa \parallel oA \parallel bd$ and $do \parallel oc$

$$\angle AOC = \angle aoc = \beta$$

$$\angle aod = \pi - \beta$$

$$\angle BOC = \angle boc = \gamma$$

$$\angle bod = \pi - \gamma$$

$$\angle BOA = \angle boa = \alpha$$

$$\angle dbo = \pi - \alpha$$

Now by the knowledge of trigonometry from $\triangle obd$, we get

$$\frac{ob}{\sin(\pi-\beta)} = \frac{bd}{\sin(\pi-\gamma)} = \frac{ob}{\sin(\pi-\alpha)}$$

$$\frac{F_1}{\sin\beta} = \frac{F_2}{\sin\gamma} = \frac{F_3}{\sin\alpha}$$

SOLVED PROBLEMS

Problem 1

Figure represents a weight of 400 N supported by two cords one 6 m long and the other 8 m long with points of support 10 m apart. Find the tensions T_1 and T_2 in the cords.

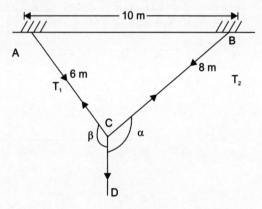

Fig. 11.9

Solution

From the knowledge of trigonometry from $\triangle ABC$.

$$AC^2 = AB^2 + BC^2 - 2AB \times BC \cos\beta.$$

$$\cos B = \frac{AB^2 + BC^2 - AC^2}{2AB.BC} = \frac{10^2 + 8^2 - 6^2}{2 \times 10 \times 8} = \frac{128}{160} = 0.8.$$

$$\angle B = 36.86°$$

$$\cos A = \frac{AB^2 + AC^2 - BC^2}{2AB.AC} = \frac{10^2 + 6^2 - 8^2}{2 \times 10 \times 6} = 0.6$$

$$\angle A = 53.13°$$

$$\angle C = 180° - (36.86° + 53.13) = 90.01°$$

$\angle BCD = \angle\alpha = 90° + 36.86 = 126.86$

$\angle ACD = \angle\beta = 90° + 53.13 = 143.13$

Applying Lami's theorem.

$$\frac{W}{\sin C} = \frac{T_1}{\sin\alpha} = \frac{T_2}{\sin\beta}$$

$$\frac{400}{\sin 90.01} = \frac{T_1}{\sin 126.86} = \frac{T_2}{\sin 143.13}$$

$$\frac{400}{1} = \frac{T_1}{0.8} = \frac{T_2}{0.6}$$

$$T_1 = 400 \times 0.8 = 320 \text{ N}$$

$$T_2 = 400 \times 0.6 = 240 \text{ N}$$

Problem 2

A smooth sphere of weight W is supported in contact with a smooth vertical wall by a string fastened to a point on its surface, the other end being attached to a point in the wall. If the length of the string is equal to the radius of the sphere, find the tension in the string and the reaction of the wall.

Solution

Since length of the string AB = radius of the sphere OC or OB.

$$OC = \frac{1}{2}OA$$

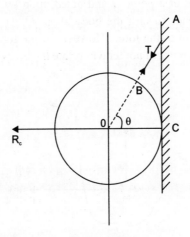

Fig. 11.10

From right angle ΔOCA

$$\cos\theta = \frac{OC}{OA} = \frac{1}{2}$$
$$\theta = 60°$$

Applying Lami's theorem to point O

$$\frac{T}{\sin 90°} = \frac{R}{\sin(90+60)}$$

$$= \frac{W}{\sin(180-60)}$$

$$T = \frac{W \sin 90°}{\sin 60°} = \frac{2W}{\sqrt{3}} \quad \text{Ans}$$

$$\therefore \text{Re} = \frac{W \sin(90° + 60°)}{\sin 60°}$$

$$= \frac{W \cos 60°}{\sin 60°}$$

$$= \frac{1}{2} \times \frac{2W}{\sqrt{3}} = \frac{W}{\sqrt{3}} \quad \text{Ans}$$

11.3 CONCEPT OF MOMENT

Consider a body nailed at a point O and acted upon by a force P as shown in figure. This force P tends to turn the body about O. This turning effect of the force P is called moment of the force. However, if the force is such that it passes through the point O, then it will not tend to rotate the body but it will tend to move

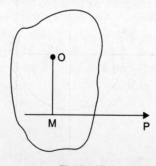

Fig. 11.11

the body in a straight line along the direction of the force. It means that turning effect i.e., moment not only depends upon the magnitude of the force but also upon the perpendicular distance from fixed point to the line of action of force.

Moment of Force and Moment of Centre

Moment of force on a body about a point or a line depends upon the following two factors:
  (1) Magnitude of the force and
  (2) Perpendicular distance from the line of action of the force to the fixed point or line of the body about which it rotates.

Moment of force for above figure is given by.

$$P \times OM$$

The point or line about which the body rotates under the influence of moment of a force is known as moment centre or fulcrum.

Sign Convention

If the turning of the body is in the same direction in which hands of a clock move, the moments are known as clockwise moments. As per the convention, these moments are taken as negative moments. If the force tends to turn the body in anticlockwise direction, the moments are known as anticlockwise moments. These are taken as positive moments.

Law of Moments

If a number of coplanar forces acting at a point be in equilibrium, the sum of clockwise moments about any point must be equal to the sum of anticlockwise moments about the same point.

Condition of Equilibrium for Bodies under Coplaner Forces

A body can only be in equilibrium if the algebraic sum of all the external forces and their moments about any point in their plane is zero.

Mathematically,
  (i) $\Sigma M = 0$
  (ii) $\Sigma V = 0$
  (iii) $\Sigma H = 0$

M = Moments, V = Vertical forces, H = Horizontal forces

Problem

A beam simply supported at both the ends carries load system as shown. Find the reactions at the two ends.

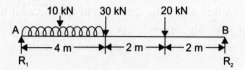

Fig. 11.12

Solution

As the load system acting on the beam is vertically downwards, so reaction at the two ends shall be vertically upward. Let R_1 and R_2 be the reactions at A and B ends respectively because beam under the load system is in equilibrium so clockwise moments about any point must be equal to anticlockwise moments about the same point.

Taking moments about R_1

$$R_2 \times 8 = 20 \times 6 + 30 \times 4 + 10 \times 4 \times 2$$

$$R_2 = \frac{120 + 120 + 80}{8}$$

$$R_2 = 40 \text{ kN.}$$

Similarly, R_1 can be found out by taking moments about point B or by applying the condition of equilibrium $\Sigma V = 0$

$$R_1 + R_2 = 20 + 30 + 40 = 90$$

By putting the value of R_2

$$R_1 = 90 - 40$$

$$R_1 = 50 \text{ N.}$$

11.4 VARIGNON'S THEOREM

Varignon's theorem states that the algebraic sum of the moments of two forces about any point in their plane is equal to the moment of their resultant about that point.

Principle of Moments

If any number of coplanar forces acting on a body having a single resultant, the algebraic sum of their moments about any point in their plane is equal to the moment of their resultant about that point.

11.5 PARTICLE DYNAMICS

In rectangular coordinates, we can express Newton's law as follows:

$$F_x = ma_x = \frac{mdVx}{dt} = \frac{md^2x}{dt^2}$$

$$F_y = ma_y = \frac{mdVy}{dt} = \frac{md^2y}{dt^2}$$

$$F_z = ma_z = \frac{mdVz}{dt} = \frac{md^2z}{dt^2}$$

If the motion is known relative to an inertial reference, we can easily solve for the rectangular components of the resultant force on the particle.

If the resultant force on a particle has the same direction and line of action at all times. The resulting motion is then confined to a straight line and is usually called rectilinear translation.

11.6 KINEMATICS OF RIGID BODIES

A particle is said to be in linear motion, if the path traced by it is a straight line. Many kinematic problems in linear motion can be solved just by using the definition of speed velocity and acceleration.

In this following topic, will be discussed:
(1) Motion with uniform acceleration
(2) Problems related to acceleration due to gravity

Motion with Uniform Acceleration

Consider the motion of a body with uniform acceleration a.

Let
  u = initial velocity
  v = final velocity
  t = time taken for change of velocity from u to v.

Acceleration is defined as rate of change of velocity. Since it is uniform, we can write

$$a = \frac{v-u}{t}$$

$$v = u + at \tag{1}$$

Displacement s is given by
  s = average velocity × time

$$= \frac{u+v}{2} \times t \tag{2}$$

Substituting the value of v from (1) into 2, we get

$$s = \frac{u+v+at}{2} t$$

$$s = ut + \frac{1}{2} at^2 \tag{3}$$

From (1)

$$t = \frac{v-u}{a}$$

putting this into 2

$$s = \left(\frac{u+v}{2}\right)\frac{(v-u)}{a}$$

$$= \frac{v^2 - u^2}{2a}$$

i.e. $v^2 - u^2 = 2as$ \hfill (4)

Thus, equations of motion of a body moving with constant acceleration are

$$\left.\begin{array}{ll} v = u + at & (a) \\ s = ut + \dfrac{1}{2}at^2 & (b) \\ v^2 - u^2 = 2as & (c) \end{array}\right\} \tag{5}$$

Acceleration Due to Gravity

Acceleration due to gravity is constant for all practical purposes when we treat motion of the bodies near earth's surface. Its value is found to be 9.81 m/sec^2 and is always directed towards centre of the earth, i.e. vertically downwards. Hence, if vertically downward motion of a body is considered, the value of acceleration a is 9.81 m/sec^2 and if vertically upward motion is considered then $a = -g = -9.81$ m/sec^2.

Example

A small steel ball is shot vertically upwards from the top of a building 25 m above the street with an initial velocity of 18 m/sec.

    (a) In what time it will reach the maximum height?
    (b) How high above the building will the ball rise?

(c) Compute the velocity with which it will strike the street and the total time for which the ball is in motion.

For upward motion

$$u = 18 \text{ m/sec}$$
$$v = 0$$
$$a = -9.81 \text{ m/sec}^2$$

and
$$s = h$$

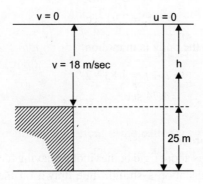

Fig. 11.13

Let t_1 be time taken to reach maximum height
From equation of motion

$$v = u + at$$
$$0 = 18 - 199.81\, t_1$$
$$t_1 = 1.83 \text{ sec.}$$

From the relation $v^2 - u^2 = 2as$

$$0 - 18^2 = 2 \times (-8.81)h$$

$$h = \frac{18^2}{2 \times 9.81} = 16.51 \text{ m}$$

∴ Total height from the ground

$$= 25 + h = 25 + 16.51$$
$$= 41.51 \text{ m}$$

Downward motion

With usual notations now

$$u = 0 \quad v = v_2 \quad s = 41.51 \text{ m} \quad a = +9.81 \text{ m/sec}^2 \quad t = t_2$$

From the relation $v^2 - u^2 = 2as$, we get

$$v_2^2 - 0 = 2 \times 9.81 \times 41.51$$
$$v_2 = 28.54 \text{ m/sec}$$

From the relation $v = u + at$, we get

$$28.54 = 0 + 9.81\, t_2$$
$$t_2 = 2.91 \text{ sec}$$

∴ Total time in which the body is in motion

$$= t_1 + t_2 = 1.83 + 2.91$$
$$= 4.74 \text{ sec.}$$

Rotation of Rigid Bodies

So far the discussion was about rigid bodies motion having rectilinear or curvilinear translation. During translation a straight line drawn on the rigid body remains parallel to its original position at any time.

Angular Motion

When a particle in a body moves from position A to B the displacement is θ. This displacement is a vector quantity since it has magnitude as well as direction. The direction is a rotation-either clockwise or counter-clockwise.

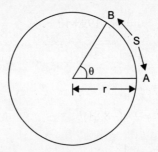

Fig. 11.14

The rate of change of angular displacement with time is called angular velocity and is denoted by ω. Thus

$$\omega = \frac{d\theta}{dt} \qquad (1)$$

The rate of change of angular velocity with time is called angular acceleration and is denoted by α.

$$\alpha = \frac{dw}{dt} = \frac{d^2\theta}{dt}$$

$$\alpha = \frac{dw}{dt} = \frac{dw}{d\theta} \times \frac{d\theta}{dt} = \frac{dw}{d\theta} w$$

$$\alpha = w\frac{dw}{d\theta} \qquad (2)$$

Relationship between angular motion and linear motion

When the particle moves from A to B the distance travelled by it is s. If r is the distance of the particle from the centre of rotation, then

$$s = r\theta$$

The tangential velocity of the particle is called linear velocity and is denoted by v then

$$v = \frac{ds}{dt} = \frac{rd\theta}{dt} \qquad (3)$$

The linear acceleration of the particle in tangential direction a_t is given by

$$a_t = \frac{dv}{dt} = \frac{rd^2\theta}{dt^2} \qquad (4)$$

Radial acceleration is then given by

$$a_r = \frac{v^2}{r} = rw^2 \qquad (5)$$

If the angular velocity is uniform, the angular distance moved in t seconds by a body having angular velocity w radians/seconds is given by

$$\theta = wt \text{ radians} \qquad (6)$$

Many times angular velocity is given in terms of revolution per minute (rpm). Since there are 2π radians in one revolution and 60 seconds in one minute, the angular acceleration w is given by

$$w = \frac{2\pi N}{60} \qquad (7)$$

N is in rpm.

Since angular velocity is uniform the time taken for one revolution T is given by equation (6)

$$2\pi = wT$$
$$T = \frac{2\pi}{w}$$

Uniformly accelerated rotation

Let us consider the uniformly accelerated motion with angular acceleration α. Then

$$\frac{dw}{dt} = \alpha$$

$$w = \alpha t + c_1$$

where c_1 is constant of integration.

If the initial velocity w_0 then

$$w_0 = \alpha \times 0 + c_1 \quad \text{or} \quad c_1 = w_0$$

$$w = w_0 + \alpha t \tag{1}$$

Again from definition of angular velocity

$$\frac{d\theta}{dt} = w = w_0 + \alpha t$$

$$\theta = w_0 t + \frac{1}{2}\alpha t^2 + c_2$$

c_2 = constant of integration

When $\theta = 0$ $t = 0$

we get

$$0 = 0 + 0 + c_2 \Rightarrow c_2 = 0$$

$$\theta = w_0 t + \frac{1}{2}\alpha t^2 \tag{2}$$

From equation (2)

$$\alpha = \frac{w \partial w}{d\theta}$$

or
$$\alpha d\theta = w dw$$

Integrating, we get

$$\alpha\theta = \frac{w^2}{2} + c_3$$

where c_3 is constant of integration.

Initially $\theta = 0$ and $w = w_0$

Hence, we get
$$\alpha \times 0 = \frac{w_0^2}{2} + c_3$$

$$c_3 = \frac{-w_0^2}{2}$$

$$\alpha\theta = \frac{w^2}{2} + \frac{(-w_0^2)}{2}$$

or
$$w^2 - w_0^2 = 2\alpha\theta \qquad (3)$$

Thus, for uniformly accelerated angular motion

$$\left.\begin{array}{ll} w = w_0 + \alpha t & \text{(i)} \\ \theta = w_0 t + \dfrac{1}{2}\alpha t^2 & \text{(ii)} \\ w^2 - w_0^2 = 2\alpha S & \text{(iii)} \end{array}\right\} \qquad (4)$$

Problem

The rotation of a flywheel is governed by the equation $w = 3t^2 - 2t + 2$ where w is in radians per second and t is in seconds. After one second from start, the angular displacement was 4 radians. Determine the angular displacement, angular velocity and angular acceleration where t = 3 seconds.

Solution

$$w = 3t^2 - 2t + 2$$

$$\frac{d\theta}{dt} = 3t^2 - 2t + 2$$

$$\theta = t^3 - t^2 + 2t + c$$

where c is constant of integration

When $\quad t = 1 \quad \theta = 4$

$$\therefore 4 = 1 - 1 + 2 + c$$

i.e., $\quad c = 2$

$$\therefore \theta = t^3 - t^2 + 2t + 2$$

When $\quad t = 3$

$$\theta = 3^3 - 2^2 + 2 \times 3 + 2 = 26 \text{ radians}$$

$$w = 3 \times 3^2 - 2 \times 3 + 2 = 23 \text{ rad/sec.}$$

Angular acceleration α is given by

$$\alpha = \frac{dw}{dt} = 6t - 2$$

$\therefore$ When $\quad t = 3$

$$\alpha = 6 \times 3 - 2 = 16 \text{ rad/sec}^2. \text{ **Ans**}$$

In circular motion.

Rotational Moment	$M = I\alpha.$
Angular Momentum	Iw
Kinetic Energy	$KE = \frac{1}{2} Iw^2$

Comparison between various terms used in linear motion and rotation.

Particulars	Linear motion	Angular motion
Displacement	s	θ
Initial velocity	u	w_0
Final velocity	v	w
Acceleration	a	α
Formulae for final velocity	$v = u + at$	$w = w_0 + \alpha t$
Formulae for displacement	$s = ut + \frac{1}{2} at^2$	$\theta = w_0 t + \frac{1}{2} \alpha t^2$
Displacement velocity and acceleration	$v^2 - u^2 = 2as$	$w^2 - w_0^2 = 2\alpha\theta$
Force causing motion	$F = ma$	$M_t = I\alpha$
Momentum	mv	Iw
Kinetic energy	$\frac{1}{2} mv^2$	$\frac{1}{2} Iw^2$

11.7 GENERAL PLANE MOTION

A body is said to have general plane motion if it possesses translation and rotation simultaneously.

Examples
(1) Wheel rolling on straight line
(2) Rod sliding against wall at one end and the floor at the other end.

Applied Mechanics 311

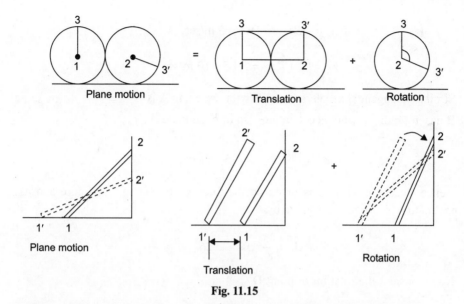

Fig. 11.15

For the analysis of general plane motion, it is convenient to split the motion into translation and pure rotation cases. The analysis for these two cases is carried out separately and then combined to get the final motion.

Problem

Wheel of radius 1 m rolls freely with an angular velocity of 5 rad/sec and with an angular acceleration of 4 rad/sec² both clockwise as shown in Fig. 13.16. Determine the velocity and acceleration of points B and D shown in the figure.

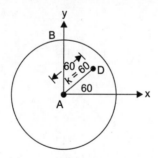

Fig. 11.16

Solution

The motion of points B and D will be split into translation, of geometric centre A and rotation about A. Translation of A is given by

$$V_A = rw = 1 \times 5 = 5 \text{ m/sec.}$$
$$a_A = r\alpha = 1 \times 4 = 4 \text{ m/sec}^2.$$

Now consider rotation of B about A. Its relative linear velocity with respect to A (normal to A) and is given by.

$$V_{B/n} = AB \times w = 1 \times 5 = 5 \text{ m/sec.}$$

$$V_B = V_A + V_{B/A} = 5 + 5 = 10 \text{ m/sec.}$$

Similarly, tangential acceleration of B with respect to A is $a_{B/A} = 1 \times 4 = 4$ m/sec^2
B has got radial inward acceleration also of magnitude

$$a_x = \frac{v_{B/A}^2}{r} = \frac{5^2}{1} = 25 \text{ m/sec}^2.$$

Hence, acceleration of B is given by the three vectors shown in figure and its inclination to horizontal is given by

$$\tan \theta = \frac{25}{8} \therefore \theta = 72.26°$$

Now consider the rotation of point D.

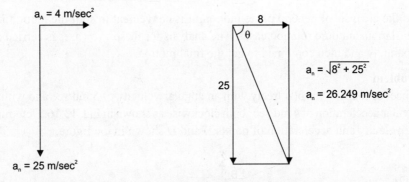

Fig. 11.17

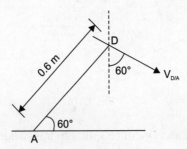

Fig. 11.18

Due to rotation about A, D has a linear velocity of $V_{D/A} = rw = 0.6 \times 5 = 3$ m/sec tangential to AD, i.e. at 60° to vertical.

$$V_{Dx} = V_A + V_{D/A} \sin 60°$$

$$= 5 + 3\sin 60° = 7.598 \text{ m/sec}$$

$$V_{Dy} = V_{D/A} \cos 60°$$

$$= 3\cos 60° = 1.5 \text{ m/sec.}$$

$$V_D = 7.745 \text{ m/sec.}$$

And its inclination to horizontal is given by

$$\tan \theta = \frac{1.5}{7.598}$$

$$\therefore \theta = 11.17°$$

Due to rotation about A, D has a tangential acceleration $= r\alpha = 0.6 \times 4 = 2.4 \text{ m/sec}^2$ and radial in word acceleration.

$$\frac{v_{D/A}^2}{r} = \frac{3^2}{0.6} = 15 \text{ m/sec}^2$$

$$v_{DA} = 0.6 \times 5 = 3 \text{ m/sec}^2$$

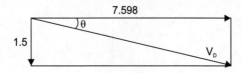

Fig. 11.19

Hence, total acceleration of D is vectorial sum of
(1) acceleration due to translation $a_A = 4 \text{ m/sec}^2$
(2) tangential acceleration $r_d = 2.4 \text{ m/sec}^2$ due to rotation
(3) radial inward acceleration of 15 m/sec^2

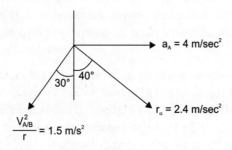

Fig. 11.20

Let a_c be acceleration of D and its inclination to horizontal be θ as shown in figure.

Then
$$a_D \sin \theta = y = 2.4 \cos 60° + 15.0 \cos 30°$$
$$= 14.190 \text{ m/sec}^2$$

$$a_D \cos \theta = 4 + 2.4 \sin 60° - 15 \cos 30°$$
$$= -1.\overline{422} \text{ m/s}^2 = 1.422 \text{ m/s}^2$$

$$a_D = \sqrt{14.190^2 + 1.422^2} = 14.261 \text{ m/s}^2$$

$$\theta = \tan^{-1} \frac{14.190}{1.422} = 84.227.$$

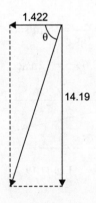

Fig. 11.21

Instantaneous Axis of Rotation

At any instant it is possible to locate a point in the plane which has zero velocity and hence plane motion of other points may be looked as pure rotation about this point. Such a point is called instantaneous centre and the axis passing through this point and at right angles to the plane of motion is called instantaneous axis of rotation.

Consider the rigid body shown in figure which has plane motion.

At any instant point B has velocity V_B. Now locate a point C perpendicular to the direction r_B at B at distance V_B. Now plane of B can be split into translation of C and rotation about C . $\therefore V_B = V_C + r_B \omega$ and direction is at right angles to CB. If we make $r_B = V_{B/\omega}$, then from the relation above we get,

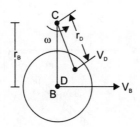

Fig. 13.22

$$V_B = V_c + \frac{V_B}{\omega}\omega$$
$$V_c = 0$$

Thus, if point C is selected at a distance $\frac{V_B}{\omega}$ along the perpendicular to the direction of velocity at B, the plane motion of B reduced to pure rotation about C. Hence, C is instantaneous centre. If D is any point on the rigid body, its velocity will be given by

$$V_D = V_c + r_D\omega$$
$$= r_D\omega \text{ since } V_c = 0.$$

and its velocity will be at right angles to C_D.

Thus, if the instantaneous centre is located, motion of all other points at that instant can be found by pure rotation case about B.

Methods of locating instantaneous centre.

Instantaneous centre can be located by any one of the following methods.

(1) If the angular velocity ω and linear velocity V_B are known instantaneous centre is located at a distance $\frac{V_B}{\omega}$ along the perpendicular to the direction of V_B at B.

(2) If the linear velocities of two points of the rigid body are known, say V_B and V_D drop perpendiculars to them at B and D. The intersection point is instantaneous centre.

Example

Find the velocity of B shown in figure by instantaneous centre method.

$$V_A = \text{Horizontal Velocity}$$
$$V_B = \text{Vertical Velocity}$$

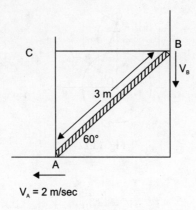

Fig. 11.23

Instantaneous centre is point C, which is obtained by dropping ⊥ to the direction V_A and V_B at points A and B respectively.

$$V_A = Ac \times w$$

$$2 = 3\sin 60° \times w$$

$$w = 0.770 \text{ rad/sec}$$

$$V_B = Bc \times w = 3\cos 60° \times 0.770$$

i.e., $\quad V_B = 1.555$ m/sec.

11.8 FREE VIBRATION

Vibration problem occurs where there are rotating or moving parts in a machinery. Apart from the machinery itself, the surrounding structure also faces the vibration hazard because of this vibrating machinery. The common examples are locomotives, diesel engines mounted on unsound foundations, whirling of shafts, etc.

Causes of Vibration
(1) Unbalanced forces in the machine. These forces are produced from within the machine itself.
(2) Dry friction between the two mating surfaces. This produces what are known as self-excited vibrations.
(3) External excitations. These excitations may be periodic, random or the nature of an impact produced external to the vibrating system.
(4) Earthquakes. These are responsible for the failure of many buildings, dams, etc.

Applied Mechanics 317

(5) Winds. These may cause the vibration of transmission and telephone lines under certain conditions.

Effects of Vibration

Negative effects
(1) Excessive stresses.
(2) Undesirable noise.
(3) Looseness of parts.
(4) Partial or complete failure of parts.

Positive effects
(1) In musical instruments
(2) Vibrating screens
(3) Shakers
(4) Stress relieving

Remedial Measures
(1) Remove the cause of vibration.
(2) Putting the screen if noise is the objection.
(3) Placing the machinery on proper type of isolators.
(4) Shock absorbers.
(5) Dynamic vibration absorbers.

Although the above methods are available to reduce vibrations at a stage where no changes in design are possible, anticipation of trouble in the original planning and design can make possible the avoidance of vibration problems at a little cost.

SOLVED PROBLEMS

Problem 1
A machine weighing 5 kN is supported by two chains attached to some point on the machine. One chain goes to the hook in the ceiling and has an inclination of 45° with the horizontal. The other chain goes to the eye bolt in the wall and is inclined at 30° to the horizontal. Make calculations for the tensions induced in the chain.

Solution
According to the problem given the figure shown in drawn. Below mentioned forces are acting on the machine.
    (i) Tension T_1 in the chain OA which goes to the hook in the ceiling.
    (ii) Tension T_2 in the chain OB which goes to the eye bolt.
    (iii) Weight of Machine W = 5 kN acting vertically downwards.
We can apply Lami's theorem since all the three forces are concurrent

318 Mechanical Engineering: Fundamentals

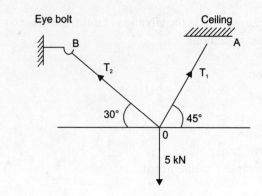

$$\frac{T_1}{\sin(90+30)} = \frac{T_2}{\sin(90+45)} = \frac{W}{\sin(180-30-45)}$$

or

$$\frac{T_1}{\sin 120} = \frac{T_2}{\sin 135} = \frac{W}{\sin 105}$$

∴

$$T_1 = W \times \frac{\sin 120°}{\sin 105°} = 5 \times \frac{0.866}{0.966} = 4.48 \text{ kN}$$

$$T_2 = W \times \frac{\sin 135}{\sin 105} = 5 \times \frac{0.707}{0.966} = 3.66 \text{ N}$$

Problem 2

A boat B is in the middle of a canal 100 m wide and is pulled by two ropes BA 150 m long and BC 100 m long by two locomotives on the banks the fully in BC being 1500 N as shown in figure. Find the pull Q in BA so that the boat moves parallel to the banks. Find also the resultant force on the boat.

Solution

Let the inclination of the ropes BC and BA be α and β with the banks

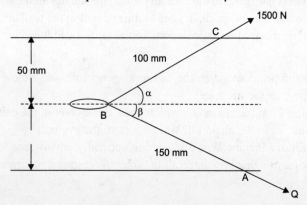

$$\sin\alpha = \frac{50}{100} = \frac{1}{2} \therefore \alpha = 30°$$

$$\sin\alpha = \frac{50}{100} = \frac{1}{3} \therefore \beta = 19.471°$$

Since the boat moves parallel to the banks, the components of the pulls in the ropes BC and BA at right angles to the bank must balance.

∴
$$Q\sin\beta = 1500\sin\alpha$$

$$Q + \frac{1}{3} = 1500 \times \frac{1}{2}$$

$$Q = 2250 \text{ N}$$

∴ Resultant force on the boat

= components of the pulls in te ropes parallel to the bank

= 1500 cos α + 2250 cos β

= 1500 cos 30° + 2250 cos 19.471° = 3420.36 N

Problem 3
Figure shows two spheres A and B resting in a smooth trough Draw the free body diagrams of A and B separately showing all the forces acting on them. Radius of A = 250 mm and that of B= 200 mm.

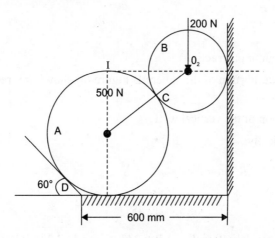

Solution
For the given configuration of the spheres.

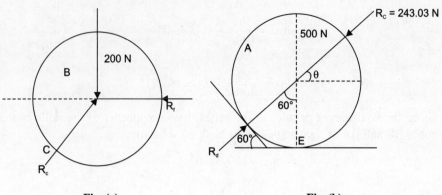

Fig. (a)   **Fig. (b)**

$$GE = 250 \tan 30° = \frac{250}{\sqrt{3}} \text{ mm}$$

$$EH = 600 - \frac{250}{\sqrt{3}} = 455.662 \text{ mm}$$

$$Q_2 F = 200 \text{ mm}$$

$IO_2 = 455.662 - 200 = 255.662$

$$\cos \theta = \frac{IO_2}{O_1 O_2} = \frac{255.662}{450} = 0.56814$$

∴ $\theta = 55° 22' 45''$

Let R_c be the common reactions

Consider the equilibrium of the sphere B Figure (a) shows the free body diagram for this sphere.

Let R_f = Reaction of the vertical wall

Resolving vertically

$$R_c \sin = 200$$

∴ $$R_c = \frac{200}{\sin 55' 22' 45''} = 243.03 \text{ N}$$

Resolving horizontally $R_f = R_c \cos \theta = 243.03 \cos 55" 22' 45" = 138.08$ N Now consider the equilibrium of the sphere A

The forces acting on the sphere are

Applied Mechanics 321

  (i) Weight of the sphere B = 500 N
 (ii) Normal Reaction R_c = 243.03 N at C
(iii) Normal reaction R_e at E
(iv) Normal reaction R_d at D.

Figure (b) shows the free body diagram

Resolving horizontally R_d cos 30° = R_c cos q = 243.03 cos 55" 22 ' 45"

$$R_d = 159.435 \text{ N}$$

Resolving vertically R_e = 243.03 sin q − R_d cos 60° = 243.03 sin 55" 22' 45"

$$- 159.435 \cos 60°$$

$$= 200 - 79.72 = 120.28 \text{ N}$$

Problem 4

A stone after falling 4 seconds from rest breaks a glass pan and in breaking it loses 25% of its velocity. How far will it fall in the next second? Take g = 9.81 metre/sec².

Solution

Velocity acquired in 4 seconds in falling from rest

$$= v = v \_ gt = gt = 9.81 \times 4 \text{ m/sec}$$

∴   v = 39.24 m/sec

since the stone loses 25% of its velocity, its velocity after breaking the glass pane

$$= 39.24 \times 3/4 = 29.43 \text{ m/sec.}$$

∴ Distance by which the stone will fall in the next one second.

$$= s = ut + \frac{1}{2} gt^2$$

$$= (29.43 \times 1) \times \left(\frac{1}{2} \times 9.81 \times 1^2 \right) = 34.335 \text{ m}$$

Problem 5

A stone dropped into well is heard to strike the water after 4 seconds. Find the depth of the well, if the velocity of sound is 350 metres/sec.

Solution

Let h = depth of the well in metres
Let time taken by the stone to reach the water = t_1 sec.

$$h = \frac{1}{2} g t_1^2$$

Time taken by sound to move from water surface to top of well

$$= \frac{h}{350} = \frac{\frac{1}{2}gt_1^2}{350}$$

Total time $= t_1 + \frac{\frac{1}{2}gt_1^2}{350} = 4$

$g = 9.8$ metres/sec.²

$$t_1 + \frac{\frac{1}{2}9.81\,t_1^2}{350} = 4$$

$$7t_1^2 + 500t_1 - 2000 = 0$$

$$t_1 = 3.798 \text{ sec}.$$

$$h = \frac{1}{2}gt_1^2 = \frac{1}{2} \times 9.81(3.789)^2 = 70.68 \text{ metres}$$

EXERCISE

1. Three marks A, B and C spaced at a distance of 100 m are made along a straight road. A car starting from rest and accelerating uniformly passes the mark A and takes 10 seconds to reach the mark B and further 8 seconds to reach the mark C. Make calculations for
   (a) The magnitude of the acceleration of the car.
   (b) The velocity of the car at A.
   (c) The velocity of the car at B.
   (d) The distance of the mark A from the starting point

   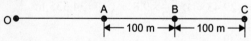

   Ans $U_a = 8.61$ m/s
   $a = 0.278$ m/s²
   $U_b = 11.388$ m/s
   $s = 133.3$ m.
2. What do you understand by free body diagram? (MRIU Dec 2009)
3. Explain free vibration and their causes, effects and remedial measures? (MRIU Dec. 2009)

4. A particle falls from rest and in the last second of its motion it passes 70 metres. Find the height from which it fell and the time of its fall. Take g = 9.8 m/sec^2
5. Explain Newton's laws of motion?
6. State and prove parallelogram law of forces?
7. State and prove Lami's theorem?
8. Define the following
   (a) Moment
   (b) Law of Moment
   (c) Varignon's theorem

OBJECTIVE TYPE QUESTIONS

1. Forces are called concurrent when their lines of action meet in
   (a) one point
   (b) two points
   (c) plane
   (d) perpendicular planes
2. Forces are called coplanar when all of them acting on body lie in
   (a) one point
   (b) one plane
   (c) different planes
   (d) perpendicular planes
3. Effect of a force on a body depends upon
   (a) magnitude
   (b) direction
   (c) position or line of action
   (d) all of the above
4. A force is completely defined when we specify
   (a) magnitude
   (b) direction
   (c) point of application
   (d) all of the above
5. If a number of forces act simultaneously on a particle it is possible
   (a) not a replace them by a single force
   (b) to replace them by a single force
   (c) to replace them by a single force through C.G.
   (d) to replace them by a couple
6. If two equal forces of magnitude P act at angle $\theta°$ their resultant will be
   (a) $\dfrac{P}{2}\cos\theta/2$
   (b) $2P\sin\dfrac{\theta}{2}$

(c) $2P\tan\dfrac{\theta}{2}$

(d) $2P\cos\dfrac{\theta}{2}$

7. Which of the following do not have identical dimensions?
   (a) Momentum and impulse
   (b) Torque and energy
   (c) Torque and work
   (d) Moment of force and angular momentum
8. The forces, which meet at one point, but their lines of action do not lie in a plane are called
   (a) coplanar non-concurrent forces
   (b) non-coplanar concurrent forces
   (c) non-coplanar non concurrent forces
   (d) intersecting forces
9. The magnitude of two forces, which when acting at right angles produce resultant force of $\sqrt{10}$ kg and when acting at 60° produce resultant of $\sqrt{13}$ kg. These forces are
   (a) 2 and $\sqrt{6}$
   (b) 2 and 1 kg
   (c) $\sqrt{5}$ and $\sqrt{5}$
   (d) 2 and 5
10. A number of forces acting at a point will be in equilibrium if
    (a) their total sum is zero
    (b) two resolved parts in two directions at right angles are equal.
    (c) Sum of resolved parts in any two perpendicular directions are both zero
    (d) all of them are inclined equally.
11. Two non-colinear parallel equal forces acting in opposite direction
    (a) balance each other
    (b) constitute a moment
    (c) constitute a couple
    (d) Constitute a moment of couple
12. According to law of triple of forces
    (a) Three forces acting at a point will be in equilibrium
    (b) Three forces acting at a point can be represented by a triangle each side being proportional to force.
    (c) If three forces acting upon a particle are represented in magnitude and direction by the sides of a triangle taken in order, they will e in equilibrium
    (d) If three forces acting at a point are in equilibrium, each force is proportional to the sine of the angle between the other two.

13. If two forces each equal to P in magnitude act at right angles, their effect may be neutralised by a third force acting along their bisector in opposite direction whose magnitude is equal to
    (a) 2P
    (b) $\dfrac{P}{2}$
    (c) $\sqrt{2}P$
    (d) $\dfrac{P}{\sqrt{2}}$

14. The resultant of two forces P and Q inclined at angle θ will be inclined at following angle w.r.t. P
    (a) $\dfrac{\theta}{2}$
    (b) $\tan^{-1}\dfrac{Q\sin\theta}{P+Q\cos\theta}$
    (c) $\tan^{-1}\dfrac{Q\sin\theta}{Q+P\cos\theta}$
    (d) $\tan^{-1}\dfrac{Q\sin\theta}{Q+P\sin\theta}$

15. According to Lami's theorem
    (a) Three forces acting at a point will be in equilibrium
    (b) Three forces acting at a point can be represented by triangle, each side being proportional to force.
    (c) If three forces acting upon a particle are represented in magnitude and direction, by the sides of a triangle, taken in order, they will being equilibrium.
    (d) If three forces acting at a point are in equilibrium, each force is proportional to the sine of the angle between the other two.

16. On a ladder resting on smooth ground and leaning against vertical wall, the force of friction will be
    (a) towards the wall at its upper end
    (b) away from the wall at its upper end
    (c) perpendicular to the wall at its upper end
    (d) zero at its upper end

17. A particle moves along a straight line such that distance (x) transversed in t seconds is given by $x = t^2(t-4)$, the acceleration of the particle will be given by the equation
    (a) $3t^2 - t$

(b) $3t^2 + 2t$
(c) $6t - 8$
(d) $6t - 4$

18. A body moves from rest with a constant acceleration of 5 m per sec. The distance covered in 5 sec is most nearly
    (a) 38 m
    (b) 62.5 m
    (c) 96 m
    (d) 124 m.

19. Two identical rollers of equal weight are supported as shown in figure the maximum reaction will occur at
    (a) point A
    (b) point B
    (c) point C
    (d) point D

20. A ladder AB of weight W kg is supported as shown in figure. The reaction at A and B and be equal to
    (a) W, W
    (b) $W, 0$
    (c) $\dfrac{W}{2}, \dfrac{W}{2}$
    (d) $\dfrac{W}{2}, 0$

21. A body is resting on a plane inclined at angle of 30° to horizontal what force would be required to slide it down, if the coefficient of friction between body and plane is 0.3
    (a) zero
    (b) 1 kg
    (c) 5 kg
    (d) would depend on weight of body

22. A ball is through up. The sum of kinetic and potential energies will be maximum at
    (a) ground
    (b) highest point
    (c) in the centre while going up.
    (d) at all the points.

23. The angular velocity of a particle changes from 69 to 71 rpm in 30 seconds. Its angular acceleration in rev/min is equal to.
    (a) 1
    (b) 2
    (c) 4
    (d) 8

24. A ball of mass 1 kg moving with a velocity of 2 m/sec collides directly on a stationary ball of man 2 kg and come to rest after impact. The velocity of the second ball of ten impact will be.
   (a) zero
   (b) 0.5 m/sec
   (c) 1.0 m/sec
   (d) 2.0 m/sec.

ANSWERS

1. a	2. b	3. d	4. d	5. b
6. d	7. d	8. b	9. c	10. c
11. c	12. c	13. c	14. b	15. d
16. c	17. c	18. b	19. c	20. b
21. a	22. d	23. d	24. d	

UNIT 12

Steam Turbines, Condensers, Cooling Towers, I.C. Engines and Gas Turbine

Upon completing this chapter, you should be able to:
(1) Explain impulse and reaction turbine working.
(2) Different types of compounding in impulse turbine.
(3) How steam condenser increases the efficiency of turbines.
(4) Different types of steam condenser.
(5) Cooling ponds and cooling towers.
(6) Terms used with I.C. engines.
(7) Types of efficiencies related with I.C. engines.
(8) Parts of I.C. engines.
(9) Working of 2-stroke and 4-stroke engines.
(10) Comparison of Otto, diesel and dual cycle.
(11) Working principle of gas turbine.

12.1 INTRODUCTION OF STEAM TURBINES

Steam turbine is a prime mover which converts the heat energy of steam (at high pressure and temperature) into mechanical work. The power so produced may be utilized in the various fields of industry such as electricity generation, transport, in driving of pumps, fans and compressors, etc. The principle of operation of any turbine depends on Newton's second law of motion. The motive power in a turbine is obtained by the change in momentum of a high velocity jet impinging on a curved blade. The steam from the boiler is expanded in a passage or nozzle, where, due to fall in pressure of steam, thermal energy of steam is converted into kinetic energy of steam resulting in the emission of a high velocity jet of steam which impinges on the moving vanes or blades mounted on a shaft, here it undergoes a change in the direction of motion which gives rise to a change in momentum and therefore, a force. This constitutes the driving force of the machine.

12.2 CLASSIFICATION OF TURBINES

The steam turbines can be classified in the following ways:
(1) On the basis of the direction of flow of fluid relative to the rotor
    (a) Axial flow turbines
    (b) Radial flow turbines
    (c) Mixed flow turbines

(2) On the basis of the way in which the rate of change of angular momentum of fluid is achieved.
    (a) Impulse turbine
    (b) Reaction turbine

(3) According to the method of compounding
    (a) Velocity compounded turbine.
    (b) Pressure compounded turbine.
    (c) Pressure-velocity compounded turbine.

(4) According to the pressure of inlet steam
    (a) High pressure turbine.
    (b) Low pressure turbine.

(5) According to the exhaust condition of steam
    (a) Condensing turbine.
    (b) Non-condensing turbine.

(6) According to the number of stages
    (a) Single stage turbine.
    (b) Multi-stage turbine.

(7) According to the position of shaft
    (a) Vertical shaft turbine.
    (b) Horizontal shaft turbine.

(8) According to the field of service i.e. stationary, variable speed, locomotive, ship, industrial etc.

12.3 IMPULSE TURBINE

It works on the principle of impulse where the kinetic energy of fluid jet is used to exert a force on a set of moving blades. Total enthalpy drop takes place in the nozzle or in the fixed blades. Thus, steam pressure remains constant while it flows through the moving blades. Kinetic energy is converted into mechanical power. Figure 12.1 shows the flow of the steam through impulse turbine, pressure and velocity variation has been also shown. Examples of such types of turbine are De-Laval, Curties, Rateau etc.

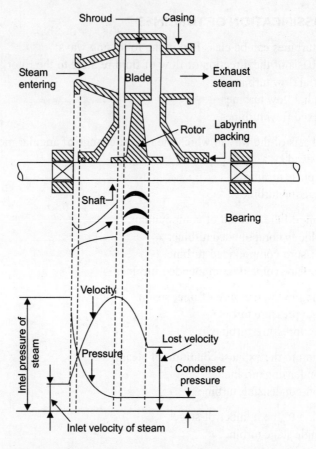

Fig. 12.1: Diagrammatic arrangement of a simple impulse turbine

The impulse turbine has two principal characteristics:
(i) It requires nozzles so that the pressure drop of steam takes place in the nozzles. The steam enters the turbine with a high velocity. The pressure in the turbine remains constant because whole of the pressure drop takes place in the nozzles.
(ii) The velocity of the steam is reduced as some of the kinetic energy in the steam is used up in producing work on the turbine shaft.

The main parts of the impulse turbine include a rotor nozzle, blades and casing. The casing forms the outer cover of the turbine and the nozzles are fixed in it. The moving blades are fixed on the rotating element, known as rotor. The rotor is coupled with the shaft of the turbine from which the useful torque is obtained. During working, the steam is first passed through a set of convergent-divergent nozzles, where the pressure energy is converted into kinetic energy. The high-speed steam jet strikes the rotor blades and due to continuous change in angular momentum, the torque is developed. The small rotor gives a very high rotational speed which can only be utilized to drive generators with large reduction gearing arrangements.

12.4 IMPULSE-REACTION TURBINE

This turbine is based on the principle that a small pressure drop of steam takes place in the moving blade which results in increase in kinetic energy of steam. This kinetic energy gives rise to a reaction in the direction opposite to that of added velocity. Thus, the gross propeller force or driving force is the vector sum of impulse and reaction force.

Expansion of steam takes place on both the blades. The reaction force due to pressure and other force due to change in momentum on moving blade provides the motive power. Therefore, entire pressure drop is achieved gradually and continuously over a series of guide blades (called fixed blade also) and moving blades in succession. This is the major difference between impulse and reaction turbine that pressure remains constant while flowing over moving blades of impulse turbine and it gradually decreases while flowing over the moving blades of reaction turbine. The example of such turbines is the Parson turbine. Figure (12.2) shows the flow of steam over fixed and moving blades. The pressure and velocity variation

Difference between impulse and reaction turbines

Impulse turbine	Reaction turbine
(i) In impulse turbine, the fluid is expanded completely in the nozzle and pressure remains constant during its passage through the moving blades.	(i) In reaction turbine, the fluid is partially expanded in the fixed blades and remaining expansion takes place in moving blades.
(ii) In impulse turbine, when the steam glides over the moving blades, the relative velocity of steam either remains constant or reduces slightly due to friction.	(ii) In reaction turbine, the steam is continuously expanding, relative velocity increases in the fixed blade.
(iii) Impulse turbines blades are of the plate or profile type and are symmetrical.	(iii) Reaction turbines blades are of aerofoil section and asymmetrical.
(iv) Impulse turbine blades are only in action when they are in front of the nozzle.	(iv) Reaction turbine blades are in action all the time.
(v) Impulse turbine have the same pressure on the two sides of the moving blades.	(v) Different pressure exist on the two sides of the moving blades of a reaction turbine.
(vi) Number of stages required for pressure drop is small for the same power.	(vi) Number of stages required will be less.
(vii) The steam velocity and blade speed for an impulse turbine are high.	(vii) The steam velocity and blade speed for a reaction turbine are low as compared with those of impulse turbine.
	(viii) The variation of diagram efficiency with blade speed ratio is more flate for reaction turbine than for impulse turbine.

332 *Mechanical Engineering: Fundamentals*

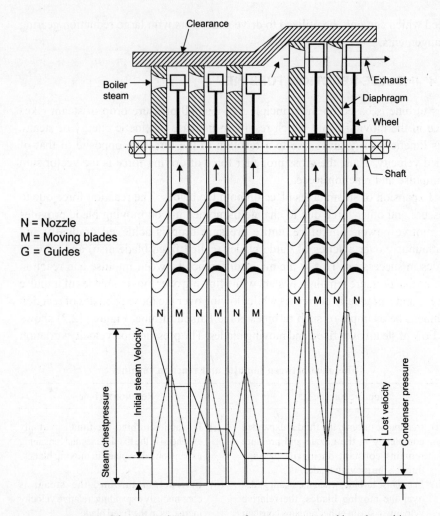

Fig. 12.2: Diagrammatic arrangement of pressure compound impulse turbine.

has also been shown. Expansion of steam takes place in both fixed and moving blades.

12.5 COMPOUNDING OF IMPULSE TURBINE

The first commercial steam turbine was developed by De-Laval in which steam enters the steam chest and expands through a group of nozzles placed around the periphery of a wheel having single row of blades. If the steam expands from very high boiler pressure (about 100 to 130 bar) in only one stage, the velocity of steam of jet will be more than 1500 m/sec and accordingly turbine speed will be as high as 30,000 rpm. This speed is excessively high. The electrical generators cannot be coupled directly and at same time, centrifugal stresses set up in the rotor will be

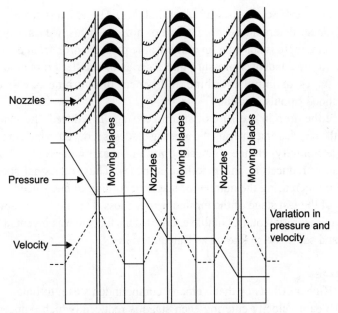

Fig. 12.3: Pressure-compounding

very high. A gearbox would be required to reduce the speed which reduces plant efficiency. It is, therefore, necessary to incorporate some improvements in the simple turbine to achieve high performance. This is done by making use of more than one set of nozzles, blades, rotors in series so that either the jet steam pressure or the jet velocity is absorbed by the turbine in stages. This process is called Compunding of Impulse Turbine. There are three important types of compunding method:
  (i) Velocity Compounding
  (ii) Pressure Compounding
  (iii) Pressure-velocity Compounding

So, the necessity of compounding is:
  (1) For reducing the rotational speed of the impulse turbine to practical limits.
  (2) Carry over loss or leaving velocity losses of the steam are reduced.

The velocity of steam at exit result in an energy loss called carry overloss or leaving velocity loss.

Pressure Compounded Impulse Turbine

In this type of turbine, the compounding is done for pressure of steam only i.e. to reduce the high rotational speed of the turbine, the whole expansion of steam is arranged in a number of steps by employing a number of simple impulse turbine in a series on the same shaft. Each of these simple impulse turbine consists of one set of nozzles and one row of moving blades and is known as a stage of the turbine and thus the turbine consists of several stages. The exhaust from each row of moving blades enters the succeeding set of nozzles. Thus, we can say that in this arrangement, there is splitting up of the whole pressure drop from the steam chest pressure to the condenser pressure into a series of smaller pressure drops across several stages of impulse turbine and hence, this turbine is called Pressure Compounded Impulse turbine.

Since the drop in pressure of steam per stage is reduced, the steam velocity leaving the nozzles and entering the moving blades is reduced, which in turn reduces the blade velocity. Hence, for economy and maximum work, shaft speed is significantly reduced to suit practical purposes. Thus, rotational speed may be reduced by increasing the number of stages according to one's need. The leaving velocity of the last stage of the turbine is much less (leaving loss is not more than 1 to 2% of the initial total available energy). This turbine was invented by Prof. L. Rateau and so, it is also known as Rateau turbine.

Advantages
(1) Higher efficiency than velocity compounded steam turbine.
(2) Steam velocity entering each stage is reduced, which reduces the blade velocity, thus reduced shaft speed which suits practical purposes.

Velocity Compounded Impulse Turbine
In velocity compounding, all the expansion of the steam takes place in a single row of nozzles. The high steam leaving the nozzles passes on to the first row of moving blades, where its velocity is only partially reduced. The steam leaving the first row of moving blades passes into a row of fixed blades which are mounted in the turbine casing. These fixed blades are needed to change the direction of the steam between one set of moving blades and the next. The steam velocity is again reduced partially in the second row of moving blades, as some of the kinetic energy in the steam is used up in producing work on the turbine shaft. Only a part of the velocity of the steam is used up in each row of blades, so it results in a slow turbine. Compared to the simple impulse turbine, the leaving velocity is small being about a percent of initial total available energy of steam. This type of turbine is also called Curtis turbine.

Advantages
(1) Length of the turbine is short which will raise the whirling speed of shaft.
(2) Because of low pressure, leakages are less in the turbine.

Pressure-Velocity Compounded Impulse Turbine
It is a combination of pressure and velocity compounding. The arrangement in the figure is for two rotors. There are two wheels or rotors and only two rows of moving blades are attached on each rotor because two-row wheels are more efficient

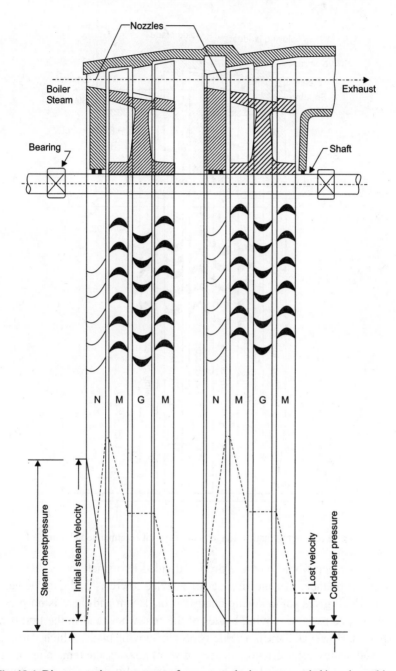

Fig. 12.4: Diagrammatic arrangement of pressure-velocity compounded impulse turbine.

than three-row wheels. The steam on passing through each wheel or rotor reduces in velocity, i.e., drop in velocity is achieved by the many rows of moving blades and is velocity compounded. The whole pressure drop takes place in the two sets

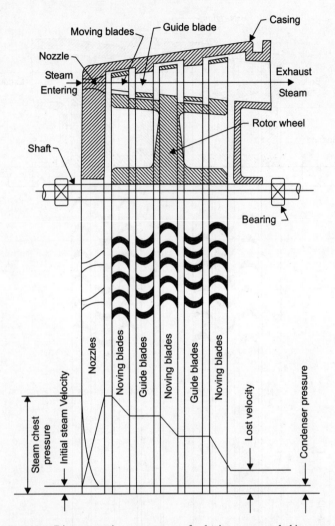

Fig. 12.5: Diagrammatic arrangement of velocity compounded impuse

of nozzles, i.e. the whole pressure drop is divided into small drops, hence, it is pressure-compounded. In the first set of nozzles, there is a slight decrease in pressure which gives some kinetic energy to the steam and there is no pressure drop in the two rows of moving blades of the first wheel and in the first row of fixed blades. However, there is a velocity drop in moving blades which also occurs in the fixed blades due to friction. In second set of nozzles, the remaining pressure drop takes place, but the velocity here increases and the drop in velocity takes place in the moving blades of the second wheel or rotor. Because of its lower efficiency, it is rarely used nowadays.

12.6 STEAM CONDENSER

Definition: Condenser is defined as a closed vessel in which exhaust steam is condensed by cooling water and vacuum is maintained resulting in an increase in work done and efficiency of the steam turbine plant and use of condensate as feed water to the boiler.

There exists a relation between temperature and pressure. If the back or exhaust steam pressure is low, the exhaust temperature is low. Thus, by lowering the back pressure, the temperature at which heat is rejected is reduced and in turn, efficiency is increased.

It is a well known fact that in a non-condensing plant, the back pressure should be either higher or equal to atmospheric pressure, otherwise the steam cannot exhaust to atmosphere. In order to reduce the back pressure below atmosphere for increasing the work done and efficiency, the steam from engine or turbine has to be exhausted in a closed vessel where it will be condensed by cooling water in a closed vessel. This enables expansion of steam to a lower back pressure. This result was expected because 1 kg of steam at 0.1 bar occupies 19.9 m³ whereas after condensation it will occupy only 0.001016 m³ of liquid. Thus, this enormous shrinkage accomplishes two practical results, firstly the vacuum in the closed vessel and secondly, little work is needed to discharge the condensate from vacuum to positive pressure. This closed vessel in which steam condenses is called condenser. Condensation may occur in an open vessel but pressure will not decrease, hence it does not serve the purpose.

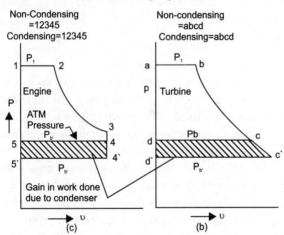

Fig. 12.6: Effect of condenser on p-v diagram

TYPES OF CONDENSER

Jet Condensers

In jet condensers, there is a direct contact between the steam to be condensed and the cooling water and the heat exchange is by direct conduction.

338 *Mechanical Engineering: Fundamentals*

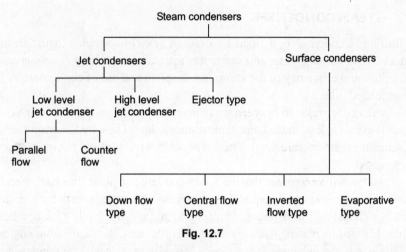

Fig. 12.7

Water is introduced in the form of spray or jet and the steam suddenly condenses, the mixture of condensate and cooling water is continuously taken out. The condensate in these condensers cannot be recovered for use as feed water to the boilers, if the cooling water is not pure and free from harmful impurities. Because of loss of condensate and high power requirement for jet condenser pump, it is rarely used in modern steam power plants. These condensers may be sub-classified as:

(i) Low level jet condenser (counter flow type)
(ii) Low level jet condenser (parallel flow type)
(iii) High level jet condenser
(iv) Ejector condenser

(i) Low level jet condenser (counter flow type)
In this type of jet condenser, the cooling water is drawn up in the condenser cell by vacuum head created in the shell by the air pump. Due to friction and velocity head, the water cannot be raised corresponding to vacuum head, hence some allowance is given. Since there is no pump to pump the water from cooling pond to condenser shell, the level of cooling pond is kept at almost the same level as the delivery pipe to the shell. Two or more perforated water trays are incorporated in the cell to break up the water into small jets or sprays. Thus, the cooling water enters at the top, cascades downwards through a series of perforated trays. The exhaust steam and any mixed air enters at the bottom portion of the shell and ascends through the falling sprays of water. Consequently, the steam is condensed and the air cools down. A separate suction pump located at the top removes the air. The capacity of air pump is reduced due to cooled air. The mixture of cooling water and condensate is pumped by an extraction pump from the bottom of the shell to the hot well. A boiler feed pump pumps the feed water from the hot well to the boiler and the surplus water overflows to the cooling pond. In this condenser, there is no under cooling of condensate. Due to mixing of condensate and cooling

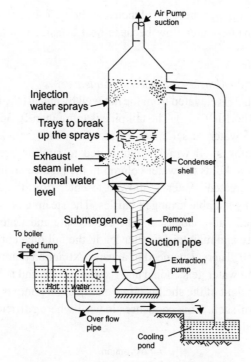

Fig. 12.8: Low level jet condenser (Counter flow)

water, this type of condenser is used only where pure water is cheap and available in plenty.

(ii) Low level jet condenser (Parallel flow type)

As the name suggests, the flow of steam and cooling water in this condenser is in the same direction. The diagram is self-explanatory for its working. This condenser

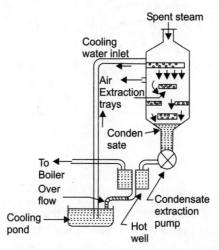

Fig. 12.9: Low level jet condenser (Parallel flow)

is used in small power units because of their simplicity. The parallel flow condenser is less efficient than counter flow because heat transfer is more in the counter flow type.

(iii) High level jet condenser (Barometric condenser)

The condensing shell is elevated from the hot well by at least the barometric height of water column (i.e.10.363 m). The tail pipe is more than 10.363 m long and one end immersed in a water vessel open to atmosphere and the other end connected with the condensing shell which is subjected to suction pressure. Thus, the atmosphere pressure holds the water column in the pipe equal to the vacuum (i.e. suction pressure) created. Water is allowed to enter from the top and is broken into fine streams by suitable arranged baffles. The steam enters from the bottom of condensing shell and condensers. The condensate and water go down to the hot well by gravity and maintain a water leg in the tail pipe corresponding to the vacuum in the shell. There is no need of a water extraction pump. The air released from the steam and water gets cooled and rises to the top and is removed by an air pump. Since the height of the shell is large, an injection pump is installed to pump water to the top of shell. This condenser is used where sufficient head required for tail pipe is available.

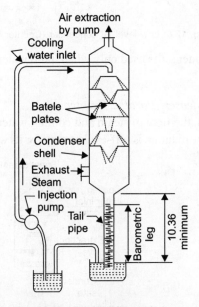

Fig. 12.10: High level jet condenser

(iv) Ejector Condenser

The main principle of the ejector condenser is that the momentum of flowing water removes the mixture of the condensate and the cooling water against the atmospheric pressure.

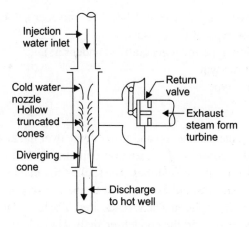

Fig. 12.11: Ejector condenser

In these condensers, the mixing of exhaust steam and cooling water takes place in a series of combining cones. The kinetic energy of the steam is utilized to drain the condensate into the hot well against the pressure of atmosphere. These condensers consist of a vertical tube with converging diverging cones as shown in figure. The cooling water enters the condenser shell from the top and passes through a series of convergent guide cones. It results an increase in the velocity and decrease in the pressure. The decrease in pressure causes suction of steam into the shell through ports and provides through mixing. The mixing causes condensation of steam. The condensation also improves the vacuum in the shell. In the converging diverging cones, in the divergent portion, the kinetic energy is partially converted into pressure energy, so as to obtain a pressure greater than the atmospheric pressure. This pressure enables the mixture of condensate and water to be discharged automatically in the hot well. The condenser is fitted with a non-return valve to prevent sudden rush of water from the hot well into the turbine, in the event of a failure in supply of injection water. These condensers require no injection pump.

Surface Condensers (Cooling water in tubes and steam surrounding it)

In surface condensers, there is no direct contact between the steam to be condensed and the cooling water. There is a wall imposed between them through which heat is transferred by conduction and convection. The temperature of the condensate may be higher than cooling water at exit because the circulating water and cooling water do not mix. The condensate is recovered as feed water to the boiler. Both cooling water and condensate are separately withdrawn. The recovery is very important for marine engines which carry a limited amount of pure water. Marine installations are, therefore, equipped with surface condenser. The only disadvantage of these condensers is high initial cost, but it can be compensated by the saving in the running cost.

342 *Mechanical Engineering: Fundamentals*

These condensers may further be classified into four sub-categories:
(i) Down flow type of condensers.
(ii) Central flow type (or regenerative condensers).
(iii) Inverted flow type.
(iv) Evaporative type

(i) Down Flow

In these, the steam enters at the top and flows down over the tubes through which water is circulated. It consists of a cast iron cylindrical shell containing a large number of small diameter parallel tubes. The shell is covered by coverplates at the two ends. The nest of brass tubes are fixed to the perforated tube plates at the ends. So, it can be easily replaced whenever desired. The tube plates are sandwiched between two water boxes and the condenser shell. The cold water enters into the water box from bottom plate. The water box on left hand side carries a divider which causes the circulating water to make two passes of the condenser tubes before being discharged through the outlet. Meanwhile, the steam enters at the top and flows downwards. It comes in contact with the comparatively cold outside surfaces of the tubes and gets condensed. The condensate is collected by the condensate extraction pump connected at the bottom. A dry air suction pump pipe is also provided near the bottom. As the steam flow normal to the direction of flow of cooling water inside the tubes. It is also called cross-surface condenser.

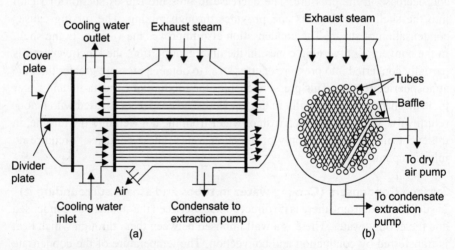

Fig. 12.12: Down flow steam condenser

(ii) Central Flow

Also called regenerative condenser. In these condensers, an air extraction pump is placed at the centre of the tube nest. The steam flowing radially towards the centre passes over the central periphery of the tubes. The condensate is extracted at the bottom by an extraction pump.

(iii) Inverted Flow

In these condensers, the steam entering the bottom rises up due to air suction and then again flows down following a path near the outer surface of the condenser. The air suction pump is placed at the top while the condensate extraction pump is provided at the bottom. The steam that is condensed collects at the bottom of shell and is extracted by condensate pump.

(iv) Evaporative Condenser

This type of condenser is based on the principle of evaporation of cooling water under a small partial pressure. The steam to be condensed enters at the top of grilled pipes kept vertical. The cooling water pumped from a cooling pond, to a horizontal header having nozzles, sprays the water over the grilled pipes. The water while descending down forms a thin film over the grilled pipes as it falls from one level to another. A natural or forced air circulation causes rapid evaporation of the water film causing condensation of steam in the pipes. The vapour of the cooling water passes off with the heated air. The remainder of the cooling water at increased temperature falls into the cooling pond and by addition of requisite quantity of make up cold water, the temperature of water is resorted to the original value for reuse.

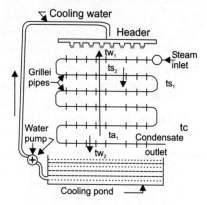

Fig. 12.13: Evaporative Condenser

VACUUM EFFICIENCY

The purpose of an air pump on condensers is to remove air from the condenser so that the total pressure in the corresponding to the condensate temperature.

The vacuum efficiency is a measure of the degree of perfection in achieving the aim of maintaining a desired vacuum in the condenser. The vacuum efficiency is defined as the ratio of actual vacuum to ideal vacuum in the condenser.

$$\text{Vacuum efficiency } \eta_{vacuum} = \frac{\text{Actual vacuum as recorded by vacuum gauge}}{\text{Ideal vacuum}}$$

$$\eta_{vacuum} = \frac{\text{Barometric pressure - actual total pressure}}{\text{Barometric pressure - ideal pressure}}$$

Ideal vacuum means the vacuum due to steam alone when air is absent. Vacuum efficiency depends upon the effectiveness of air cooling and the rate at which it is removed by air pump.

CONDENSER EFFICIENCY

The purpose of an ideal condenser is to remove only the latent heat so that the temperature of condensate equals the saturation temperature corresponding to condenser pressure. It means there should be no undercooling of condensate. Maximum temperature achieved by cooling water will be equal to condensate temperature. The condenser efficiency is then defined as the ratio of actual rise in the temperature of cooling water to the maximum possible rise in the temperature of the cooling water.

Let t_s = saturation temperature corresponding to condenser pressure °C
t_0 = exit temperature of cooling water
t_i = inlet temperature of cooling water

The condenser efficiency

$$\eta_{cond} = \frac{\text{Temperature rise of cooling water.}}{\text{Saturated temperature corresponding to condenser pressure - inlet temperature of cooling water}}$$

$$\eta_{cond} = \frac{\text{Actual rise in temperature of cooling water}}{\text{Maximum possible rise in temperature of cooling water}}$$

$$= \frac{t_0 - t_i}{t_s - t_i}$$

12.7 COOLING WATER SUPPLY

For the condensation of steam into condensate to be used as a feed water to the boiler, cooling water has to be supplied to the condenser. If there is abundant source of cooling water like river or pond, there is no problem. But if the supply of water is limited with high cost, it is necessary to use cooling towers for water cooled condensers.

The method commonly used to overcome a condition of limited cooling water supply is to repeatedly cool and circulate it through the following means.

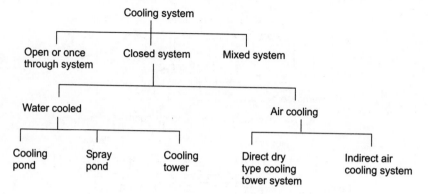

Fig. 12.14

Cooling Pond

The simplest system of removing heat from the cooling water consists of cooling it in an open pond. The effectiveness of this method depends upon a very large surface area of the pond and hence it is used mostly for small condensers only. In this system, sufficient amount of water is lost by evaporation and windage. The factors which affect the rate of heat dissipation from a cooling pond are area and depth of pond, temperature of water entering the pond, wind velocity, atmospheric temperature, shape and size of water spray nozzles and relative humidity. Cooling ponds are of two types namely:

(i) Non-directed flow type
(ii) Directed flow type

Spray Pond

For a given cooling capacity, the size of pond in this case is much less than that of open pond. Schematic diagram of spray pond is shown in Figure 12.15. Hot water coming out from powerhouse is sprayed into the atmosphere, through many nozzles. Tiny particles of sprayed water lose the sensible heat to air and gets cooled and are finally collected into a reservoir from which cold water is supplied to the power house for reuse. In this case, cooling achieved is more effective.

12.8 COOLING TOWERS

The place where acquisition of land is very expensive, we may use cooling towers for cooling purposes. A cooling tower requires smaller area than a spray pond. It is an artificial device used to cool the hot cooling water coming out of condenser effectively. The cooling tower is a semi-enclosed device made of wooden, steel or concrete structure and corrugated surfaces or troughs or baffles or perforated trays are provided inside the tower for uniform distribution and better automization of water in the tower. The hot water coming out from the condenser falls down in radial sprays from a height and the atmospheric air enters from the base of the

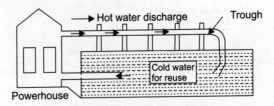
Non-directed flow type cooling pond

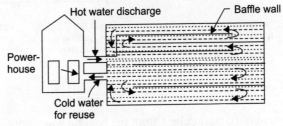
Directed flow type cooling pond

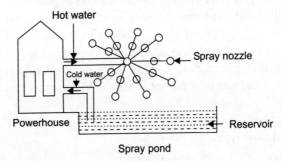
Spray pond

Fig. 12.15

tower. The partial evaporation of water takes place which reduces the temperature of circulating water. This cooled water is collected in the pond at the base of the tower and pumped into the condenser. Draft eliminators are provided at the top of the tower to prevent the escaping of water particles with air.

According to the method of air-circulation, cooling towers are classified as:
  (i) Natural draught type cooling tower
 (ii) Mechanical draught type cooling tower
    (a) Forced draught type
    (b) Induced draught type
(iii) Hyperbolic cooling tower

(i) Natural Draught Cooling Tower
In this, hot water from condenser is pumped at the top where water sprays through a series of spray nozzles. Then water falls over decks. The decks also increase the amount of wetted surface in the tower and breaks up the water into droplets. The air flowing across in transverse direction cools the falling water. These towers are used for small capacity power plants such as diesel power plants.

(ii) Forced Draught Cooling Tower

In this tower, draught air fan is installed at the bottom of tower. The hot water from the condenser enters the nozzles. This water is sprayed over the tower filling slats and the rising air cools the water. The entrained water is removed by eliminators located at the top.

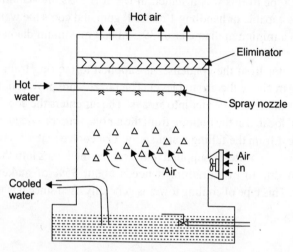

Fig. 12.16: Forced draught cooling tower

(iii) Induced Draught Cooling Tower

The difference here is in the supply of air. The draught fans installed at the top of tower draw air through the tower. The hot water is allowed to pass through the tower below the eliminators. The air moving in the upward direction cools the

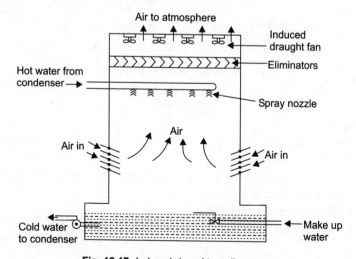

Fig. 12.17: Induced draught cooling tower

down coming hot water particles issued from spray nozzles. Some percentage (1%) of total water goes into air in the form of water vapour.

(iv) Hyperbolic Cooling Tower

It is usually made of steel reinforced cement concrete to withstand high wind pressure. First type of this was installed in US at Big-Sandy station of Kentucky Power Co. It is capable of handling 120×10^3 gpm and cools the water from 43°C to 30°C. It has a minimum diameter of 39.5 m and maximum diameter of 74.5 m and 400 metres high.

The hot water from the condenser is supplied to the ring troughs which are placed at 8–10 m above the ground level. The nozzles are provided on the bottom side of troughs to break up water into sprays. The air enters the cooling tower just above the pond located at the bottom, from the air openings provided, rises upwards and absorbs heat from the falling water spray. The cooled water is collected in the pond. Some makeup water is supplied to overcome the losses into the atmosphere along with air due to evaporation. It needs about 35% of makeup water for compensating. This type of cooling tower is generally used since it is very efficient.

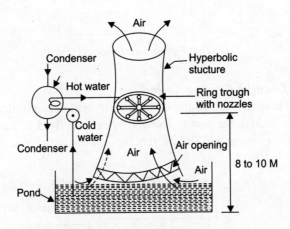

Fig. 12.18: Hyperbolic cooling tower

Advantages

(1) No air fans are required, so power cost and auxillary equipments are totally eliminated.
(2) Hyperbolic structure creates its own draught assuring efficient operation even when there is no wind.
(3) Operation and maintinance cost are reduced.
(4) Ground fogging is avoided in hyperbolic towers.

Drawbacks
(1) Its initial cost is considerably very high.
(2) Its performance varies with the seasonal changes in DBT (Dry bulb temperatur and RH (Relative humidity) of air.

12.9 INTERNAL COMBUSTION ENGINE

Introduction

The Internal Combustion engine (i.e. I.C. engine) is a heat engine, in which the energy of fuel-air mixture is released by combustion in the engine cylinder itself. The heat energy, thus, liberated in the engine cylinder increases the pressure and temperature of the cylinder gas (i.e. the products of combustion) and subsequent expansion of this gas converts the heat energy into mechanical work.

Some important applications of the I.C. engines are:
 (i) I.C engines are used in all road vehicles, such as automobiles, trucks, tractors, etc.
 (ii) Used in railroad, aviation and marine.
 (iii) Used in lawn movers, motor boats, concrete mixing equipment, etc.
 (iv) These engines serve as a standby or substitute power plant. Moreover, they are widely used in running mills.

Classification of I.C. Engine

The internal combustion engines are generally classified on the basis of the following:
 (a) Number of strokes per cycle
   (i) Four-stroke engine: An engine in which the cycle is completed in four strokes of the piston.
   (ii) Two-stroke engine
 (b) Nature of thermodynamic cycle
   (i) Otto cycle (Constant volume combustion)
   (ii) Diesel cycle (Constant pressure combustion)
   (iii) Partly constant volume and pressure combustion (Dual or mixed cycle)
 (c) Ignition system
   (i) Spark ignition (S.I. engine): These engines in which a spark plug is used for combustion.
   (ii) Compression ignition (C.I. engine): In these, auto ignition of the fuel takes place when it is injected at high pressure in a cylinder.
 (d) Fuel used
   (i) Gas engines: Gaseous fuel is used in these engines. e.g. CNG, LPG.
   (ii) Petrol engines: These engines use highly volatile liquid fuel such as petrol.

350 *Mechanical Engineering: Fundamentals*

(iii) Oil engines: These engines use less volatile liquid fuel, such as diesel oil, kerosene, heavier residual fuel.

(iv) Bifuel engines: In these engines, gas is used as a basic fuel while liquid fuel is used for starting purposes.

(e) Cylindrical arrangements

(i) Inline engines: These engines contain back of cylinders with their axis parallel. They transmit power to a single crankshaft.

(ii) V engines: In this, the cylinders are arranged in the form of the letter V. In other words, these engines contain two banks of cylinders set at an angle and attached to the same crankcase and with the same crankshaft.

(iii) Opposed cylinder: These engines consist of two or more cylinders on opposite sides of a common crankshaft.

(iv) Opposed piston: When a single cylinder houses two pistons, each of which drives a separate crankshaft, it is called an opposed piston.

(v) Radial engines: In these engines, the cylinders are arranged radially like the spokes of a wheel and are connected to a single crankshaft. These engines are used in aircraft.

(f) Cooling system

(i) Water cooled engines: In these engines, the cylinder is cooled by water.

(ii) Air cooled engine: In air cooled engine, the cylinder wall is cooled by air.

(g) Field of application

(i) Stationary engines: These engines are used as small and medium capacity electric power plants and to drive pumping units in agriculture.

(ii) Mobile engines: These engines are used in motor vehicles, aeroplanes, ships, locomotives, etc.

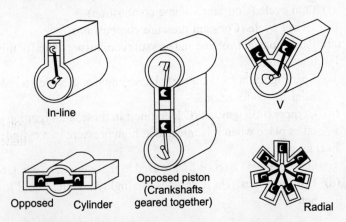

Fig. 12.19: Classification of engines by cylinder arrangement

(h) Fuel supply system
- (i) Carburettor engines: In these engines (petrol engines), the air and fuel is properly mixed in a carburettor before induction in the cylinder.
- (ii) Air injection engines: In some cases of diesel engines, the fuel is supplied under pressure to the engine cylinder by using compressed air.
- (iii) Airless or solid or mechanical injection system: In these diesel engines, the fuel is injected with mechanical or pump injection.

(i) Method of control
- (i) Quality control engines: In these, the composition of the mixture is changed by admitting more or less fuel according to the load.
- (ii) Quantity control engines: In these engines, the composition of mixture remains constant but the quantity of the mixture is changed.
- (iii) Quality and quantity control engines: Here both the quality and quantity of the mixture is changed to affect control.

(j) Lubrication system
- (i) Wet sump lubrication
- (ii) Dry sump lubrication
- (iii) Pressurized lubrication

A detailed classification of heat engines in tabular form can be given as below:

Terms used with I.C. engine:

(a) Bore: The inside diameter of the cylinder is called bore.

(b) Stroke: It is the distance travelled by piston when it moves from TDC to BDC or vice versa. The stroke length is double the crank radius.

(c) Top dead centre (TDC): When the engine is placed vertically, the farthest position of piston from crankshaft is termed TDC. It is also the position when the piston is closest to cylinder head, it is termed Inner Dead Centre (IDC).

(d) Bottom dead centre (BDC): When the engine is placed vertically, the closest position of piston from the crankshaft is termed BDC. It is also the position

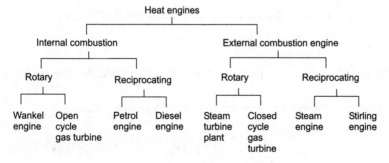

Fig. 12.20

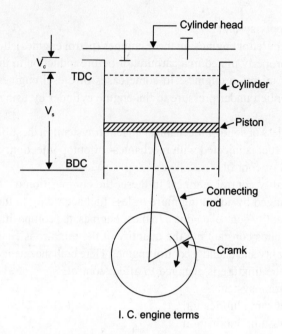

Fig. 12.21: I.C. engine terms

when the piston is farthest from the cylinder head. In case of horizontal engine, it is termed Outer Dead Centre (ODC).

(e) **Clearance volume:** The volume occupied by cylinder when the piston is at TDC is called Clearance Volume.

(f) **Swept volume:** It is the difference of volume occupied by cylinder when the piston is at BDC and volume occupied by the cylinder when the piston is at TDC. It can be calculated by multiplying cylinder cross-sectional area by stroke length.

(g) **Cylinder volume and compression ratio:** The volume of working fluid in the cylinder, when the piston is at the bottom dead centre is referred to as cylinder volume and it equals the clearance volume plus swept volume.

Cylinder volume = Clearance volume + Swept volume

$$= V_c + V_s$$

The ratio of cylinder volume to clearance volume is called the compression ratio.

Compression ratio = Cylinder volume / Clearance volume

$$= \frac{V_c + V_s}{V_c} = 1 + V_s / V_c$$

Higher the compression ratio, better will be the performance of an engine.

The compression ratio varies from (5/1) to (10.5/1) in S.I. engine and from 12 to 25 in I.C. engines.

Mean Effective Pressure

Mean Effective Pressure (mep) is defined as the average pressure acting on the piston which will produce the same output as is done by the varying pressure during a cycle. That is

area of indicator loop = area of rectangular abcd

The height of the rectangle then represents the mean effective pressure:

$$\text{mep} = \frac{\text{Work done per cycle}}{\text{Swept volume}}$$

$$= \frac{\text{Area of indicator loop}}{\text{length of loop}}$$

$W_{net} = P_m \times V_s$

Or in other words, the ratio of network done during the cycle to swept volume is called mean effective pressure.

P_m = Work done per cycle/Swept volume

Mean effective pressure is used as a parameter to compare the performance of reciprocating engines of equal size. An engine that has a large volume of mep will deliver more network and will, thus, perform better.

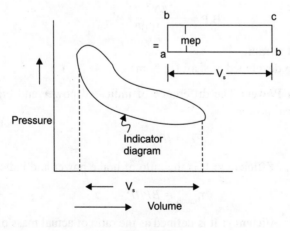

Fig. 12.22

Performance of I.C. Engines

For valuation of engine performance, certain basic parameters are chosen. Some of them are as follows:

(a) Indicated power: The indicated power is the power actually developed inside the engine cylinder. It is measured with the help of an indicator diagram obtained from an instrument known as indicator.

In measuring the IP, the first step is to calculate the indicated mean effective pressure from the indicator diagram. The indicated mean effective pressure is given by:

$$P_m = \frac{\text{Area of indicator diagram in cm}^2 \times \text{spring scale in bar/cm}}{\text{Length of diagram in cm}}$$

I.P = Work given by cycle × number of cycles/sec × number of cylinder
    = $P_m \times V_s \times N/60 \times k \times n$

n = Number of cylinders P_m = Indicated mean effective pressure (kPa)
L = Length of stroke (m) A = Area of piston (m²)
k = 1/2 for 4-stroke engine k = 1 for 2-stroke engine
N = Number of revolutions of crankshaft/minute (rpm)

1 H.P = 746 watt

In M.K.S system it was called Indicated Horse Power (IHP).

$$\text{IHP} = \frac{nP_m \cdot LANk}{4500} \quad [P_m \text{ is in bars}]$$

(b) Brake Power (B.P): The brake power for an engine is available at the crankshaft or clutch shaft. The brake power is less than indicated power.

$$\text{B.P} = \frac{2\pi NT}{50 \times 1000} \text{kW}$$

N = speed in rpm
T = torque developed at shaft (in N-m)

(c) Frictional Power: The difference of indicated power and brake power is termed frictional power.

$$FP = IP - BP$$

(d) Mechanical Efficiency: It is the ratio of brake power and indicated power.

$$\eta_{\text{mech}} = BP/IP$$

(e) Volumetric efficiency: It is defined as the ratio of actual mass of air inducted by the engine on the intake stroke to the theoretical mass of air that should have

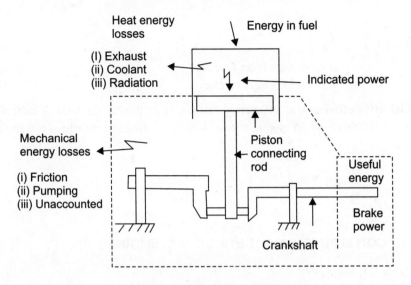

Fig. 12.23: Energy flow in an I.C. engine

been inducted by filling the piston displacement volume with air at atmospheric pressure and temperature. Thus,

$$\eta_v = \frac{m_a}{m_c}$$

m_a = actual mass of air inducted per intake stroke
m_c = theoretical mass of air to fill the piston displacement volume under atmospheric conditions

(f) Thermal Efficiency: It is ratio of indicated power to energy supplied by the fuel. It is of two types:
(i) Indicated thermal efficiency

$$\eta_{thermal} = \frac{\text{Work done per cycle}}{\text{Heat added per cycle}} = \frac{\text{I.P.}}{m_f \times c}$$

m_f = mass of fuel supplied in kg/sec
c = lower calorific value of the fuel (kJ/kg)

(ii) Brake thermal efficiency = $\eta_{thermal} = \dfrac{\text{B.P.}}{m_f \times c}$

(g) Specific Output: It is defined as the brake power per unit piston displacement.

$$\text{S.P output} = \frac{\text{B.P.}}{A*L} \ (kW/m^3)$$

$$A = \frac{\pi}{4} d^2$$

d = bore diameter (metre)
L = stroke length (metre)

(h) Relative efficiency or efficiency ratio: It is defined as the ratio of indicated thermal efficiency of an engine and air standard efficiency on which the engine works

$$\eta_{rel} = \frac{\eta_{indicated}}{\eta_{cycle}}$$

12.10 CONSTRUCTIONAL DETAIL OF I.C. ENGINES

4-strok S.I. engine

(i) **Cylinder block:** It forms the basic framework of a multicylinder engine. It houses the engine cylinders in which the piston reciprocates. In cylinder block, there are also the passages for circulation of cooling water and lubrication oil.

(ii) **Cylinder:** It is a hollow cylinder, one side of which is closed by a head known as cylinder head. It is supposed to be the heart of the engine inside which the fuel is burnt and power is developed due to reciprocating motion of the piston. The cylinder is made of cast iron for automobile engines. For aircraft engines, thin steel hardened inside against wear resistance is used. The inside diameter of the cylinder is known as bore.

(iii) **Cylinder head:** It covers one side of the cylinder and usually contains both the valves, valve ports and fuel injector or spark plug. It is made of cast iron.

(iv) **Cylinder liner:** A cylinder is equipped with cylinder lines so that if the wear and tear that takes place only the linear is damaged. This is easily replaceable as compared to the cylinder. It is also made of cast iron.

(v) **Piston:** The piston is a cylindrical casting of iron, aluminium alloy, or steel with one end closed by a crown, which forms the lower surface of the combustion chamber. The outer periphery of each piston is accurately machined to a running fit in the cylinder bore and is provided with several grooves into which the piston rings are fitted. The main function of the piston is to transmit the force created by combustion process to the connecting rod.

(vi) **Piston rings:** The function of the piston rings is to prevent any leakage of gas past the piston and to prevent the wear of the piston. Whatever the wear that takes place, it takes place only in piston rings which is less costly to replace than the piston. They have a slot cut so that they can be

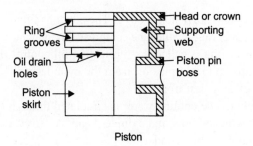

Piston

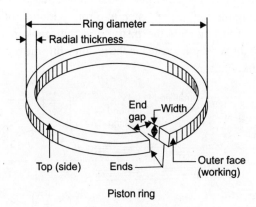

Piston ring

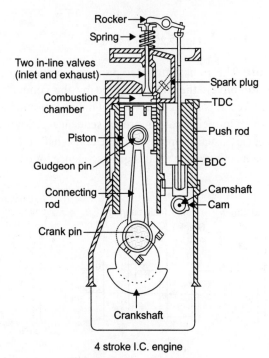

4 stroke I.C. engine

Fig. 12.24: 4 I.C. engine

easily inserted in the piston grooves. The piston is fitted with at least three rings. The upper ring is called the compression ring which prevents leakage and the lower ring is called the lubricating oil control ring and the third which scraps up the surplus oil from the cylinder wall. The piston rings are made of grey cast iron.

(vii) **Water jacket:** The cooling water is circulated through the water jackets surrounding the cylinder and cylinder head. Actauly, the water jacket is cast integral with the cylinder.

(viii) **Connecting rod:** The function of the connecting rod is to convert the reciprocating motion of the piston into rotary motion of crankshaft. The connecting rod connects the piston and the crankpin. Smaller end is attached to the wrist pin located in the piston while the big end is connected to the crankpin.

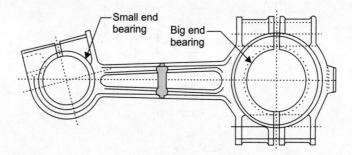

Fig. 12.25: Connecting rod

(ix) **Crankshaft:** The crankshaft is the principal rotating part of the engine. It serves to convert the forces applied by the connecting rods into a rotational force. It controls the motion of the pistons and must be designed to cause them to reciprocate in proper sequence. Alloy steel forgings or carbon steel are usually employed in the crankshaft. It receives the power from the connecting rod in the designed sequence for onward transmission

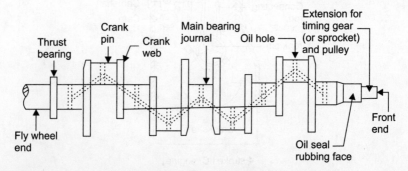

Fig. 12.26: Crankshaft

to clutch and subsequently to the wheels. It also drives the camshaft which actuate the valve of the engine.

(x) **Valves:** There are two valves for every cylinder. One is an inlet valve which admits air or a mixture of air and fuel in the suction stroke. The other is the exhaust valve through which the product of combustion after doing work on the piston escapes to the atmosphere. The valve mechanism consists of cams, camfollower, push rod, rocker arms and spring.

(xi) **Camshaft:** The camshaft provides the means to operate the inlet and exhaust valves with the help of camfollowers, push rods, and rocker arms. It actuates the opening and controls the period before closing for both the inlet as well as exhaust valves. It also provides a drive for the ignition distributor.

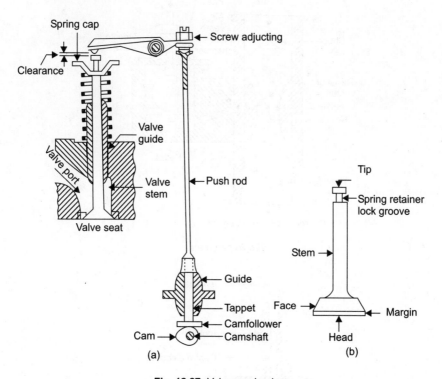

Fig. 12.27: Valve mechanism

(xii) **Manifolds:** The piping which connects the inlet ports of the various cylinders to a common air intake for the engine is called the inlet manifold. Similarly, the piping which connects the exhaust ports to a common exhaust system and sends the exhaust products to the muffler is called the exhaust manifold.

(xiii) **Fly wheel:** It is fitted with the main shaft to work as a reservoir of mechanical energy. It diminishes the cyclic variations in speed. It acquires energy during power stroke and releases during other stroke and maintain fairly constant output torque. It is made up of cast iron.

(xiv) **Carburettor:** This is a device employed in S.I engines to atomise, vaporize, and mix the fuel with air in required proportions at all loads and speed before entering the engine cylinder.

(xv) **Spark Plug:** It is mounted in the cylinder head to initiate combustion of compressed charge. The spark plug creates spark after regular intervals for a minute period of the order of milliseconds. The spark timing is controlled by cams provided in the ignition system. A very high voltage surge (upto 25000 volts) is required for ignition purposes.

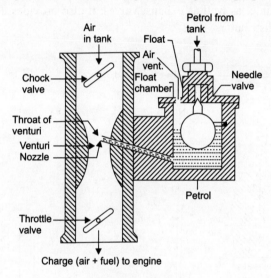

Simple carburettor

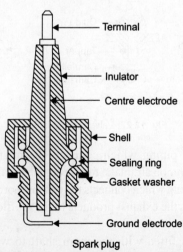

Spark plug

Fig. 12.28

12.11 CONSTRUCTIONAL DETAILS OF 2-STROKE SI ENGINES

In this case, crank case plays a vital role in compression process. The valves are replaced by ports.

Crankcase

The crankcase is attached to the cylinder block or it is cast integral with the block. It is shaped simply like a box, its roof being formed by the lower deck of the cylinder block. The bottom of the crankcase walls is flanged to strengthen the casing and to provide a machined joint face for the crankcase. Compression of the charge crank is supported in the crankcase through bearings called main bearings. The construction of the crankcase is designed to provide very high rigidity because it must provide reactions for the heavy forces set up due to gas pressures in the cylinders.

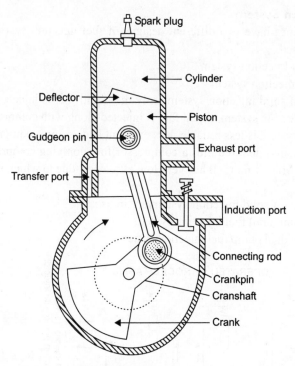

Fig. 12.29: 2 Stroke S.I. engine

Ports

Ports are open passages for flow of gas or charge without any valve controlled mechanism. It comprises three ports:
   (i) Intake port
  (ii) Exhaust port
 (iii) Transfer port

Deflector

The deflector is provided above the piston crown to divert the flow of exhaust gases to the exhaust port. The term scavenging refers to the process of removal of the burnt gases from the cylinder at the end of exhaust stroke, by deflecting the fresh charge across the cylinder.

Other components are similar to that of S.I. engines.

12.12 CONSTRUCTIONAL DETAILS OF C.I. ENGINES

Here since the compression of only air is taking place while the highly pressurized fuel is being injected at the end of the compression stroke. So, the main change is present in the fuel supply system. Here the spark plug is replaced by fuel injector and the carburettor is replaced by fuel pump. Rest of the components remain almost the same with certain changes in design.

Fuel Injection System

The C.I. engines have two different designs of fuel injection system which are commonly used.
  (i) Solid Injection system
  (ii) Air Injection system

In case of solid injection system, only the liquid fuel is injected, whereas in case of air injection system, liquid fuel is injected along with compressed air. The air injection system is less reliable, less efficient and requires compressed air. As the air injection system requires a compressor for supplying compressed air at a pressure of 7 MPa or more. It has very low efficiency, due to which it has become obsolete.

Two types of solid injection system are in use:
  (i) Common rail fuel injection system
  (ii) Individual pump fuel injection system

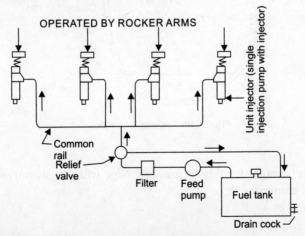

Fig. 12.30: Common rail fuel injection system

In a common rail fuel injection system, a single injection pump with injector called unit injector, is employed on each cylinder. The unit injectors are operated by rocker arms and springs similar to the engine valves. A linkage connects the control rocks of all the unit injectors, so that fuel injection in all the cylinders may be equal and simultaneously controlled. The fuel is taken from the fuel tank by the feed pump and is supplied at a low pressure through a filter, to all the unit injectors. This avoids the high pressure fuel lines necessary in the individual pump system. Any excess fuel from the relief valve is returned to the fuel tank.

Individual Pump Fuel Injection System: Fuel is drawn from the fuel tank by means of a fuel feed pump which is obtained from the injection pump camshaft. Generally, the plunger type or the diaphragm type of fuel feed pumps are employed in automobiles. The fuel is then passed through a filter and hence to the fuel injection pump. The fuel injection pump then injects definite quantity of fuel into individual cylinders, in turn according to firing order through injectors fitted on them. The injection pump is gear driven from the engine camshaft so that it is driven at half the engine speed. Contained in the injection pump on its side, is a governor which provides automatic speed control, relative to any set position of the accelerator pedal. Any excess fuel after lubrication of injector nozzle is returned to the fuel tank.

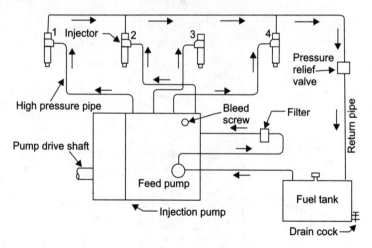

Fig. 12.31: Individual pump fuel injection system

Fuel Feed Pump

This feeds the fuel from the fuel tank to the injection pump at a rate which varies with the engine requirements. It is driven from an eccentric or cam on the injection pump camshaft or the engine camshaft. The feed pump is usually a part of the injection pump assembly. Some are mounted on the outside of the injection pump housing, while others are enclosed in the housing.

Fuel Injection Pump

The function of a fuel injection pump is to deliver accurately metered quantity of fuel under high pressure, at the correct instant and in the correct sequence to the injector fitted on each engine cylinder. The injection pressure ranges from 7 to 30 MPa. The injection pump is driven from the engine's timing gears and its output is controlled by the driver through accelerator pedal. As the volume of fuel to be metered for each injection is very small and frequency of injection quite high, the pump has to be manufactured to very precision. For an idea in a 4-stroke 4-cylinder diesel engine, at maximum speed of 6000 rpm about 150 mm³ of fuel has to be metered and injected 20 times in a second. In a 2-stroke engine, number of injections per second are twice this value. These are generally jerk pumps but sometimes, they can be distributor type pumps.

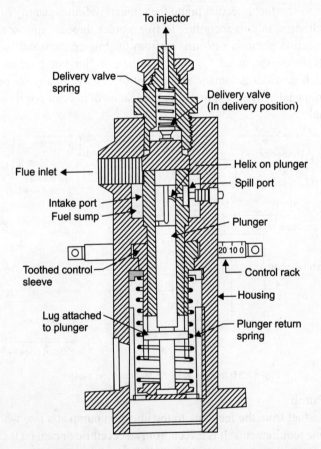

Fig. 12.32: Single cylinder jerk pump type fuel injection pump

Fuel Injector

This is also known as nozzle, atomizer or fuel valve. Its function is to inject the fuel in the cylinder in properly atomised form and in proper quantity. It consists of mainly two parts, i.e., the nozzle and the nozzle holder. A spring loaded spindle in the nozzle holder keeps the nozzle valve pressed against its seat, till the fuel supplied by fuel injection pump through inlet passage lift the nozzle against the spring force.

The other components like cylinder block, cylinder head, piston, piston rings, connecting rod, gudgeon pin, crank pin, crankshaft, camshaft, valves and flywheel perform the same function as they perform in S.I. engine.

12.13 WORKING OF INTERNAL COMBUSTION ENGINES

Working of 4-stroke S.I. or Petrol Engine

Suction Stroke: Suppose that the piston is very near to the top dead centre position (TDC). During this stroke, the inlet valve is opened and the discharge valve is closed and the piston moves towards BDC creating vacuum inside the cylinder. Since the inlet manifold is connected to carburettor on one side where the pressure is atmospheric, this excess pressure will help the air to rush towards the cylinder. In carburettor, the fuel gets mixed with air and it provides (fuel + air) mixture to the cylinder. On the PV diagram, it has been shown by line a-1.

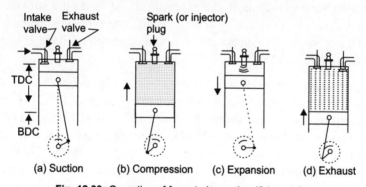

Fig. 12.33: Operation of four-stroke engine (S.I. and C.I.)

Compression stroke (1-2): During a compression stroke, both the valves are closed and the piston moves from bottom to top dead centre position. The charge or air is compressed upto a compression ratio. There is a reduction in volume of the charge of the air which results in an increase of pressure and temperature of the cylinder contents. In petrol engines, the compression ratio varies from 5 to 10.5 and pressure and temperature at the end of compression are 7 to 14 bar and 250 to 300°C respectively shortly before the piston reaches TDC, charge is ignited (2-3) by a spark from spark plug.

Expansion Stroke: During power stroke, both the valves are closed. The power stroke includes combustion of fuel and expansion of the products of combustion. During combustion, the chemical energy of the fuel is released and there is a rise in temperature and pressure of the gas. The temperature of the gases increased between 1800–2300°C and the pressure to 30–40 bar. The high pressure and the high temperature of the products of combustion thus obtained pushes the piston outward from TDC to BDC position for expansion stroke. This reciprocating motion of the piston is converted into rotary motion by the crankshaft, connecting rod and crank mechanism.

Exhaust Stroke: When 7/8th of the power stroke is completed, the exhaust valve begins to open and the inlet valve remains closed. The piston moves from BDC to TDC, thrusting the gases to go out of the cylinder through exhaust valve which is open. In the end of exhaust stroke, piston reaches to TDC and the cycle gets completed.

Working of Four Stroke Diesel Engine
Here the diesel engine will be used for working.

Suction Stroke (a-1): Air at atmospheric pressure is drawn into the cylinder when the piston moves from TDC to BDC. During this, the inlet valve remains open while the exhaust valve remains closed.

Compression Stroke (1–2): Both valves remain closed during this stroke and the piston moves from BDC to TDC. For diesel engines, the compression ratio is 12 to 20 and the pressure and temperature at the end of compression are 28 to 59 bar and 600°C to 700°C respectively.

Working or Expansion or Power Stroke : In C.I. engine (mechanical airless injection), one or more jets of fuel, compressed to a pressure of 105-210 bar by an injector pump are injected into the combustion chamber by a fuel nozzle at the end of the compression stroke. The injected fuel is vaporized in the combustion chamber when the fuel vapour is raised to self-ignition temperature, the combustion then starts automatically and there is a sudden rise of pressure at approximately constant volume. The hot gases inside the cylinder now expand and thrust the piston to move towards BDC. Thus, power is obtained by piston and flywheel during this stroke. Both valves remain closed during this stroke. At the end of expansion stroke, the heat is rejected at constant volume condition process (4-1).

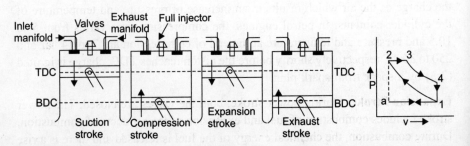

Fig. 12.34

Exhaust (1-a): Exhaust valve gets opened and the piston moves from BDC to TDC. During this, intake remains closed. The upward moving piston thrusts the gases to move out of the cylinder.

Working of Two-stroke Petrol Engine

The working cycle is completed in two strokes of the piston, or in one revolution of the crankshaft as against two crankshaft revolutions in a four-stroke cycle engine. The preparatory strokes (suction and exhaust) are combined with the working strokes (compression and expansion).

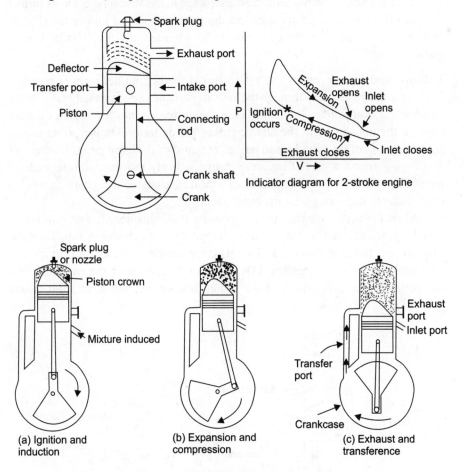

Fig. 12.35: Working of two-stroke cycle engine

Ignition and induction: The piston occupies the almost TDC position towards the end of the compression stroke. The compressed charge is being ignited by providing a spark. The combustion of fuel occurs and thermal energy is released. There occurs a rise both in the pressure and temperature of combustion products.

At the same time a partial vacuum (pressure lower than atmosphere) exists in the crankcase and fresh charge is being inducted into the crankcase through the inlet port, which is uncovered by the piston.

Expansion and compression: The high pressure gases push the piston down, expansion takes place and power is developed. With downward movement of the piston, the charge in the crankcase gets compressed by the underside of the piston to a pressure of about 1.4 bar absolute. After completion of about 80% of expansion stroke, the piston uncovers the exhaust port. Some of the combustion products which are still above atmospheric pressure escape to the atmosphere. On its further downward motion, the piston uncovers the transfer port and allows the slightly compressed charge from the crankcase to be admitted into the cylinder via the transfer part.

Exhaust and transference: The piston lies at the bottom dead centre position. The expanded gases escap through the exhaust port and simultaneously the slightly compressed charge from the crankcase is being forced into the engine cylinder through the transfer port. The charge strikes the deflector. On the piston crown rises to the top of the cylinder and pushes out most of the burnt gases. During this scavenging action, a part of the fresh charge is likely to leave with the exhaust gases. The cylinder is completely filled with the fresh charge, although it is somewhat diluted due to its mixing with the burnt gases.

When the piston moves upward from its BDC position, it first covers the transfer port and stops the flow of fresh charge into the cylinder. A little later, the exhaust port too gets covered and actual compression of the charge begins and continues till the piston reaches TDC position. The cycle of the engine is thus completed within two strokes of the piston (one up and one down) and in one revolution of the crankshaft.

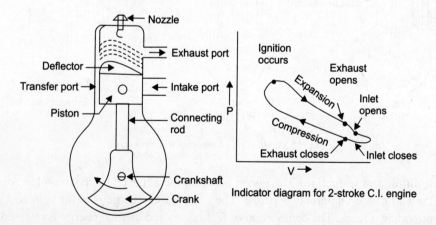

Fig. 12.36

Comparison of two-stroke and four-stroke engines

Four-Stroke	Two-Stroke
(i) The cycle is completed in four strokes of the piston or in two revolutions of the crankshaft. Thus, one power stroke is obtained in every two revolutions of the crankshaft.	(i) The cycle is completed in two strokes of the piston or in one revolution of the crankshaft. Thus, one power stroke is obtained in each revolution of crankshaft.
(ii) More heavy and bulky engine for the same power developed.	(ii) Light and compact engine for the power developed.
(iii) A heavier flywheel is needed because one power stroke is available for every two revolutions of flywheel.	(iii) Lighter flywheel is needed because power is obtained in each revolution.
(iv) Lesser cooling and lubrication is required.	(iv) Greater cooling and lubrication is required.
(v) The engine contains valves and valve mechanism.	(v) There are no valves, only ports are fitted.
(vi) It has higher initial cost.	(vi) It is cheaper in initial cost.
(vii) Volumetric efficiency is more due to more time of induction.	(vii) Volumetric efficiency is less due to lesser time of induction.
(viii) Higher thermal efficiency (almost double of 2-stroke engine).	(viii) Lower thermal efficiency.
(ix) Used where fuel economy is the prime concern.	(ix) Used where smooth power cycle is required.

Working of two-stroke Diesel Engine

The working is similar to that of the two stroke petrol engine except that air is compressed during compression stroke and fuel is being injected after the end of compression stroke instead of spark plug ignition.

Advantages of two-stroke engine over four-stroke

(1) Higher power per cycle.
(2) Simple in design.
(3) Easy to start.
(4) Lesser size of flywheel.
(5) Higher mechanical efficiency.
(6) Less space and lighter foundation.

Disadvantages of two-stroke engine over four-stroke

(1) Overheating of piston.
(2) Lower Compression ratio.
(3) Noisy exhaust.
(4) Lesser thermal efficiency.
(5) More lubrication.
(6) Higher fuel consumption.
(7) Less charge sucked in.

Comparison of S.I. and C.I. engines

S.I. engines	C.I. engines
(i) It works on Otto cycle.	(i) It works on diesel cycle.
(ii) Compression ratio is kept 5 to 10.5. Upper limit of compression ratio is fixed by anti-knock quality of fuel.	(ii) Compression ratio is from 12 to 25. Upper limit is limited by thermal and mechanical stresses developed in the cylinder material.
(iii) Fuel is supplied with air through carburettor.	(iii) Fuel is supplied by fuel injector.
(iv) Engine has generally high speed.	(iv) Low engine speed compared to S.I. engine.
(v) Low maintainance cost but high running cost.	(v) Higher maintenance cost but low running cost.
(vi) Spark plug is required for ignition of fuel.	(vi) Injector is required to inject fuel but no spark plug is needed to initiate combustion.
(vii) Low thermal efficiency.	(vii) High thermal efficiency.

12.14 AIR STANDARD CYCLES

In order to compare the efficiencies of various cycles a hypothetical efficiency called the standard efficiency is calculated on the following assumptions.

(1) The working substance is pure dry air which in addition is assumed to be a perfect gas.
(2) The specific heat remains constant at all temperatures.
(3) To eliminate the effect of calorific value of fuel the heat is supplied by a hot reservoir and is rejected to a cold reservoir. The hot reservoir or the cold reservoir is brought in contact with the end of the cylinder which is assumed perfectly conducting for these operations. The hot reservoir and the cold reservoir are of large capacities and their temperatures do not change during the heat interchange. Further, heat interchange causes no increase or decrease in the number of molecules.
(4) No account is taken how the heat is supplied or rejected.
(5) There is no dissociation at higher temperatures.
(6) Apart from intentional changes in heat, no heat is either gained or lost during the cycle.
(7) No mechanical losses like friction, etc. occur in the working, i.e. all processes are assumed to be reversible.
(8) There are no suction or exhaust strokes.

Both the source and sink are of infinite capacity so that when the heat is transferred from or to them, their temperatures remain unchanged. From second law of thermodynamics, only a portion of heat supplied to a system is converted into work and the remainder has to be rejected to an outside sink if Q_1 and Q_2

represents the heat addition and heat rejection respectively, then assuming no friction and heat losses

Work done = $Q_1 - Q_2$

Thermal efficiency = $\dfrac{\text{work done}}{\text{heat input}} = \dfrac{Q_1 - Q_2}{Q_1}$

A thermodynamic cycle working with air as the working fluid is called an air standard cycle and the thermal efficiency of an air standard cycle is called air standard efficiency. The approach and concept of ideal air cycle helps to:
(1) Indicate the ultimate performance, i.e. to determine the maximum ideal efficiency of a specific thermodynamic cycle.
(2) Study qualitatively the influence of different variables on the performance of an actual engine.
(3) Evaluate one engine relative to another.

Otto Cycle or Constant Volume Cycle

The cycle was presented by Beau De Rochas (1862) and was successfully applied by a German scientist Nikolous. A. Otto to produce a successful four-stroke cycle engine in 1876, that was far superior to an I.C. engine previously built. The thermodynamic cycle is operated with iso chroic (constant volume) heat addition, and consists of two adiabatic processes and two constant volume changes.

The sequence of processes is as follows:
Process (1–2): Reversible adiabatic compression.
Process (2–3): Constant volume heat addition.
Process (3–4): Reversible adiabatic expansion.
Process (4–1): Constant volume heat rejection.

Thermal Efficiency

Consider unit mass of air undergoing the cyclic change, heat supplied during process 2–3.

$$Q_1 = C_v (T_3 - T_2)$$

Heat rejected during process 4–1

$$Q_2 = C_v (T_4 - T_1)$$

Work done = heat supplied – heat ejected.

$$= C_v (T_3 - T_2) - C_v (T_4 - T_1)$$

Thermal efficiency $\eta = \dfrac{\text{Work done}}{\text{Heat supplied}}$

$$= \frac{C_v(T_3 - T_2) - C_v(T_4 - T_1)}{C_v(T_3 - T_2)}$$

Since C_v remains constant and is independent of temperature

$$\eta = \frac{(T_3 - T_2) - (T_4 - T_1)}{T_3 - T_2}$$

$$= 1 - \frac{T_4 - T_1}{T_3 - T_2}$$

The expression for thermal efficiency in terms of compression ratio and adiabatic exponent γ can be set up by expressing all the temperatures in terms of lowest temperature T_1 in the cycle.

At point 2 $V_2 = \dfrac{V_1}{r}$ where r is the compression ratio

$$p_2 = p_1 \left(\frac{V_1}{V_2}\right)^\gamma = p_1 r^\gamma$$

$$T_2 = T_1 \left(\frac{V_1}{V_2}\right)^{\gamma-1} = T_1 r^{\gamma-1}$$

At point 3. $\qquad V_3 = V_2 = \dfrac{V_1}{r}$

$$P_3 = P_2 \alpha = p_1 \alpha r^\gamma$$

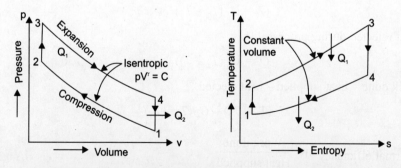

Fig. 12.37: Air-Standard Otto cycle on p-V diagram and T-s diagram

where α is the explosion pressure ratio.

$$T_3 = \frac{T_2 p_3}{p_2} = T_2\alpha = T_1\alpha r^{\gamma-1}$$

At point 4 $V_4 = V_1$

$$p_4 = p_3\left(\frac{V_3}{V_4}\right)^\gamma = p_3\left(\frac{V_2}{V_1}\right)^\gamma = \frac{p_3}{r^\gamma} = \frac{p_1 r^\gamma \alpha}{r^\gamma} = p_1\alpha$$

$$T_4 = T_3\left(\frac{V_3}{V_4}\right)^{\gamma-1} = T_1\alpha r^{\gamma-1} \times \frac{1}{r^{\gamma-1}} = T_1\alpha.$$

When these temperature values are substituted in expression, we wet

$$\eta = 1 - \frac{T_4 - T_1}{T_3 - T_2} = 1 - \frac{T_1\alpha - T_1}{T_1\alpha r^{\gamma-1} - T_1 r^{\gamma-1}}$$

$$= 1 - \frac{T_1(\alpha-1)}{T_1 r^{\gamma-1}(\alpha-1)}$$

$$= 1 - \frac{1}{r^{\gamma-1}}$$

Thus, thermal efficiency of an Otto cycle engine depends upon compression ratio r and adiabatic exponent γ i.e. on the nature of working medium.

Figure shows the variation of air standard efficiency of Otto cycle with compression ratio. The curve tends to become rather flat at higher compression ratios which means that though the efficiency is still increasing, the rate of increase starts diminishing.

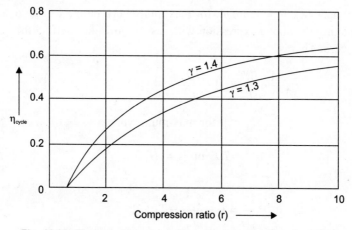

Fig. 12.38: Thermal efficiency of Otto cycle as a function of r and γ.

CONDITION FOR MAXIMUM WORK

The ambient conditions determine the minimum temperature T_1 while the maximum temperature T_3 is determined by the design conditions and metallurgical considerations of the piston and cylinder for unit mass of air

work output $w = C_v(T_3 - T_2) - C_v(T_4 - T_1)$

$$\frac{w}{C_v} = T_3 - T_2 - T_4 + T_1 \tag{1}$$

From adiabatic compression process 1–2 and adiabatic expansion process 3–4, we have

$$T_2 = T_1 \left(\frac{V_1}{V_2}\right)^{\gamma-1} = T_1 r^{\gamma-1}$$

$$T_4 = T_3 \left(\frac{V_3}{V_4}\right)^{\gamma-1} = T_3 \left(\frac{V_2}{V_1}\right)^{\gamma-1} = \frac{T_3}{r^{\gamma-1}}$$

$$= \frac{T_3}{T_2/T_1} = \frac{T_1 T_3}{T_2}$$

Substituting $T_4 = \dfrac{T_1 T_3}{T_2}$ in expression (i), we get

$$\frac{W}{C_v} = T_3 - T_2 - \frac{T_1 T_3}{T_2} + T_1$$

The intermediate temperature T_2 for maximum work output can be obtained by differentiating the above expression with respect to T_2 and setting the derivative equal to zero. That is,

$$\frac{1}{C_v}\frac{dW}{dT_2} = -1 + \frac{T_1 T_3}{T_2^2}$$

$= 0$ for maximum work.

$$T_2^2 = T_1 T_3 \text{ or } T_2 = \sqrt{T_1 T_3}$$

Similarly,

$$\frac{W}{C_v} = T_3 - \frac{T_1 T_3}{T_4} - T_4 + T_1$$

$$\frac{1}{C_v}\frac{dW}{dT_4} = \frac{T_1 T_3}{T_4^2} - 1 = 0 \text{ for maximum work}$$

$$T_4 = \sqrt{T_1 T_3}$$

Thus, for maximization of work $T_2 = T_4 = \sqrt{T_1 T_3}$, i.e. the intermediate temperatures T_2 and T_4 must be equal.

The corresponding maximum work is

$$W_{max} = C_v(T_3 - T_2 - T_4 + T_1)$$

$$= C_v(T_3 - \sqrt{T_1 T_3} - \sqrt{T_1 T_3} + T_1)$$

$$= C_v(T_3 + T_1 - 2\sqrt{T_1 T_3})$$

Diesel Cycle (Constant Pressure Cycle)

The diesel cycle was conceived and developed by Rudolph Diesel in 1893 to the point where he was able to begin commercial production of the corresponding diesel engine in his own factory.

The operations are as follows:

Process (1–2): Reversible adiabatic compression.
Process (2–3): Constant pressure heat supply.
Process (3–4): Reversible adiabatic expansion.
Process (4–1): Constant volume heat rejection.

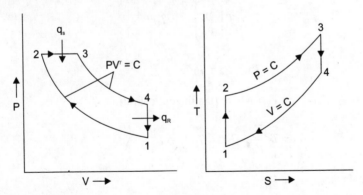

Fig. 12.39

Thermal Efficiency

Consider unit mass of air undergoing the cyclic change.

Heat supplied during process (2–3) $Q_1 = C_p (T_3 - T_2)$

Heat rejected during process (4–1) $Q_2 = C_v (T_4 - T_1)$

Work done = heat supplied − heat rejected.
$$= C_p(T_3 - T_2) - C_v(T_4 - T_1)$$

Internal efficiency $\eta = \dfrac{\text{work done}}{\text{heat supplied}}$

$$= \dfrac{C_p(T_3 - T_2) - C_v(T_4 - T_1)}{C_p(T_3 - T_2)}$$

$$= 1 - \dfrac{1(T_4 - T_1)}{\gamma(T_3 - T_2)} \quad (2)$$

r = compression ratio

ρ = cut-off ratio

γ = adiabatic exponent.

At point 2 $\quad V_2 = \dfrac{v_1}{r}$

$$p_2 = p_1\left(\dfrac{v_1}{v_2}\right)^\gamma = p_1 r^\gamma$$

$$T_2 = T_1\left(\dfrac{v_1}{v_2}\right)^{\gamma-1} = T_1 r^{\gamma-1}$$

At point 3 $\quad p_3 = p_2 = p_1 r^\gamma$

$V_3 = V_2 \rho$ where ρ is the cut off ratio

$$T_3 = T_2 \dfrac{V_3}{V_2} = T_2 \rho = T_1 r^{\gamma-1} \rho$$

At point 4 $\quad V_4 = V_1$

$$p_4 = p_3\left(\dfrac{V_3}{V_4}\right)^\gamma = p_3\left(\dfrac{v_3}{v_1}\right)^\gamma = p_3\left(\dfrac{V_3/V_2}{V_1/V_2}\right)^\gamma$$

$$= p_3\left(\dfrac{\rho}{r}\right)^\gamma$$

$$T_4'' = T_3 \left(\frac{V_3}{V_4}\right)^{\gamma-1} = T_3 \left(\frac{\rho}{r}\right)^{\gamma-1}$$

$$= \frac{T_1 r^{\gamma-1} \rho \rho^{\gamma-1}}{r^{\gamma-1}} = T_1 \rho^\gamma$$

when these values are placed in equation 2

$$\eta = 1 - \frac{1}{y} \frac{T_1 p' - T_1}{T_1 r^{\gamma-1} \rho - T_1 r^{\gamma-1}}$$

$$= 1 - \frac{1}{y} \frac{T_1(\rho^\gamma - 1)}{T_1 r^{\gamma-1}(\rho - 1)}$$

$$= 1 = \frac{1}{r^{\gamma-1}} \left[\frac{\rho^\gamma - 1}{y(\rho - 1)}\right]$$

Apparently the efficiency of Diesel cycle depends upon the compression ratio (r) and cut of ratio (ρ) and hence upon the quantity of heat supplied.

Figure shows the air standard efficiency of diesel cycle for various cut-off ratio. If we consider.

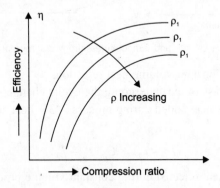

Fig. 12.40: Diesel cycle efficiency for various cut-off ratios

$$\text{factor } D = \left[\frac{\rho^\gamma - 1}{y(\rho - 1)}\right]$$

It shows that with the increase in the cut-off ratio ρ, the value of D increases, that implies that for a Diesel engine at constant compression ratio the efficiency would increase with decrease in ρ and in the limit $\rho \to 1$ the efficiency would become $1 - \frac{1}{r^{\gamma-1}}$ since D is always greater. than one, the Diesel cycle is always less efficient than a corresponding Otto cycle having the same compression.

The Dual or Limited Pressure Cycle

This is a cycle in which the addition of heat is partly at constant volume and partly at constant pressure.

Thermal efficiency for dual cycle.
The cycle consists of five process:
Process (1–2): Reversible adiabatic compression.
Process (2–3): Constant volume heat addition.
Process (3–4): Constant pressure heat addition.
Process (4–5): Reversible adiabatic expansion.

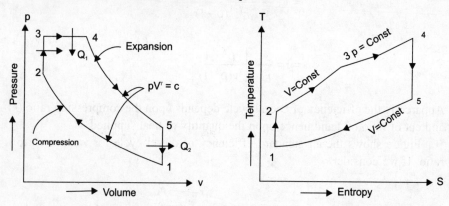

Fig. 12.41: Pressure-volume and temperature-entropy diagram for air standard cycle

Process (5–1): Constant volume heat rejection.
Consider unit mass of air undergoing the cyclic change.
Heat added Q_1 = heat added during process 2–3 + heat added during process 3–4

$$= C_v (T_3 - T_2) + C_p (T_4 - T_3)$$

Heat rejected Q_2 = heat rejected during constant volume process 5–1

$$= C_v (T_5 - T_1)$$

Work done = heat supplied – heat rejected

$$= C_v (T_3 - T_2) + C_p (T_4 - T_3) - C_v (T_5 - T_1)$$

Thermal efficiency $\eta = \dfrac{\text{Work done}}{\text{Heat supplied}}$

$$= \frac{C_v(T_3 - T_2) + C_p(T_4 - T_3) - C_v(T_5 - T_1)}{C_v(T_3 - T_2) + C_p(T_4 - T_3)}$$

$$= 1 - \frac{T_5 - T_1}{(T_3 - T_2) + \gamma(T_4 - T_3)} \qquad (3)$$

Compression ratio $r = \dfrac{V_1}{V_2}$ explosion ratio $\alpha = \dfrac{p_3}{p_2}$

cut-off ratio $\rho = \dfrac{V_4}{V_3}$ adiabatic exponent $= \gamma$

At point 2 $V_2 = \dfrac{V_1}{r}$ $p_2 = p_1 \left(\dfrac{V_1}{V_2}\right)^{\gamma} = p_1 r^{\gamma}$

$$T_2 = T_1 \left(\frac{V_1}{V_2}\right)^{\gamma-1} = T_1 r^{\gamma-1}$$

At point 3 $V_3 = V_2 = \dfrac{V_1}{r}$

$$p_3 = p_2 \alpha = p_1 \alpha r^{\gamma}$$

$$T_3 = T_2 \frac{p_3}{p_2} = T_2 \alpha = T_1 \alpha r^{\gamma-1}$$

At point 4 $V_4 = V_3 \rho$ $p_4 = p_3 = p_1 \alpha r^{\gamma}$

$$T_4 = T_3 \frac{V_4}{V_3} = T_3 \rho = T_1 \alpha \rho^{\gamma-1}$$

At point 5 $V_5 = V_1$

$$P_5 = P_4 \left(\frac{V_4}{V_5}\right)^{\gamma} = P_4 \left(\frac{V_4}{V_1}\right)^{\gamma} = P_3 \left(\frac{V_4/V_3}{V_1/V_3}\right)^{\gamma}$$

$$= p_3 \left(\frac{V_4/V_3}{(V_1/V_2)}\right) = p_3 \left(\frac{\rho}{r}\right)^{\gamma}$$

$$T_5 = T_4\left(\frac{V_4}{V_5}\right)^{\gamma-1} = T_4\left(\frac{\rho}{r}\right)^{\gamma-1}$$

$$= \frac{T_1 \alpha r^{\gamma-1} \times \rho^{\gamma-1}}{r^{\gamma-1}} = T_1 \alpha \rho^\gamma$$

When these values are substituted in equation 3, we get

$$\eta = 1 - \frac{T_1 \alpha \rho - T_1}{(T_1 \alpha r^{\gamma-1} - T_1 r^{\gamma-1}) + y(T_1 \alpha \rho r^{\gamma-1} - T_1 \alpha r^{\gamma-1})}$$

$$= 1 - \frac{1}{r^{\gamma-1}}\left[\frac{T_1(\alpha\rho^\gamma - 1)}{T_1(\alpha-1) + \gamma T_1(\alpha\rho - \alpha)}\right] = 1 - \frac{1}{r^{y-1}}\left[\frac{\alpha\rho^\gamma - 1}{(\alpha-1) + \gamma\alpha(\rho-1)}\right]$$

Since the value of pressure ratio $\alpha > 1$ the value of η_{dual} will increase for given value of ρ and γ. Thus the efficiency of a dual cycle is intermediate between those of Otto and Diesel cycle having the same compression ratio.

If α and $\rho = 1$ it becomes Otto cycle

$\alpha = 1$ it becomes Diesel cycle

Comparison of Otto, Diesel and Dual Cycle

The comparison of Otto, Diesel and dual cycles can be made on the basis of compression ratio, maximum pressure, maximum temperature, heat input, work output, etc.

(a) For the same compression ratio and same heat input

The efficiency of Otto cycle depends upon compression r and is given by

$$\eta_{otto} = 1 - \frac{1}{r^{y-1}} \quad (1)$$

The efficiency of Diesel cycle depends upon compression ratio r, the cut-off ratio ρ and is prescribed by the relation

$$\eta_{Diesel} = 1 - \frac{1}{r^{\gamma-1}}\left[\frac{\rho^\gamma - 1}{\gamma(\rho-1)}\right] \quad (2)$$

for fixed compression ratio r, the efficiency of Diesel cycle depends upon the factor

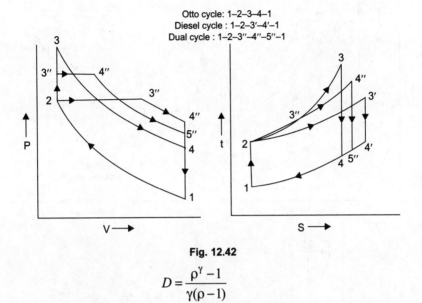

Fig. 12.42

$$D = \frac{\rho^\gamma - 1}{\gamma(\rho - 1)}$$

For adiabatic exponent $\gamma = 1.4$, the value of D for different cut off ratios is

$\rho = 1.5$	2.0	2.5	3.0
$K = 1.092$	1.17	1.24	1.31

The value of factor *K* is thus always greater than unity. A close examination of expression (1) and (2) then reveals that the Diesel cycle is always less efficient than the corresponding Otto cycle having the same compression ratio.

When compression ratio is kept constant process (1–2) remains the same for all the three cycles. But process (2–3) which shows the heat addition is different for these cycles. If same heat is transferred in all three cycles, the temperature attained is maximum for otto cycle, than dual cycle and least by Diesel cycle. Since we know the work done during the cycle is proportional to the area inside the bonded region. The area is maximum for Otto cycle and minimum for Diesel cycle. Thus, for same heat input, efficiency of Otto cycle will be maximum while that of Diesel cycle will be minimum.

$$\eta_{otto} > \eta_{dual} > \eta_{diesel}$$

for the same compression ratio and same heat input.

(b) For same maximum pressure and same heat input

For the same maximum pressure points 3, 3′ and 3″ must be on some pressure line and for the same heat input the area 2 – 3 – ad –, 2′ – 3′ – c – d – 2′ and 2″ 3″ 4″ b – d – 2″ should be equal. It is obvious from figure that heat rejected by Otto cycle 1 – 4 – a – d – 1 is more than 1 – 5″ – b – d – 1 and diesel 1 – 4′ – c – d – 1

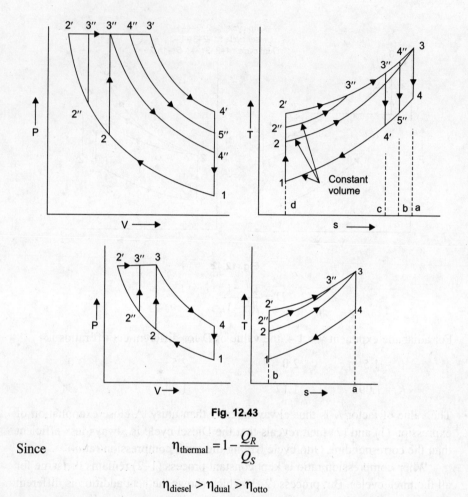

Fig. 12.43

Since
$$\eta_{\text{thermal}} = 1 - \frac{Q_R}{Q_S}$$

$$\eta_{\text{diesel}} > \eta_{\text{dual}} > \eta_{\text{otto}}$$

(iii) For the same heat rejection

It is clear from the figure that same amount of heat is being rejected by all the three cycles. But maximum heat is supplied in Diesel cycle (area $2' - 3 - a - b - 2'$) and minimum in Otto cycle (area $2 - 3 - a - b - 2$) while for dual cycle it is in between the two (area $2' - 3'' - 3 - a - b - 2''$)

Since
$$\eta_{\text{thermal}} = 1 - \frac{\text{Heat rejected}}{\text{Heat supplied}}$$

Hence,
$$\eta_{\text{diesel}} > \eta_{\text{dual}} > \eta_{\text{otto}}$$

12.15 WORKING PRINCIPLE OF A GAS TURBINE

A simple gas turbine consists of: (1) compressor (2) combustion chamber: (3) a turbine. The air is compressed in the compressor heated by the combustion process to raise its pressure and temperature, expanded in turbine and finally the expanded products of combustion are rejected through the exhaust.

Types of Gas Turbine

The gas turbine are of two types:
  (1) open cycle gas turbine
  (2) closed cycle gas turbine.

In the open cycle, fresh air is drawn into the compressor from the atmosphere, heat is added by combustion of fuel in the air itself and the spent up combustion products are rejected to the atmosphere. The arrangement thus necessitate continuous replacement of the working medium. The quantity and speed of the

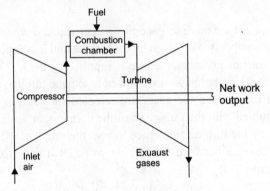

Open cycle gas turbine

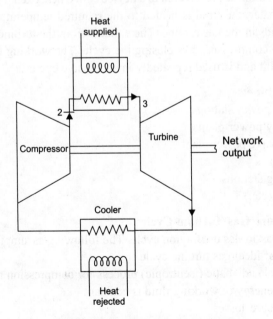

Fig. 12.44: Closed cycle gas turbine unit

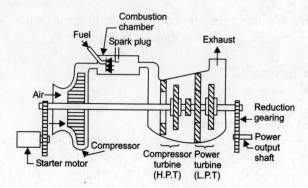

Fig. 12.45: Ope cycle gas turbine

working fluid required are more so generally a centrifugal or an axial compressor is employed. The compressor is driven by the turbine and is coupled to the turbine shaft. In the combustion chamber, the heat is supplied by the fuel sprayed into air stream, then the resulting hot gases are allowed to expand through a turbine. After passing through the turbine the gases are rejected into atmosphere since the compressor is coupled with the turbine shaft the compressor absorbs some of the power produced by the turbine and hence lowers the efficiency. The net work is therefore the difference between the turbine work (W_T) and work required by compressor to derive it W_c.

In a closed system, the compressed working fluid which may be air, hydrogen, helium or carbon dioxide is delivered to a device called heat exchanger or simply a heater. In the heater, the fluid is heated to the required temperature and the hot fluid then expands in the gas turbine. The exhaust from the turbine is cooled and again, fed to the compressor, thus closing the cycle. The working fluid thus does not leave the plant and is used repeatedly in successive cycles.

Uses of gas turbine:
 (1) Central power stations
 (2) Stand by power plant
 (3) Locomotive power plant
 (4) Jet engines
 (5) Pumping stations
 (6) Marine (etc.)

Constant Pressure Gas Turbine Cycle
It is also known as Joules or Brayton cycle. The following assumptions are made for the analysis of ideal gas turbine cycle.
 (1) Reversible adiabatic (isentropic) process for compression and expansion
 (2) Kinetic energy of working fluid is constant
 (3) No pressure losses.

(4) The working fluid considered to be ideal gas with constant specific heats.
(5) Counter flow heat exchanger with 100% efficiency.

In the ideal diagram of Joule cycle the cycle is assumed to be a closed one and heat addition or rejection is only the heat transfer process. Figure shows the representation of ideal Brayton cycle on the P-V and T-S diagram.

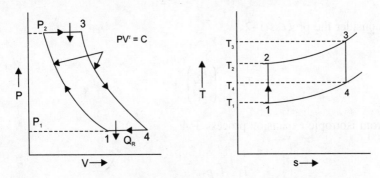

Fig. 12.46: Brayton cycle

The various processes are

1–2: Isentropic compression in the compressor thus raising pressure and temperature from p_1, T_1, to p_2, T_2

2–3: Addition of heat at constant pressure raising temperature from T_2 to T_3

3–4: Isentropic expansion of air from high pressure and temperature to low pressure and temperature and thus doing work.

4–1: Rejection of heat at constant pressure to restore the original state of fluid.

For example, consider 1 kg of working fluid.

From 1st law for steady flow neglecting ΔKE and ΔPE

$$\delta Q - \delta W = dh$$

During heat addition or heat rejection no work is done, hence

$$Q = \Delta h$$

∴ Heat added = $Q_A = Q_{2-3} = h_3 - h_2 = C_p(T_3 - T_2)$

Heat rejected = $Q_R = Q_{4-1} = h_4 - h_1 = C_p(T_4 - T_1)$

Net work $= Q_A - Q_R$

$= C_p[(T_3 - T_2) - (T_4 - T_1)]$

$$\eta_{thermal} = \frac{\text{Net work}}{\text{heat added}} = \frac{C_p[(T_3 - T_2) - (T_4 - T_1)]}{C_p(T_3 - T_2)}$$

$$\eta_{thermal} = 1 - \frac{T_4 - T_1}{T_3 - T_2} \tag{1}$$

Now consider the process 1–2

$$T_2 = T_1 \left(\frac{P_2}{P_1}\right)^{\frac{\gamma-1}{\gamma}}$$

And from isotropic expansion process 3–4

$$\frac{T_3}{T_4} = \left(\frac{p_3}{p_4}\right)^{\frac{\gamma-1}{\gamma}}$$

But $p_2 = p_3$ and $p_1 = p_4$

Gas turbine versus recirocating I.C. engines

(1)	Mechanical efficiency	Less prone to friction loss because less members are involved in transmission of power so high mechanical efficiency (95% to 97%)	Large number of pairs of sliding or bearing members so more friction less mechanical efficiency 85 to 95%
(2)	Expansion	Full adiabatic expansion of the products of combustion	No full expansion of the products of combustion
(3)	Balancing	Accurately balanced	Not fully balanced
(4)	Knocking or detonation	Nil	Present
(5)	Size	Large power from units of comparatively small size and weight specific weight of 10 kg/kW	Less power 55 kg/H.P. for a diesel engine
(6)	Fuels	Cheap and readily available fuels such as benzene powdered coal and heavy graded hydrocarbons	Costly fuels used
(7)	Pressure	Low operating pressure so joints and pipes can be easily designed	High operating pressure so design part is tough

$$\therefore \frac{T_2}{T_1} = \frac{T_3}{T_4} = \left(\frac{p_2}{p_1}\right)^{\frac{\gamma-1}{\gamma}}$$

or
$$\frac{T_1}{T_2} = \frac{T_4}{T_3} = \frac{T_4 - T_1}{T_3 - T_2}$$

Put this value of $\frac{(T_4 - T_1)}{T_3 - T_2}$ in equation (1), we get

$$\eta_{thermal} = 1 - \frac{T_1}{T_2} = 1 - \frac{T_4}{T_3}$$

Get r_p = pressure ratio = $\frac{p_2}{p_1}$

$$\frac{T_2}{T_1} = \left(\frac{p_2}{p_1}\right)^{\frac{\gamma-1}{\gamma}} = (r_p)^{\frac{\gamma-1}{\gamma}}$$

$$\frac{T_1}{T_2} = \frac{1}{(r_p)^{\frac{\gamma-1}{\gamma}}}$$

$$\eta_{thermal} = 1 - \frac{1}{(r_p)^{\frac{\gamma-1}{\gamma}}} \quad \text{where } \gamma = \frac{C_p}{C_v} = 1.4 \text{ for air.}$$

Hence, we can say that the thermal efficiency progressively increases with increasing value of pressure ratio.

Disadvantages of gas turbine:
(1) Lower value of thermal efficiency because a considerable portion of the power produced in the turbine is spent up in running the compressor.
(2) Difficult to start
(3) Initial cost of gas turbine plant is very high due to costly materials and high research, development and manufacturing cost, but in long span of time the cost may be compensated by running the turbine on full load for long duration.
(4) The efficiency of the turbine is very low at part load conditions.
(5) The decreasing size of the turbine worsen the features such as specific power, acceleration, fuel economy, relative cost etc.

EXERCISE

1. Explain the following cycles.
   (a) Otto
   (b) Diesel
   (c) dual
2. Explain the following terms as applied to I.C. engines
   (a) Bore   (b) Stroke
   (c) Clearance volume.   (d) Swept volume
   (e) Compression ratio   (f) cult off ratio.
   (g) Expansion ratio   (h) Indicated Horse power.
   (i) Brake horse power.
3. What is a C.I engine? Why it has more compression ratio as compared to S.I engine.
4. Explain construction and working of 4-stroke diesel engine.
   (MRIU Dec 2009)
5. For the same compression ratio show that efficiency of otto cycle is greater than that of diesel cycle.
6. Explain the working principle of gas turbine.
7. What do you mean by steam condenser? What are its function?
8. Differentiate between surface and jet condenser.
9. Explain the necessity of compounding in steam turbine?
10. What is the function of cooling towers in modern steam power plants.
11. Differentiate between natural and artificial draught cooling towers.
12. What is the function of cooling pond.

OBJECTIVE TYPE QUESTIONS

1. In the impulse turbine the steam expands
   (a) in the nozzle
   (b) in the blades
   (c) partly in the nozzle and partly in the blades
   (d) neither in the nozzle nor in the blades
2. Parson's turbine is
   (a) simple reaction turbine
   (b) simple impulse type turbine
   (c) velocity compounded type turbine
   (d) pressure compounded type turbine
3. In reaction turbine the expansion of steam as it flows over blades represents
   (a) throttling process
   (b) free expansion process
   (c) isothermal expansion
   (d) adiabatic process

4. De-level turbine is a
   (a) simple reaction turbine
   (b) simple Impulse turbine
   (c) velocity compounded impulse turbine
   (d) pressure compounded impulse turbine
5. Reteau turbine is a
   (a) simple reaction turbine
   (b) simple impulse turbine
   (c) pressure compounded turbine
   (d) velocity compounded impulse turbine
6. Velocity compounding involves
   (a) expansion of steam in stages
   (b) recovery of genetic energy of steam leaving first set of blades in subsequent rows of blades
   (c) velocity and pressure equalization at different stages
   (d) increased velocity after each stage due to expansion of steam
7. For a given horsepower an impulse turbine compared to a reaction turbine has
   (a) less rows of blade
   (b) more rows of blade
   (c) equal rows of blade
   (d) it can be anything
8. In a reaction turbine, a stage is represented by
   (a) each row of blades
   (b) number of entries of steam
   (c) number of casings
   (d) number of exists of steam
9. In Impulse turbine
   (a) The steam is expanded in nozzles and there is no fall in pressure as the steam passes over the rotor blades
   (b) Steam is directed over bucket like blades which propels the rotor
   (c) expansion of steam takes place as it passes through the moving blades on the rotor as well as through the guide blades fixed to the casing
10. In an impulse turbine the energy supplied to the blades per kg of steam equals to
    (a) work done by steam
    (b) sum of kinetic energy and potential energy at inlet
    (c) reaction energy of steam
    (d) kinetic energy of jet at entrance per kg of steam
11. The pressure velocity compounding of steam turbine results in
    (a) shorter turbine for a given total pressure drop
    (b) large turbine for a given pressure drop
    (c) large number of stages
    (d) lesser friction losses

12. Combining impulse stages in series results in
    (a) increase of speed
    (b) decrease of speed
    (c) speed remains unaffected
    (d) unpredictable speed effect
13. In reaction turbines
    (a) the steam is expanded in nozzles and there is no fall in pressure as the steam passes over the rotor blades
    (b) steam is directed over bucket-like blades which propels the rotor
    (c) expansion of steam takes places as it passes through the moving blades on the rotor as well as through the guide blades fixed to the casing
    (d) steam pressure remains constant
14. In Impulse reaction turbine, the pressure drops
    (a) In fixed nozzles
    (b) In moving blades
    (c) In fixed blades
    (d) In both fixed and moving blades
15. A cooling tower is to be installed in a place where dry bulb temperature and wet bulb temperature are almost constant throughout such a proportion is
    (a) excellent
    (b) not desirable
    (c) can be considered
    (d) other data are required to determine the same
16. Cooling effect in a cooling tower can be speed by
    (a) increasing air velocity over the wet surfaces
    (b) lower the barometric pressure
    (c) reducing humidity of air
    (d) all of the above
17. Cooling tower are installed where
    (a) water is available in plenty
    (b) water is scarce
    (c) for very big plants
    (d) for very small plants
18. Approach of cooling tower means
    (a) difference in temperature of hot water entering and cold water leaving
    (b) difference in temperature of the cold water leaving the cooling tower and the wet bulb temperature of surrounding air
    (c) difference in temperature of the cold water and atmospheric temperature
    (d) amount of heat thrown away by the cooling tower in kcal/hr
19. In cooling tower, water is cooled by the process of
    (a) condensation
    (b) fusion
    (c) evaporation
    (d) sublimation

20. Compression ratio of I.C. engine is
    (a) The ratio of volume of air in cylinder before compression stroke and after compression stroke.
    (b) ratio of pressure after compression and before compression
    (c) swept volume/cylinder volume
    (d) cylinder volume/swept volume
21. Scavenging air in diesel engine means
    (a) air used for combustion sent under pressure
    (b) forced air for cooling cylinder
    (c) burnt air containing products of combustion
    (d) air used for forcing burnt gases out of engine's cylinder during the exhaust period.
22. If the intake air temperature of I.C. engine increases, its efficiency will
    (a) increase
    (b) decrease
    (c) remain same
    (d) unpredictable
23. For the same compression ratio
    (a) Otto cycle is more efficient than diesel
    (b) Diesel cycle is more efficient than otto
    (c) both Otto and Diesel cycles are equally efficient
    (d) compression ratio has nothing to do with efficiency
24. The process of breaking up of a liquid into fine droplets by spraying is called
    (a) vaporisation
    (b) carburation
    (c) ionization
    (d) atomisation
25. Diesel fuel, compared to petrol is
    (a) less difficult to ignite
    (b) more difficult to ignite
    (c) highly ignitable
    (d) none of the above

ANSWERS

a. a	2. a	3. d	4. b	5. c
6. b	7. b	8. a	9. a	10. d
11. a	12. b	13. c	14. d	15. b
16. d	17. b	18. b	19. c	20. a
21. d	22. b	23. a	24. d	25. b

Steam Tables

The preparation of steam tables has been made possible bystudying experimentally the steam properties t different presure and temperature conditions of its formation. The experimental data was then utilized to formulate suitale equations of state in different regions and subsequently obtain comprehensive data on steam properties.

The steam tables gives values of specific volume, enthalpy and entropy for saturated liquid and dry saturated vapour tabulated against pressure (p) or corresponding saturation temperature (t_s). Steam table make it possible to rapidly find the basic characteristics of steam according to the given parameters. [see table (1) and (2)]

Table (1) Properties of saturated water and steam (temperature base)

Temperature	Absolute pressure bar	Specific volume		Specific enthalpy			Specific entropy		
t	p	v_f	v_g	h_f	h_{fg}	h_g	s_f	s_{fg}	s_g
20	0.023	0.001	57.791	83.96	2454.1	2538.1	0.297	8.371	8.668
40	0.07384	0.001	19.523	167.57	2406.7	2574.3	0.572	7.685	8.257

Table (2) Properties of saturated water and steam (pressure base)

Absolute pressure bat	Saturation temperature °C	Specific volume m³/kg		Specific enthalpy kJ/kg			Specific entropy kJ/kg K.		
p	t_s	v_f	v_g	h_f	h_{fg}	h_g	s_f	s_{fg}	s_g
20	212.4	0.001177	0.0996	908.8	1890.7	2799.5	2.447	3.894	6.341
40	250.4	0.001	0.0498	1087.3	1714.1	2801.4	2.796	3.274	6.070

Values of specific volume, enthalpy and entropy depend not only on pressure but also on temperature or degree of superheat. A typical structure is given below.

Properties of superheated steam

p (bar)	t_s			200	400	600
10	$t_g = 0.1944$		v	.2060	0.3066	0.4011
$t_s = 179.9°C$	$h_g = 2778.1$		h	2828.9	3263.9	3697.9
	$s_g = 6.587$		s	6.694	7.465	8.029

MOLLIER DIAGRAM

The solutions toproblems dealing with steam vapour as working substance are easily solved by certain charts/diagrams which have been prepared by having data from steam tables on the basis of this available data. Enthalpy-Entropy chart is prepared which is also known as Mollier diagram. The important feature of this diagram are:

(1) The specific enthalpy is plotted along y-axis while the specific entropy is laid along the x-axis.
(2) The diagram is divided into two regions by the saturation curve below the saturation curve lies the wets team region which gives the values of steam properties pertaining to wet steam. The region of superheated steam lies above the saturation curve and indicates values pertaining to superheated steam.
(3) From these two regions passes the constant pressure line and constant temperature lines.
(4) Lines of constant dryness fraction are drawn in the wet region.

The chart makes it possible to rapidly determine the steam parameters with an accuracy sufficient for practical purposes and give solution to many thermodynamic problems relating to a change in state of steam.

SYMBOLS

S.No.	Term	Symbol	Unit
1.	Temperature	t	°C K
2.	Saturation temperature	t_s	°C
3.	Pressure	p	bar
4.	Specific volume of saturated liquid (fluid)	v_f	m³/kg
5.	Specific volume of sturated steam (vapour gas)	v_g	m³/kg
6.	Specific enthalpy of saturated liquid (fluid)	h_f	kJ/kg
7.	Specific enthalpy of evaporation (Latent heat)	h_{fg}	kJ/kg
8.	Specifc enthalpy of saturated steam (vapour, gas)	h_g	kJ/kg
9.	Specific entropy of saturated liquid (fluid)	s_f	kJ/kgK
10.	Specific entropyof evaporation	S_{fg}	kJ/kgK
11.	Specific entropy of steam (vapour, gas)	s_g	kJ/kgK

1 bar = 10^5 N/m² = 10^5 Pa = 100 KPa = 0.1 MPa

SUBSCRIPTS

f - refers to a property of the saturated liquid
g - refers to property of the saturated vapour
fg - refers to change of phase at constant pressure
s - refers to the saturated state

Table 1: Properties of Saturated Water and Steam (Temperature Base)
(Values based on datum temperature at 0°C)

Temperature °C	Absolute pressure	Specific volume m³/kg		Specific enthalpy kJ/kg			Specific entropy kJ/kg K.		
t	p	v_f	v_g	h_g	h_{fg}	h_g	s_f	s_{fg}	s_g
0.01	0.006113	0.00100	206.14	0.01	2501.3	2501.4	0	9.1562	9.1562
1	0.006567	0.001	192.577	4.16	2499	2503.2	0.015	9.115	9.13
2	0.007056	0.001	179.889	8.37	2496.7	2505	0.031	9.073	9.104
3	0.007577	0.001	168.132	12.57	2494.3	2506.9	0.046	9.032	9.078
4	0.008131	0.00100	157.232	16.78	2491.9	2508.7	0.061	8.99	9.051
5	0.008721	0.001	147.12	20.98	2489.6	2510.6	0.076	8.95	9.026
6	0.009349	0.001	137.734	25.2	2487.2	2512.4	0.091	8.909	9
7	0.010016	0.001	129.017	29.39	2484.8	2514.2	0.106	8.869	8.975
8	0.010724	0.001	120.917	33.6	2482.5	2516.1	0.121	8.829	8.95
9	0.011477	0.001	113.386	37.8	2480.1	2517.9	0.136	8.789	8.925
10	0.012276	0.001	106.379	42.01	2477.7	2519.8	0.151	8.75	8.901
11	0.013123	0.001	99.857	46.2	2475.4	2521.6	0.166	8.711	8.877
12	0.014022	0.001	93.784	50.41	2473	2523.4	0.181	8.672	8.853
13	0.14974	0.001	88.124	54.6	2470.7	2525.3	0.195	8.633	8.828
14	0.05983	0.001	82.848	58.8	2468.3	2527.1	0.21	8.595	8.805
15	0.017051	0.001	77.926	62.99	2465.9	2528.9	0.225	8.557	8.782
16	0.018181	0.00100	73.333	67.19	2463.6	2530.8	0.239	8.519	8.758
17	0.019376	0.00100	69.044	71.38	2461.2	2532.6	0.254	8.482	8.736
18	0.020640	0.00100	65.038	75.58	2458.8	2534.4	0.268	8.444	8.712
19	0.021975	0.00100	61.293	79.77	2456.5	2536.2	0.282	0.407	8.689
20	0.02339	0.00100	57.791	83.96	2454.1	2538.1	0.297	8.371	8.668
21	0.02487	0.00100	54.514	83.96	2454.1	2539.9	0.311	8.334	8.645
22	0.02645	0.00100	51.447	92.33	2449.4	2541.7	0.325	8.298	8.623

Contd. ...

Table 1: (Contd.)

t	p	v_f	v_g	h_g	h_{fg}	h_g	s_f	s_{fg}	s_g
23	0.02810	0.00100	48.574	96.52	2447.0	2543.5	0.339	8.262	8.601
24	0.02985	0.00100	45.883	100.70	2444.7	2545.4	0.353	8.226	0.579
25	0.03169	0.00100	43.360	104.89	2442.3	2547.2	0.367	8.190	8.558
26	0.03363	0.00100	40.994	109.07	2439.9	2449.0	0.381	8.155	8.536
27	0.03567	0.00100	40.994	109.07	2437.6	2550.8	0.395	8.120	8.515
28	0.03782	0.00100	36.690	117.43	2435.2	2552.6	0.409	8.085	8.494
29	0.04008	0.00100	34.773	121.61	2432.8	2554.5	0.423	8.051	8.453
30	0.04246	0.00100	32.894	125.79	2430.5	2556.3	0.437	8.016	8.453
31	0.04496	0.00100	31.165	129.97	2428.1	2558.1	0.451	7.982	8.433
32	0.04759	0.00101	29.540	134.15	2425.7	2559.9	0.464	7.948	8.412
33	0.05034	0.00101	28.011	138.33	2423.4	2561.7	0.478	7.915	8.393
34	0.05324	0.00101	26.571	142.50	2421.0	2563.5	0.492	7.881	8.373
35	0.05628	0.00101	25.216	146.68	2418.6	2565.3	0.505	7.848	8.353
36	0.05947	0.00101	23.940	150.86	2416.2	2567.1	0.519	7.815	8.334
37	0.06281	0.00101	22.737	155.03	2413.9	2568.9	0.532	7.782	8.314
38	0.06632	0.00101	21.602	159.21	2411.5	2570.7	0.546	7.749	8.295
39	0.06999	0.00101	20.533	163.39	2409.1	2572.5	0.559	7.717	8.276
40	0.07384	0.00101	19.523	167.57	2406.7	2574.3	0.572	7.685	8.257
41	0.07786	0.00101	18.570	171.74	2404.3	2576.1	0.586	7.652	8.238
42	0.08208	0.00101	17.671	175.91	2404.9	2577.9	0.599	7.621	8.220
43	0.08649	0.00101	16.821	180.10	2399.5	2579.6	0.612	7.589	8.201
44	0.09111	0.00101	16.018	184.27	2397.2	2581.4	0.626	7.557	8.183
45	0.09593	0.00101	15260	188.45	2394.8	2583.2	0.639	7.526	8.165
46	0.10101	0.001010	14.540	192.61	2392.4	2585.0	0.652	7.495	8.147
47	0.10622	0.001011	13.861	196.82	2390.0	2586.8	0.665	7.464	8.129
48	0.11101	0.001011	13.223	201.00	2387.6	2588.6	0.678	7.433	8.111
49	0.11753	0.001012	12.612	205.12	23852	2590.3	0.691	7.403	8.094

Contd. ...

Table 1: (Contd.)

t	p	v_f	v_g	h_f	h_{fg}	h_g	s_f	s_{fg}	s_g
50	0.12354	0.001012	12.034	209.31	2382.7	2592.1	0.704	7.372	8.076
51	0.12961	0.001013	11.499	213.4	2380.5	2593.9	0.716	7.344	8.060
52	0.13632	0.001013	10.972	217.7	2377.9	2595.6	0.731	7.312	8043
53	0.14293	0.001014	10.488	221.8	2375.7	2597.5	0.742	7.284	8.026
54	0.15021	0.001014	10.013	226.0	2373.1	2599.1	0.755	7.253	8.008
55	0.15741	0.001015	9.579	230.2	2370.8	2601.0	0.768	7.225	7.993
56	0.16530	0.001015	9.149	234.4	2368.2	2602.6	0.781	7.194	7.975
57	0.17313	0.001016	8.760	238.5	2366.0	2604.5	0.793	7.166	7.959
58	0.18172	0.001016	8.372	242.8	2363.4	2606.2	0.806	7.136	7.942
59	0.19016	0.001017	8.021	246.9	2361.1	2608.0	0.818	7.109	7.927
60	0.19941	0.001017	7.671	251.1	2358.5	2609.6	0.831	7.078	7.909
61	0.20861	0.001018	7.353	255.3	2356.1	2611.4	0.844	7.015	7.895
62	0.21863	0.001018	7.037	259.5	2353.6	2613.1	0.856	7.022	7.878
63	0.22855	0.001019	6.749	263.6	2351.3	2614.9	0.868	6.995	7.863
64	0.23934	0.001019	6.463	267.9	2348.7	2616.5	0.881	6.965	7.846
65	0.25009	0.001020	6.202	272.0	2346.4	2618.4	0.893	6.939	7.832
66	0.261 72	0.001020	5.943	276.2	2343.7	2619.9	0.906	6.910	7.816
67	0.273 34	0.001021	5.706	280.4	2341.4	2621.8	0.918	6.884	7.802
68	0.285 90	0.001022	5.471	284.6	2338.8	2623.4	0.930	6.855	7.785
69	0.29838	0.001022	5.256	288.8	2336.4	2625.2	0.943	6.828	7.771
70	0.311 92	0.001023	5.042	293.0	2333.8	2626.8	0.955	6.800	7.755
71	0.32535	0.001 024	4.846	297.2	2331.4	2628.6	0.967	6.775	7.742
72	0.339 58	0.001 024	4.656	301.3	2329.0	2630.3	0.979	6.748	7.727
73	0.35434	0.001 025	4.474	305.5	2326.5	2632.0	0.991	6.721	7.712
74	0.36964	0.001 025	4.300	309.7	2324.0	2633.7	1.003	6.695	7.698
75	0.38549	0.001 026	4.131	313.9	2321.5	2635.4	1.015	6.668	.7.683
76	0.40191	0.001 027	3.976	318.1	2318.9	2637.0	1.027	6.641	7.668
77	0.41891	0.001 027	3.824	322.3	2316.4	2638.7	1.039	6.616	7.655

Contd. ....

Table 1: (Contd.)

t	p	v_f	v_g	h_g	h_{fg}	h_g	s_f	s_{fg}	s_g
78	0.436 52	0.001028	3.680	326.5	2313.9	2640.4	1.051	6.591	7.642
79	0.45474	0.001029	3.541	330.7	2311.4	2642.1	1.063	6.564	7.627
80	0.47360	0.001029	3.409	334.9	2308.8	2643.7	1.025	6.537	7.612
81	0.493 11	0.001030	3.283	339.1	2306.3	2645.4	1.087	6.512	7.599
82	0.51329	0.001031	3.162	343.3	2303.7	2647.0	1.098	6.487	7.585
83	0.53416	0.001031	3.046	347.5	2301.2	2648.7	1.111	6.461	7.572
84	0.55573	0.001 032	2.935	351.7	2298.6	2650.3	1.123	6.436	7.559
85	0.578 03	0.001033	2.829	355.9	2296.1	2652.0	1.134	6.411	7.545
86	0.60108	0.001033	2.727	360.1	2293.5	2653.6	1.146	6.387	7.533
87	0.624 89	0.001 034	2.630	364.3	2291.0	2655.3	1.158	6.361	7.519
88	0.64947	0.001035	2.536	368.5	2288.3	2656.8.	1.169	6.336	7.505
89	0.67487	0.001035	2.447	372.7	2285.8	2658.5	1.181	6.312	7.493
90	0.70144	0.001 036	2.361	376.9	2283.2	2660.1	1.192	6.287	7.479
91	0.728 15	0.001037	2.279	381.1	2280.6	2661.7	1.204	6.263	7.467
92	0.756 06	0.001038	2.200	385.3	2278.0	2663.3	1.216	6.238	7.454
93	0.78489	0.001038	2.124	389.6	2275.4	2665.0	1.227	6.215	7.442
94	0.81461	0.001039	2.052	393.7	2272.7	2666.4	1.238	6.190	7.428
95	0.84526	0.001040	1.982	398.0	2270.1	2668.1	1.250	6.167	7.417
96	0.87685	0.001041	1.915	402.2	2267.3	2669.5	1.261	6.143	7.404
97	0.90944	0.001041	1.851	406.4	2264.9	2671.3	1.273	6.119	7.392
98	0.943 01	0.001042	1.789	410.6	2262.3	2672.9	1.283	6.095	7.378
99	0.97761	0.001 043	1.730	414.8	2259.6	2674.4	1.296	6.072	7.368
100	1.013 50	0.001044	1.673	419.0	2257.0	2676.0	1.307	6.048	7.355
105	1.208 2	0.001048	1.419	440.1	2243.7	2683.8	1.363	5.933	7.296
110	1.432 7	0.001052	1.210	461.3	2230.2	2691.5	1.418	5.820	7.238
115	1.690 6	0.001056	1.037	482.5	2216.5	2699.0	1.473	5.711	7.183
120	1.985 3	0.001061	0.892	503.7	2202.6	2706.3	1.527	5.602	7.129

Contd. ...

Table 1: (Contd.)

t	p	v_f	v_g	h_f	h_{fg}	h_g	s_f	s_{fg}	s_g
125	1.9853	0.001065	0.771	524.0	2188.5	2713.5	1.581	5.496	7.077
130	2.7011	0.001070	0.669	546.3	2174.2	2720.5	1.635	5.392	7.027
135	3.1318	0.001075	0.582	567.7	2159.6	2727.3	1.687	5.291	6.978
140	3.6132	0.001080	0.509	589.1	2144.7	2733.9	1.739	5.191	6.930
145	4.1543	0.001085	0.446	610.6	2129.6	2740.2	1.791	5.093	6.884
150	4.7580	0.001091	0.393	632.2	2114.3	2746.5	1.842	4.996	6.838
155	5.433	0.00110	0.3464	653.8	2097.4	2751.2	1.892	4.900	6.792
160	6.178	0.00110	0.3071	675.5	2082.6	2758.1	1.943	4.807	6.750
165	7.007	0.00111	0.2725	697.2	2064.8	2762.0	1.992	4.713	6.705
170	7.917	0.00111	0.2428	719.2	2019.5	2768.7	2.042	4.624	6.666
175	8.924	0.00121	0.2165	741.1	2030.7	2771.8	2.091	4.531	6.622
180	10.021	0.00113	0.1940	763.2	2015.0	2778.2	2.139	4.446	6.585
185	11.233	0.00113	0.1739	785.3	1995.1	2780.4	2.187	4.355	6.542
190	12.544	0.00114	0.11;)65	807.6	1978.8	2786.4	2.236	4.272	6.598
195	13.987	0.00115	0.1408	829.9	1957.9	2787.8	2.283	4.182	6.465
200	15.538	0.00116	0.1274	852.5	1940.7	2793.2	2.331	4.101	6.432
205	17.243	0.00116	0.1150	875.0	1918.8	2793.8	2.378	4.013	6.391
210	19.062	0.00117	0.1044	897.8	1900.7	2798.5	2.425	3.934	6.359
215	21.060	0.00118	0.0946	920.6	1877.7	2798.3	2.471	3.846	6.317
220	23.182	0.00119	0.0862	943.6	1858.5	2802.1	2.518	3.768	6.286
225	25.501	0.00120	0.0784	966.9	1834.3	2801.2	2.564	3.682	6.246
230	27.951	0.00121	0.0716	990.2	1813.8	2804.0	2.610	3.605	6.215
235	30.632	0.00122	0.0652	1013.8	1788.5	2802.3	2.656	3.519	6.175
240	33.444	0.00123	0.0598	1037.3	1766.5	2803.8	2.706	3.442	6.148
245	36.523	0.00124	0.0547	1061.6	1740.0	2801.6	2.748	3.358	6.106
250	39.734	0.00125	0.0513	1085.3	1716.2	2801.5	2.793	3.280	6.073
255	43.246	0.00126	0.0459	1110.2	1688.5	2798.7	2.839	3.197	6.036
260	46.882	0.00128	0.0422	1134.4	1662.5	2796.9	2.884	3.118	6.002

Contd. ....

Table 1: (Contd.)

t	p	v_f	v_g	h_g	h_{fg}	h_g	s_f	s_{fg}	s_g
265	50.877	0.00129	0.0387	1159.9	1633.6	2793.5	2.931	3.035	5.966
270	54.993	0.00130	0.0356	1184.5	1605.2	2789.7	2.975	2.955	5.930
275	59.496	0.00132	0.0327	1210.8	1574.7	2785.5	3.022	2.873	5.895
280	64.120	0.00133	0.0302	1236.0	1543.6	2779.6	3.067	2.790	5.857
285	69.186	0.00135	0.0277	1263.2	1511.3	2774.5	3.115	2.707	5.822
290	74.364	0.00137	0.0256	1289.1	1477.1	2766.2	3.159	2.623	5.782
295	80.037	0.00138	0.0235	1317.3	1442.5	2759.8	3.208	2.539	5.717
300	85.812	0.00140	0.0217	1344.0	1404.9	2749.0	3.253	2.451	5.704
305	92.14	0.00142	0.0199	1373.4	1367.7	2741.1	3.303	2.366	5.669
310	98.56	0.00145	0.0183	1401.3	1326.0	2727.3	3.349	2.274	5.623
315	105.61	0.00147	0.0168	1432.1	1285.5	2717.6	3400	2.136	5.586
320	112.74	0.00150	0.0155,	1461.5	1238.6	2700 1	3.448	2.088	5.536
330	128.45	0.00156	0.0130	1525.3	1140.6	2665.9	3.551	1.891	5.442
340	14.'5.86	0.00164	0.0108	1594.2	1027.9	2662.0	3.659	1.676	5.335
350	165.13	0.00174	0.0088	1670.6	893.4	2563.9	3.778	1.433	5.211
360	186.51	0.00189	0.0069	1760.5	720.5	2481.0	3.915	1.138	5.053
370	210.32	0.00221	0.0049	1890.5	441.6	2332.1	4.111	0.686	4.797
374.14	220.9	0.003 15	0.00315	2099.3	0.0	2099.3	4.4298	0	4.4298

Table 2: Properties of Saturated Water and Steam (Pressure Base)
(Values based on datum temperature at 0°C)

Absolute Pressure Bar	Temperature °C	Specific Volume m³/kg		Specific enthalpy kJ/kg			Specific Entropy kJ/kg K			Absolute Pressure kPa
		Saturated liquid (water)	Saturated vapour (steam)	Saturated liquid (water)	Evaporation (latent heat)	Saturated vapour (steam)	Saturated liquid (water)	Evaporation	Saturated vapour (steam)	
P	t	v_f	v_g	h_f	h_{fg}	h_g	s_f	s_{fg}	s_g	p
0.006113	0.01	0.00100	206.136	0.01	2501.3	2501.4	0.000	9.156	9.156	0.6113
0.007	1.89	0.00100	181.225	7.91	2496.9	2504.8	0.029	9.077	9.106	0.7
0.008	3.77	0.00100	159.675	15.81	2492.5	2508.3	0.057	9.000	9.057	0.8
0.009	5.45	0.00100	142.789	22.89	2488.5	25.11.4	0.083	9.931	9.014	0.9
0.010	6.98	0.00100	129.208	29.30	2484.9	2514.2	0.106	8.869	8.976	1.0
0.011	8.37	0.00100	118.042	35.17	2481.6	2516.8	0.127	8.814	8.941	1.1
0.012	9.66	0.00100	108.696	40.58	2478.6	2519.1	0.146	8.763	8.909	1.2
0.013	10.86	0.00100	100.755	45.60	2475.7	2521.3	0.164	8.716	8.880	1.3
0.014	11.98	0.00100	93.922	50.31	2473.1	2523.4	0.180	8.673	8.853	1.4
0.015	13.03	0.00100	87.980	54.71	2470.6	2525.3	0.196	8.632	8.828	1.5
0.016	14.02	0.00100	82.763	58.87	2468.3	2527.1	0.210	8.594	8.804	1.6
0.017	14.95	0.00100	78.146	62.80	2466.0	2528.8	0.224	8.559	8.762	1.8
0.019	16.69	0.00100	70.337	70.10	2461.9	2532.0	0.249	8.493	8.742	1.9
0.020	17.51	0.001001	67.004	73.48	2460.0	2533.5	0.261	8.463	8.724	2.0
0.025	21.09	0.001002	54.340	88.45	2451.6	2540.0	0.32	8.333	8.645	2.5
0.030	24.10	0.001003	45.670	101.00	2444.5	2545.5	0.354	8.224	8.578	3.0
0.035	26.69	0.001003	39.510	111.85	2438.4	2550.3	0.391	8.133	8.523	3.5
0.040	28.98	0.001004	34.803	121.45	24.32.9	2554.4	0.423	8.053	8.476	4.0
0.045	31.03	0.001005	31.155	130.00	2428.2	2558.2	0.451	7.983	8.434	4.5
0.050	32.90	0.001005	28.195	137.80	2423.7	2561.5	0.476	7.920	8.396	5.0

Contd. ....

Table 2: (Contd.)

p	t	v_f	v_g	h_f	h_{fg}	h_g	s_f	s_{fg}	s_g	p
0.055	34.60	0.001006	25.778	144.90	2419.6	2565.5	0.500	7.862	8.362	5.5
0.060	36.18	0.001006	23.741	151.50	2415.9	2567.5	0.521	7.810	8.331	6.0
0.065	37.65	0.001007	22.070	157.65	2412.4	2570.1	0.541	7.762	8.303	6.5
0.070	39.03	0.001007	20.530	163.40	2409.1	2572.5	0.559	7.718	8.277	7.0
0.075	40.32	0.001008	19.242	168.80	2406.0	2574.8	0.576	7.676	8.259	7.5
0.080	41.54	0.001008	18.104	173.90	2403.1	2577.0	0.593	7.637	8.230	8.0
0.085	42.69	0.001009	17.101	178.70	2400.3	2579.0	0.608	7.600	8.208	8.5
0.090	43.79	0.001009	16.203	183.30	2397.7	2581.0	0.622	7.566	8.188	9.0
0.095	44.84	0.001010	15.401	187.70	2395.2	2582.9	0.636	7.533	8.169	9.5
0.10	45.81	0.001010	14.674	191.83	2392.8	2584.7	0.649	7.501	8.150	10
0.11	47.69	0.001010	13.415	199.67	2388.3	2588.0	0.674	7.443	8.117	11
0.12	49.42	0.001010	12.361	206.92	2384.1	2591.1	0.696	7.390	8.086	12
0.13	51.04	0.001010	11.465	213.67	2380.2	2593.9	0.717	7.341	8.058	13
0.14	52.55	0.001010	10.693	219.99	2376.6	2596.6	0.737	7.296	8.032	14
0.15	53.97	0.001010	10.022	225.94	2373.1	2599.1	0.755	7.2545	8.008	15
0.16	55.34	0.001015	9.433	231.6	2369.9	2601.5	0.772	7.214	7.986	16
0.17	56.62	0.001015	8.911	236.9	2366.8	2603.7	0.788	7.177	7.965	17
0.18	57.83	0.001016	8.445	242.0	2363.8	2605.8	0.804	7.141	7.945	18
0.19	58.98	0.001017	8.027	246.8	2361.0	2607.8	0.818	7.108	7.926	19
0.20	60.09	0.001017	7.650	251.4	2558.3	2609.7	0.832	7.077	7.909	20
0.21	61.15	0.001018	7.307	255.9	2355.8	2611.7	0.845	7.047	7.892	21
0.22	62.16	0.001018	6.995	260.1	2353.2	2613.3	0.858	7.018	7.876	22
0.23	63.14	0.001019	6.709	264.2	2351.0	2615.2	0.870	6.991	7.861	23
0.24	64.08	0.001019	6.447	268.1	2348.5	2616.6	0.882	6.964	7.846	24
0.25	64.99	0.001020	6.204	272.0	2346.3	2618.3	0.893	6.939	7.832	25
0.26	65.87	0.001020	5.980	275.6	2344.1	2619.7	0.904	6.914	7.818	26
0.27	66.72	0.001021	5.772	279.2	2342.1	2621.3	0.915	6.891	7.806	27

Contd. ....

Table 2: (Contd.)

P	t	v_f	v_g	h_f	h_{fg}	h_g	s_f	s_{fg}	s_g	P
0.28	67.55	0.001021	5.578	282.6	2340.0	2622.6	0.925	6.868	7.793	28
0.29	68.35	0.001022	5.398	286.0	2338.1	2624.1	0.935	6.847	7.781	29
0.30	69.12	0.001022	5.230	289.2	2336.1	2625.3	0.944	6.825	7.768	30
0.32	70.62	0.001023	4.922	295.5	2332.4	2627.9	0.962	6.784	7.746	32
0.34	72.03	0.001024	4.650	301.4	2328.8	2630.2	0.979	6.746	7.725	34
0.36	73.37	0.001025	4.407	307.0	2325.5	2632.5	0.996	6.710	7.706	36
0.38	74.66	0.001026	4.190	312.4	2322.3	2634.7	1.011	6.676	7.687	38
0.40	75.90	0.001027	3.993	317.6	2319.2	2636.8	1.026	6.644	7.670	40
0.42	77.06	0.001027	3.815	322.6	2316.3	2638.9	1.040	6.614	7.654	42
0.44	78.19	0.001028	3.652	327.3	2313.4	2640.7	1.054	6.584	7.638	44
0.46	79.28	0.001029	3.503	331.9	2310.7	2642.6	1.067	6.556	7.623	46
0.48	80.33	0.001029	3.366	336.3	2308.0	2644.3	1.079	6.530	7.609	48
0.50	81.35	0.001030	3.240	340.6	2305.4	2646.0	1.091	6.504	7.595	50
0.52	82.33	0.001031	3.123	344.7	2302.9	2647.6	1.103	6.478	7.581	52
0.54	83.28	0.001031	3.015	348.7	2300.5	2649.2	1.114	6.455	7.569	54
0.56	84.19	0.001032	2.914	352.5	2298.2	2650.7	1.125	6.431	7.556	56
0.58	85.09	0.001033	2.819	356.3	2295.8	2652.1	1.135	6.409	7.544	58
0.60	85.95	0.001033	2.732	359.9	2293.7	2653.6	1.145	6.388	7.533	60
0.62	86.80	0.001034	2.649	363.5	2291.4	2654.9	1.155	6.367	7.522	62
0.64	87.62	0.001034	2.571	366.9	2289.4	2656.3	1.165	6.346	7.511	64
0.66	88.42	0.001035	2.498	370.3	2287.3	2657.6	1.174	6.326	7.500	66
0.68	89.20	0.001036	2.429	373.C	2285.2	2658.8	1.183	6.307	7.490	68
0.70	89.96	0.001036	2.365	376.8	2283.3	2660.1	1.192	6.288	7.480	70
0.72	90.70	0.001037	2.303	379.9	2281.4	2661.3	1.201	6.270	7.471	72
0.74	91.43	0.001037	2.245	382.9	2279.5	2662.4	1.209	6.253	7.462	74
0.76	92.14	0.001038	2.189	385.9	2277.7	2663.6	1.217	6.235	7.452	76
0.78	92.93	0.001038	2.137	388.9	2275.8	2664.7	1.225	6.219	7.444	78
0.80	93.51	0.001039	2.087	391.7	2274.1	2665.8	1.233	6.202	7.435	79

Contd. ...

Table 2: (Contd.)

P	t	v_f	v_g	h_f	h_{fg}	h_g	s_f	s_{fg}	s_g	p
0.85	95.15	0.001040	1.972	398.6	2269.8	2668.4	1.252	6.163	7.415	85
0.90	96.71	0.001041	1.869	405.2	2265.7	2670.9	1.270	6.125	7.395	90
0.95	98.20	0.001042	1.777	411.4	2261.8	2673.2	1.287	6.090	7.377	95
1.00	99.63	0.001043	1.694	417.5	2258.0	2675.5	1.303	6.057	7.360	100
1.0132	100.0	0.001044	1.673	419.1	2256.9	2676.6	1.307	6.048	7.365	101.32
1.05	101.0	0.001044	1.618	423.2	2254.4	2677.6	1.318	6.025	7.343	105
1.1	102.3	0.001045	1.549	428.8	2250.9	2679.7	1.333	5.994	7.327	0.11 ↓ MPa
1.2	104.8	0.001047	1.428	439.3	2244.2	2683.5	1.361	5.937	7.298	0.12
1.3	107.1	0.001049	1.325	449.1	2238.0	2687.1	1.387	5.884	7.271	0.13
1.4	109.3	0.001051	1.237	458.4	2232.0	2690.4	1.411	5.835	7.246	0.14
1.5	111.4	0.001053	1.159	467.1	2226.5	2693.6	1.434	5.789	7.223	0.15
1.6	113.3	0.001054	1.091	475.4	2221.1	2696.5	1.455	5.747	7.202	0.16
1.7	115.2	0.001056	1.031	483.2	2216.0	2699.2	1.475	5.706	7.181	0.17
1.8	116.9	0.001058	0.977	490.7	2211.2	2701.8	1.494	5.668	7.162	0.18
1.9	118.6	0.001059	0.929	497.3	2206.5	2704.3	1.513	5.631	7.144	0.19
2.0	120.2	0.001061	0.886	504.7	2201.9	2706.7	1.530	5.597	7.127	0.20
2.1	121.8	0.001062	0.846	511.3	2197.6	2708.9	1.547	5.564	7.111	0.21
2.2	123.3	0.001063	0.810	517.6	2193.4	2711.0	1.563	5.532	7.095	0.22
2.3	124.7	0.001065	0.777	523.7	2189.3	2713.1	1.578	5.502	7.080	0.23
2.4	126.1	0.001066	0.747	529.6	2185.4	2715.0	1.593	5.473	7.066	0.24
2.5	127.4	0.001067	0.719	535.4	2181.5	2716.9	1.607	5.446	7.053	0.25
2.6	128.7	0.001069	0.693	540.9	2177.8	2718.7	1.621	5.419	7.040	0.26
2.7	130.0	0.001070	0.669	546.3	2174.2	2720.5	1.634	5.393	7.027	0.27
2.8	131.2	0.001071	0.646	551.4	2170.7	2722.1	1.647	5.368	7.015	0.28
2.9	132.4	0.001072	0.625	556.5	2167.3	2723.8	1.660	5.343	7.003	0.29

Contd. ....

Table 2: (Contd.)

P	t	v_f	v_g	h_f	h_{fg}	h_g	s_f	s_{fg}	s_g	P
3.0	133.5	0.001073	0.606	561.5	2163.8	2725.3	1.672	5.320	6.992	0.30
3.1	134.7	0.001074	0.588	566.3	2160.6	2726.8	1.684	5.297	6.981	0.31
3.2	135.8	0.001075	0.570	571.0	2157.3	2728.3	1.695	5.275	6.970	0.32
3.3	136.8	0.001076	0.554	575.5	2154.2	2729.7	1.706	5.254	6.960	0.33
3.4	137.9	0.001078	0.539	580.0	2151.1	2731.1	1.717	5.233	6.950	0.34
3.5	138.9	0.001079	0.524	584.3	2148.1	2732.4	1.728	5.213	6.941	0.35
3.6	140.0	0.001080	0.514	588.6	2145.1	2733.7	1.738	5.193	6.931	0.36
3.7	140.8	0.001081	0.498	592.8	2142.2	2735.0	1.748	5.174	6.922	0.37
3.8	141.8	0.001082	0.485	596.8	2139.4	2736.2	1.758	5.155	6.913	0.38
3.9	142.7	0.001083	0.474	600.8	2136.6	2737.4	1.767	5.137	6.904	0.39
4.0	143.6	0.001084	0.463	604.7	2133.8	2738.5	1.777	5.119	6.896	0.40
4.1	144.5	0.001085	0.452	608.6	2131.1	2739.7	1.786	5.102	6.888	0.41
4.2	145.4	0.001086	0.442	612.4	2128.4	2740.8	1.795	5.085	6.880	0.42
4.3	146.3	0.001086	0.432	616.1	2125.8	2741.9	1.804	5.068	6.872	0.43
4.4	147.1	0.001087	0.423	619.7	2123.2	2742.9	1.812	5.052	6.864	0.44
4.5	147.9	0.001088	0.414	623.3	2120.6	2743.9	1.821	5.036	6.857	0.45
4.6	148.7	0.001089	0.406	626.8	2118.2	2744.9	1.829	5.020	6.849	0.46
4.7	149.5	0.001090	0.397	630.2	2115.7	2745.9	1.837	5.005	6.842	0.47
4.8	150.3	0.001091	0.390	633.6	2113.2	2746.8	1.845	4.990	6.835	0.48
4.9	151.1	0.001092	0.382	636.9	2110.8	2747.8	1.853	4.975	6.828	0.49
5.0	151.9	0.001093	0.375	640.2	2108.5	2748.7	1.861	4.961	6.821	9.50
5.2	153.3	0.001094	0.361	646.7	2103.8	2750.5	1.876	4.932	6.808	0.52
5.4	154.8	0.001096	0.349	652.9	2099.3	2752.1	1.890	4.905	6.795	0.54
5.6	156.2	0.00L097	0.337	658.9	2094.8	2753.8	1.904	4.879	6.783	0.56
5.8	157.5	0.001099	0.326	664.8	2090.5	2755.3	1.918	4.853	6.771	0.58
6.0	158.9	0.001101	0.316	670.6	2086.3	2756.8	1.931	4.829	6.760	0.60
6.2	160.1	0.001102	0.306	676.2	2082.1	2758.3	1.944	4.805	6.749	0.62

Contd. ...

Table 2: (Contd.)

P	t	v_f	v_g	h_f	h_{fg}	h_g	s_f	s_{fg}	s_g	P
6.4	161.4	0.001104	0.297	681.6	2078.0	2759.6	1.956	4.782	6.738	0.64
6.6	162.6	0.001105	0.288	686.9	2074.0	2761.0	1.969	4.759	6.728	0.66
6.8	163.8	0.001107	0.281	692.1	2070.1	2762.2	1.981	4.737	6.718	0.68
7.0	165.0	0.001108	0.273	697.2	2066.3	2763.5	1.992	4.716	6.708	0.70
7.2	166.1	0.001109	0.266	702.2	2062.5	2764.7	2.004	4.695	6.699	0.72
7.4	167.2	0.001111	0.259	707.1	2058.8	2765.9	2.014	4.675	6.689	0.74
7.6	168.3	0.001112	0.252	711.8	2055.2	2767.0	2.025	4.655	6.680	0.76
7.8	169.4	0.001113	0.246	716.5	2051.6	2768.1	2.036	4.635	6.671	0.78
8.0	170.4	0.001115	0.240	721.1	2048.0	2769.1	2.046	4.617	6.663	0.80
8.2	171.5	0.001116	0.235	725.6	2044.6	2770.2	2.056	4.598	6.654	0.82
8.4	172.5	0.001117	0.230	730.1	2041.1	2771.2	2.066	4.580	6.646	0.84
8.6	173.5	0.001119	0.225	734.4	2037.7	2772.1	2.076	4.562	6.638	0.86
8.8	174.4	0.001120	0.220	738.6	2034.4	2773.0	2.085	4.545	6.630	0.88
9.0	175.4	0.001121	0.215	742.8	2031.1	2773.9	2.095	4.528	6.623	0.90
9.2	176.3	0.001122	0.211	747.0	2027.8	2774.8	2.104	4.511	6.615	0.92
9.4	177.2	0.001124	0.206	751.0	2024.7	2775.7	2.113	4.495	6.608	0.94
9.6	178.1	0.001125	0.202	755.0	2021.5	2776.5	2.122	4.479	6.601	0.96
9.8	179.0	0.001126	0.198	758.9	2018.4	2777.3	2.130	4.463	6.593	0.98
10.0	179.9	0.001127	0.194	762.8	2015.3	2778.1	2.139	4.448	6.587	1.00
10.5	182.0	0.001130	0.186	772.2	2007.7	2779.9	2.159	4.411	6.570	1.05
11.0	184.1	0.001133	0.178	781.3	2000.4	2781.7	2.179	4.374	6.553	1.10
11.5	186.1	0.001136	0.170	790.1	1993.2	2783.3	2.198	4.340	6.538	1.15
12.0	188.0	0.001139	0.163	798.6	1986.2	2784.8	2.217	4.306	6.533	1.20
12.5	189.8	0.001141	0.157	806.8	1979.4	2786.2	2.234	4.275	6.509	1.25
13.0	191.6	0.001144	0.151	814.9	1972.7	2787.6	2.251	4.244	6.495	1.30
13.5	193.4	0.001146	0.146	822.6	1966.2	2788.8	2.268	4.214	6.482	1.35
14.0	195.0	0.001149	0.141	830.3	1959.7	2790.0	2.284	4.185	6.469	1.40

Contd. ....

Table 2: (Contd.)

P	t	v_f	v_g	h_f	h_{fg}	h_g	s_f	s_{fg}	s_g	P
14.5	196.7	0.001151	0.136	837.6	1953.5	2791.1	2.300	4.157	6.457	1.45
15.0	198.3	0.001154	0.132	844.9	1947.3	2792.2	2.315	4.130	6.445	1.50
15.5	199.9	0.001156	0.128	851.9	1941.2	2793.1	2.330	4.103	6.433	1.55
16.0	201.4	0.001159	0.124	858.8	1935.2	2794.0	2.344	4.078	6.422	1.60
16.5	202.9	0.001161	0.120	865.5	1929.4	2794.9	2.358	4.053	6.411	1.65
17.0	204.3	0.001163	0.117	872.1	1923.6	2795.7	2.372	4.028	6.400	1.70
17.5	205.8	0.001166	0.113	878.5	1917.9	2796.4	2.385	4.005	6.390	1.75
18.0	207.2	0.001168	0.110	884.8	1912.4	2797.2	2.398	3.981	6.379	1.80
18.5	208.5	0.001170	0.108	891.0	1906.8	2797.8	2.411	3.958	6.369	1.85
19.0	209.8	0.001172	0.105	897.0	1901.4	2798.4	2.423	3.936	6.359	1.90
19.5	211.1	0.001175	0.102	903.0	1896.0	2799.0	2.435	3.915	6.350	1.95
20.0	212.4	0.001177	0.996	908.8	1890.7	2799.5	2.447	3.894	6.341	2.00
21.0	214.9	0.001181	0.0950	920.2	1880.3	2800.5	2.470	3.852	6.323	2.10
22.0	217.3	0.001185	0.0907	931.1	1870.2	2801.3	2.493	3.813	6.306	2.20
23.0	219.6	0.001189	0.0869	941.8	1860.2	2802.0	2.514	3.775	6.289	2.30
24.0	221.8	0.001193	0.0833	952.1	1850.5	2802.6	2.535	3.738	6.73	2.40
25.0	224.0	0.001197	0.0800	962.1	1841.0	2803.1	2.555	3.703	6.258	2.50
26.0	226.1	0.0012u1	0.0769	971.9	1831.6	2803.5	2.574	3.669	6.243	2.60
27.0	:128.1	0.001205	0.0741	981.3	1822.4	2803.8	2.593	3.635	6.228	2.70
28.0	230.1	0.001209	0.0715	990.6	1813.4	2804.0	2.611	3.603	6.214	2.80
29.0	232.0	0.001213	0.0690	996.6	1804.5	2804.1	2.628	3.572	6.200	2.90
30.0	233.9	0.001217	0.0667	1008.5	1795.7	2804.2	2.646	3.541	6.187	3.00
31.0	235.7	0.001220	0.0645	1017.0	1787.1	2804.1	2.662	3.512	6.174	3.10
32.0	237.5	0.001224	0.0625	1025.5	1778.6	2804.1	2.679	3.483	6.161	3.20
33.0	239.2	0.001227	0.0606	1033.7	1770.2	2803.9	2.695	3.454	6.149	3.30
34.0	240.9	0.001231	0.0588	1041.8	1761.9	2803.7	2.710	3.427	6.137	3.40
35.0	242.6	0.001235	0.0571	1049.7	1753.7	2803.4	2.725	3.400	6.125	3.50
36.0	244.2	0.001238	0.0555	1057.5	1745.6	2803.1	2.740	3.374	6.114	3.60

Contd. ...

Table 2: (Contd.)

p	t	v_f	v_g	h_f	h_{fg}	h_g	s_f	s_{fg}	s_g	p
37.0	245.8	0.001242	0.0539	1065.2	1737.6	2802.8	2.755	3.348	6.103	3.70
38.0	247.4	0.001245	0.0523	1072.7	1729.7	2802.4	2.769	3.323	6.092	3.80
39.0	248.9	0.001249	0.0511	1080.1	1721.8	2801.9	2.783	3.298	6.081	3.90
40.0	250.4	0.001252	0.0498	1087.3	1714.1	2801.4	2.796	3.274	6.070	4.00
42.0	253.3	0.001259	0.0473	1101.5	1698.8	2800.3	2.823	3.227	6.050	4.20
44.0	256.0	0.001266	0.0451	1115.2	1683.8	2799.0	2.849	3.181	6.030	4.40
46.0	258.8	0.001273	0.0431	1128.6	1669.0	2797.6	2.873	3.137	6.010	4.60
48.0	261.4	0.001279	0.0412	1141.5	1654.5	2796.0	2.897	3.095	5.992	4.80
50.0	264.0	0.001286	0.0394	1154.2	1640.1	2794.3	2.920	3.053	5.973	5.00
52.0	266.4	0.001293	0.0378	1166.9	1625.7	2792.6	2.943	3.013	5.956	5.20
54.0	268.8	0.001299	0.0363	1179.0	1611.8	2790.8	2.965	2.974	5.939	5.40
56.0	271.1	0.001306	0.0349	1190.8	1598.2	2789.0	2.986	2.937	5.923	5.60
58.0	273.4	0.001312	0.0336	1202.4	1584.6	2787.0	3.007	2.899	5.906	5.80
60.0	275.6	0.001319	0.0324	1213.3	1571.0	2784.3	3.027	2.862	5.889	6.00
62.0	277.7	0.001325	0.0313	1224.9	1558.0	2782.9	3.047	2.828	5.875	6.20
64.0	279.8	0.001332	0.0302	1235.8	1544.8	2780.6	3.066	2.794	5.860	6.40
66.0	281.9	0.001338	0.0292	1246.5	1531.8	2778.3	3.066	2.760	5.845	6.60
68.0	283.9	0.001345	0.0283	1257.1	1518.8	2775.9	3.104	2.727	5.831	6.80
70.0	285.9	0.001351	0.0274	1267.0	1505.8	2772.1	3.121	2.695	5.813	7.00
72.0	287.7	0.001358	0.0265	1277.7	1493.2	2770.9	3.140	2.662	5.802	7.20
74.0	289.6	0.001365	0.0257	1287.8	1480.4	2768.2	3.157	2.631	5.788	7.40
76.0	291.4	0.001371	0.0249	1297.7	1467.8	2765.5	3.174	2.600	5.774	7.60
78.0	293.2	0.001378	0.0242	1307.5	1455.4	2762.7	3.191	2.569	5.760	7.80
80.0	295.1	0.001384	0.0235	1316.6	1441.3	2758.0	3.207	2.536	5.743	8.00
82.0	296.7	0.001391	0.0228	1326.7	1430.3	2757.0	3.224	2.510	5.734	8.20
84.0	298.4	0.001398	0.0222	1336.2	1417.8	2754.0	3.240	2.481	5.721	8.40
86.0	300.1	0.001404	0.0216	1345.4	1405.5	2750.9	3.256	2.452	5.708	8.60

Contd. ...

Table 2: (Contd.)

P	t	v_f	v_g	h_f	h_{fg}	h_g	s_f	s_{fg}	s_g	P
88.0	301.7	0.001411	0.0210	1354.7	1393.1	2747.8	3.271	2.424	5.695	8.80
90.0	303.4	0.001418	0.0205	1363.2	1378.9	2742.1	3.286	2.391	5.677	9.00
92.0	304.9	0.001425	0.0199	1372.8	1368.5	2741.3	3.302	2.367	5.669	9.20
94.0	306.5	0.001432	0.0194	1381.7	1356.3	2738.0	3.317	2.340	5.657	9.40
96.0	308.0	0.001439	0.0189	1390.6	1344.1	2734.7	3.332	2.313	5.644	9.60
98.0	309.5	0.001446	0.0185	1399.4	1331.9	2731.2	3.346	2.286	5.632	9.80
100	311.1	0.001452	0.0180	1407.6	1317.1	2724.7	3.360	2.254	5.614	10.00
105	314.7	0.001470	0.0170	1429.0	1286.4	2715.4	3.395	2.188	5.583	10.50
110	318.2	0.001489	0.0160	1450.1	1255.5	2705.6	3.430	2.123	5.553	11.00
115	321.5	0.001507	0.0151	1470.8	1224.6	2695.4	3.453	2.059	5.522	11.50
120	324.8	0.001527	0.0143	1491.3	1193.6	2684.9	3.496	1.996	5.492	12.00
125	327.9	0.001547	0.0135	1511.1	1162.2	2673.7	3.528	1.934	5.462	12.50
130	330.9	0.001567	0.0128	1531.5	1130.7	2662.2	3.561	1.817	5.432	13.00
135	333.9	0.001588	0.0121	1551.4	1098.8	2650.2	3.592	1.810	5.402	13.50
140	336.8	0.001611	0.0115	1571.1	1066.5	2637.6	3.623	1.749	5.372	14.00
145	339.5	0.001634	0.0109	1590.9	1033.5	2624.4	3.654	1.687	5.341	14.50
150	342.2	0.001658	0.0103	1610.5	1000.0	2610.5	3.685	1.625	5.310	15.00
155	344.9	0.001684	0.00981	1630.3	965.7	2596.0	3.715	1.563	5.278	15.50
160	347.4	0.001711	0.00931	1650.1	930.6	2580.6	3.746	1.499	5.245	16.00
165	349.9	0.001740	0.00883	1670.1	894.3	2564.4	3.777	1.435	5.212	16.50
170	352.4	0.001770	0.00836	1690.3	856.9	2547.2	3.808	1.370	5.178	17.00
175	354.7	0.001804	0.00793	1711.0	817.8	2528.8	3.839	1.302	5.141	17.50
180	357.1	0.001840	0.00749	1732.0	777.1	2509.1	3.871	1.233	5.104	18.00
185	359.3	0.001880	0.00708	1753.9	733.9	2487.8	3.905	1.160	5.065	18.50
190	361.5	0.001924	0.00666	1776.5	688.0	2464.5	3.939	1.084	5.023	19.00
195	363.7	0.001976	0.00625	1800.6	638.2	2438.8	3.975	1.002	4.977	19.50
200	365.8	0.002036	0.00583	1826.3	583.4	2409.7	4.014	0.913	4.927	20.00

Contd. ...

Table 2: (Contd.)

P	t	v_f	v_g	h_f	h_{fg}	h_g	s_f	s_{fg}	s_g	P
205	367.9	0.002110	0.00541	1855.0	520.8	2375.8	4.057	0.812	4.869	20.50
210	369.9	0.002207	0.00495	1888.4	446.2	2334.6	4.107	0.694	4.801	21.00
215	371.9	0.002358	0.00442	1933.0	344.9	2277.9	4.175	0.535	4.710	21.50
220	373.8	0.002742	0.00357	2022.2	143.4	2165.6	4.311	0.222	4.533	22.00
221.2	374.15	0.003170	0.00317	2107.4	0.0	2107.4	4.443	0.0	4.443	22.12

Table 3: Thermodynamic Properties of Superheated Steam (p in bar, t in .C, v in m^3/kg, h in kJ/kg and S in kJ/kg K)

Absolute pressure, (t_s)	v_g, h_g, s_g	v, h, s	\multicolumn{11}{c}{Temperature of steam t in .C}
			50	100	150	200	250	300	350	400	450	500
$p = 0.01$ bar = 1 kPa = 0.001 MPa ($t_s = 6.98.$C)	$v_g = 129.21$ $h_g = 2514.2$ $s_g = 8.976$	v h s	149.1 2594.5 9.242	172.2 2688.6 9-513	1.95.3 2783.6 9.752	218.4 2880.0 9.967	241.5 2978.7 10.163	264.5 3076.8 10.344	287.6 3178.2 10.507	310.7 3279.7 10.671	333.7 3384.4 10.815	356.8 3489.2 10.960
$p = 0.02$ bar = 2 kPa = 0.002 MPa ($t_s = 17.51.$C)	$v_g = 67.00$ $h_g = 2533.5$ $s_g = 8.724$	v h s	74.49 2594.1 9.201	86.08 2688.5 9.193	97.63 2783.7 9.433	109.2 2880.0 9.648	128.7 2977.7 9.844	132.2 3076.8 10.025	143.8 3178.3 10.188	155.3 3279.8 10.351	166.8 3384.5 10.496	178.4 3489.2 10.641
$p = 0.04$ bar = 4 kPa. = 0.004 MPa ($t_s = 28.98.$C)	$v_g = 34{:}80$ $h_g = 2544.4$ $s_g = 8.476$	v h s	37.22 2593.8 8.952	43.03 2688.3 8.873	48.81 2783.5 90.113	54.58 2879.9 9.328	60.35 2977.6 9.524	66.12 3076.8 9.705	71.89 3178.2 9.868	77.66 3279.7 9.705	'83.43' 3384.4 10.176	89.20 3489.2 10.031
$p = 0.06$ bar = 6 kPa. = 0.006 MPa	$v_g = 23.74$ $h_g = 2567.4$ $s_g = 8.331$	v h s	24.83 2593.2 8.614	26.68 2688.0 8.685	32.53 2783.4 8.925	36.38 2879.8 9.141	40.23 2977.6 9.337	44.08 3076.7 9.518	47.93 3178.1 9.681	51.77 3279.6 9.844	55.62 3384.4 9.989	59.47 3489.2 10.134
$p = $ O.O bar = 8 kPa = 0.008 MPa ($l = 41.54$°C)	$v_g = 18.10$ $h_g = 2577.0$ $s_g = 8.230$	v h s	18.59 2592.9 8.352	21.50 2687.8 8.552	24.40 2783.2 8.792	27.28 2879.7 9.00S	30.17 2977.5 9.204	33.06 S076.7 9.385	35.94 3178.1 9.548	38.83 3279.6 9.711	41.72 3384.3 9.856	44.60 3489.1 10.001
$p = 0.1$ bar = 10 kPa = O.01MPa ($l. = 45.81$°C)	$v_g = 14.57$ $h_g = 584.7$ $s_g = 8.150$	v h s	14.87 2592.6 8.175	17.20 2687.5 8.448	19.51 2783.0 8.688	21.83 2879.5 8.904	24.14 2977.3 9.100	26.45 3076.6 9.281	28.75 3178.1 9.444	31.06 3279.6 9.608	33.37 3384.3 9.753	35.68 3489.1 9.898
$p = 0.15$ bar	$v_g = 10.02$	v	100 11.51	150 13.06	200 14.61	250 16.16	300 17.71	400 20.80	500 23.89	600 26.98	700 30.07	800 33.16

Contd. ....

Table 3: (Contd.)

(t_s)	v_g, h_g, s_g	v, h, s	50	100	150	200	250	300	350	400	450	500
= 15 kPa = 0.015 MPa (t = 53.97°C)	v_g = 2599.1 h_g = 8.009 s_g =	h s	2686.9 8.261	2782.4 8.502	2879.5 8.718	2977.3 8.915	3076.5 9.096	3279.5 9.422	3489.1 9.712	3705.5 9.975	3928.7 10.217	4158.7 10.442
P = 0.2 bar = 20 kPa = 0.02 MPa (t = GO.rC)	v_g = 7.65 h_g = 2609.7 s_g = 7.909	v h s	8.858 2686.3 8.126	9.748 2782.3 8.368	10.91 2879.2 8.584	12.07 2977.1 8.781	13.22 3076:4 8.962	15.53 3279.4 9.288	17.84 3489.0 9.578	20.15 3705.4 9.842	22.45 3928.7 10.084	24.76 4158.7 10.309
p = 0.25 bar = 25 kPa z = 0.025 MPa	v_g = 6.20 h_g = 2618.3 s_g = 7.832	v h s	6.874 2685.7 8.022	7.808 2782.0 8.264	8.737 2879.0 8.481	9,665 2977.0 8.678	10.59 3076.3 8.859	12.44 3279.3 9.186	14.29 3489.0 9.476	16.14 3705.4 9.379	17.99 3928.7 9.981	19.84 4158.6 10.206
p = 0.30 bar = 30kPa. = 0.030 MPa (l. 69.rC)	v_g = 5.23 h_g = 2625.3 s_g = 7.768	v h s	5.714 2685.1 7.936	6.493 2781.6 8.179	7.268 2878.7 8.396,	8.040 2976.8 8.593	8.811 3076.1 8.774	10.35 3279.3 9.101	11.89 3488:9 9.391	13.43 3705.4 9.654	14.70 3928.7 9.897	16.51 4158.6 10.121
p = 0.35 bar = 35 kPa = 0.035 MPa (t. = 72.7°C)	v_g = 4.53 h_g = 2631.4 s_g = 7.715	v h s	4.898 2684.5 7.864	5.568 2781.2 8.107	6.233 2878.5 8.325	6.896 2976.7 8.522	7.557 3076.0 8.703	8.88 3279.3 9.030	10.20 3488.9 9.320	11.52 3705.3 9.583	12.84 3928.7 9.826	14.16 4158.6 10.050
P = 0.40 bar = 40 kPa = 0.04 MPa (t. = 75.9°C)	v_g = 3.99 h_g = 2636.8 s_g = 7.670	v h s	4.279 2683.8 7.801	4.866 2780.9 8.045	5.45 2878.2 8.263	6.03 2976.5 8.460	6.61 3075.9 8.641	7.76 3279.1 8.968	8.92 3488.8 9.258	10.07 3705.3 9.522	11.23 3928.6 9.764	12.38 4158.6 9.989
p = 0.45 bar = 45kPa = 0.045 MPa (t_s = 78.7°C)	v_g = 3.58 h_g = 2641.6 s_g = 7.631	v h s	3.803 2683. 7.745	4.325 2780.5 7.990	4.844 2878.0 8.208	5.360 2976.3 8.405	5.875 3075.8 8.587	6.903 3279.1 8.914	7.930 3488.8 9.204	8.957 3705.2 9.467	9.984 3928.6 9.709	10.99 4158.5 9.934
P = 0.50 bar	v_g = 3.240	v	3.418	3.889	4.356	4.821	5.284	6.209	7.134	8.057	8.981	9.904

Contd. ...

Table 3: (Contd.)

(t_s)	v_g, h_g, s_g	v, h, s	50	100	150	200	250	300	350	400	450	500
=50kPa = 0.05 MPa	$h_g = 2645.9$ $s_g = 7.594$	h s	2682.5 7.695	2780.1 7.940	2877.7 8.158	2976.0 8.356	3075.5 8.537	3278.9 8.864	3488.7 9.155	3705.2 9.419	3928.6 9.661	4158.5 9.886
p = 0.60 bar = 60 kPa = 0.060 MPa	$v_g = 2.732$ $h_g = 2653.6$ $s_g = 7.533$	v h s	2.844 2681.3 7.609	3.238 2779.4 7.885	3.628 2877.3 8.074	4.016 2975.8 3.272	4.402 3075.4 3.454	5.174 3278.8 8.781	5.944 3488.6 9.071	6.714 3705.1 9.334	7.484 3928.5 9.576	8.254 4158.5 9.801
(t_s = 86.0°C) p = 0.70 bar = 70 kPa = 0.070 MPa	$v_g = 2.365$ $h_g = 2660.1$ $s_g = 7.480$	v h s	2.434 2680.1 7.535	2.773 2778.6 7.783	3.108 2876.8 8.002	3.441 2975.5 8.200	3.772 3075.2 8.382	4.434 3278.7 8.709	5.095 3488.5 9.000	5.755 3705.0 9.263	6.415 3928.4 9.505	7.074 4158.4 9.732
(t_s = 90.0°C) p = 0.80 bar = 80kPa = 0.080 MPa	$v_g = 2.087$ $h_g = 2665.8$ $s_g = 7.435$	v h s	2.126 2678.8 7.470	2.425 2777.8 7.720	2.718 2876.3 7.940	3.010 2975.2 8.138	3.300 3075.0 8.320	3.879 3278.5 8.648	4.457 3488.4 8.938	5.035 3705.0 9.214	5.613 3928.4 9.444	6.190 4158.4 9.669
(t_s = 93.5°C) p = 0.90 bar = 90kPa = 0.090 MPa	$v_g = 1.869$ $h_g = 2670.9$ $s_g = 7.395$	v h s	1.887 2677.5 7.413	2.153 2777.1 7.664	2.415 2875.8 7.884	2.674 2974.8 8.083	2.933 3074.7 8.266	3.448 3278.4 8.593	3.962 3488.3 8.884	4.475 3704.9 9.147	4.989 3928.3 9.389	5.502 4158.3 9.614
(t_s = 96.7°C) P = 1.00 bar = 100 kPa = 0.1 MPa	$v_g = 1.694$ $h_g = 2675.5$ $s_g = 7.360$	v h s	1.696 2676.2 7.361	1.936 2776.4 7.613	2.172 2875.3 7.834	2.406 2974.3 8.033	2.639 3074.3 8.216	3.103 3278.2 8.544	3.565 3488.1 8.834	4.028 3704.8 9.098	4.490 3928.2 9.341	4.952 4158.3 9.565
(t_s = 99.6°C) p = 1.5 bar = 150 kPa = 0.15 MPa (t_s = 111.4°C)	$v_g = 1.159$ $h_g = 2693.6$ $s_g = 7.223$	v h s	- - -	1,285 2772.6 7.419	1.143 2872.9 7.643	1.607 2972.7 7.844	1.757 3073.1 8.027	2.067 3277.4 8.356	2.376 3487.6 8.647	2.685 3704.4 .8.911	2.993 3927.9 9.152	3.301 4158.0 9.377

Contd. ...

Table 3: (Contd.)

(t_s)	v_g, h_g, s_g	v, h, s	50	100	150	200	250	300	350	400	450	500
P = 2.0 bar = 200 kPa = 0.2 MPa (t_s = 120.2°C)	v_g = 0.886 h_g = 2706.7 s_g = 7.127	v h s	– – –	0.960 2768.3 7.279	1.080 2870.5 7.507	1.199 2971.0 7.709	1.316 3071.8 7.893	1.549 3276.6 8.222	1.781 3487.1 8.513	2.013 3704.0 8.778	2.244 3927.6 9.021	2.475 4157.8 9.245
P = 2.5 bar = 250 kPa = 0.25 MPa (t_s = 127.4°C)	v_g = 0.719 h_g = 2716.9 s_g = 7.053	v h s	– – –	0.764 2764.5 7.169	0.862 2868.0 7.401	0.957 2969.6 7.604	1.052 3070.9 7.789	1.238 3275.9 8.119	1.424 3486.5 8.410	1.610 3703.6 8.674	1.794 3927.2 8.916	1.981 4157.6 9.141
P = 3.0 bar = 300 kPa = 0.30 MPa (t_s = 133.5°C)	v_g = 0.606 h_g = 2725.3 s_g = 6.992	v h s	– – –	0.634 2761.0 7.078	0.716 2865.6 7.311	0.796 2967.6 7.517	0.875 3069.3 7.702	1.031 3275.0 8.033	1.187 3486.1 8.325	1.341 3703.2 8.590	1.496 3927.0 8.833	1.650 4157.3 9.058
P = 4.0 bar = 400 kPa = 0.40 MPa (t_s = 143.6°C)	v_g = 0.463 h_g = 2738.5 s_g = 6.896	v h s	– – –	0.471 2752.8 6.930	0.534 2860.5 7.171	0.595 2964.2 7.379	0.655 3066.8 7.566	0.773 3273.4 7.899	0.889 3484.9 8.191	1.005 3702.2 8.455	1.121 3926.4 8.699	1.237 4156.8 8.926
p = 5.0 bar = 500kPa = 0.50 MPa (is = 151.9°C)	v_g = 0.3749 h_g = 2748.7 s_g = 6.821	v h s	0.4249 2855.4 I 7.059	0.4744 2960.7 7.271	0.5226 3064.7 7.460	0.5701 3167.7 7.633	0.6173 3271.9 7.794	0.6642 3377.2 7.945	0.7109 3483.9 8.087	0.8041 3701.7 8.353	0.8969 3925.8 8.596	0.9896 4156.3 8.821
P = 6.0 bar ; = 600 kPa = 0.60 MPa (is = 158.9°C)	v_g :: 0.3157 h_g = 2726.8 s_g = 6.760	v h s	0.3520 2850.1 6.967	0.3938 2957.2 7.182	0.4344 3061.6 7.372	0.47421 3165.7 7.546	0.5137 3270.3 7.708	0.5530 3376.0 7.859	0.5920 3482.8 8.002	0.66971 3700.9 8.267	0.7471 3925.1 8.510	0.8244 4155.8 8.736
P = 7.0 bar = 700 kPa = 0.70MPa (is = 165.0°C)	v_g = 0.2729 h_g = 2763.5 s_g = 6.708	v h s	0.2999 2844.8 6.886	0.3363 2953.6 7.105	0.3714 3059.1 7.298	0.4058 3163.7 7.473	0.43971 3268.7 7.635	0.47351 3374.7 7.787	0.5070 3481.8 7.930	0.57381 3700.2 8.196	0.64021 3924.4 8.440	0.7066 4155.5 8.665

Contd. ....

Table 3: (Contd.)

(t_s)	v_g, h_g, s_g	v, h, s	50	100	150	200	250	300	350	400	450	500
P = 8.0 bar	v_g = 0.2404	v	0.2608	0.2931	0.3241	0.3544	0.3843	0.4139	0.44331	0.50181	0.5600	0.6181
= 800 kPa	h_g = 2769.1	h	2839.3	2950.1	3056.5	3161.7	3267.1	3373.4	3480.6	3699.4	3923.8	4155.0
= 0.80 MPa	s_g = 6.663	s	6.816	7.038	7.233	7.409	7.572	7.724	7.867	8.133	8.376	8.604
p = 9.0	v_g = 0.2150	v	0.2303	0.2696	0.2874	0.3144	0.3411	0.3676	0.3938	0.4459	0.4976	0.5492
= 900 kPa	h_g = 2773.9	h	2833.6	2946.3	3053.8	3159.7	3265.5	3372.1	3479.6	3698.6	3923.4	4154.5
= 0.90 MPa	s_g = 6.623	s	6.752	6.979	7.175	7.352	7.516	7.668	7.812	8.078	8.322	8.549
(t_s = 175.4°C)												
p = 10.0	v_g = 0.1944	v	0.2060	0.2327	0.2579	0.2825	0.3066	0.3304	0.3541	0.4011	0.4477	0.4943
= 1000 kPa	h_g = 2778.1	h	2828.9	2942.6	3051.2	3157.8	3263.9	3370.7	3478.5	3697.9	3922.8	4154.2
= 1 MPa	s_g = 6.587	s	6.694	6.925	7.123	7.301	7.465	7.618	7.762	8.029	8.273	8.500
(t_s = 179.9°C)												
P = 11.0	v_g = 0.1775	v	0.1860	0.2107	0.2339	0.2563	0.2783	0.3001	0.3217	0.3645	0.4068	0.4492
= 1100 kPa	h_g = 2781.7	h	2822.0	2938.8	3048.5	3155.7	3262.3	3369.4	3477.2	3697.C	3922.0	4153.6
= 1.1 MPa	s_g = 6.553	s	6.640	6.875	7.075	7.255	7419	7.573	7.717	7.984	8.228	8.456'
(t_s = 184.rC)												
p = 12.0	v_g = 0.1633	v	0.1693	0.1923	0.2138	0.2345	0.2548	0.2748	0.2946	0.3339	0.3729	0.4118
= 1200 kPa	h_g = 2784.8	h	2815.9	2935.0	3045.8	3153.6	3260.7	3368.2	2476.3	3296.3	3921.4	4153.1
= 1.2 MPa	s_g = 6.523	s	6.590	6.829	7.032	7.212	7.377	7.531	7.676	7.944	8.187	8.415
(t_s = 188.0°C)												
p = 13.0	v_g = 0.1513	v	0.1552	0.1768	0.1968	0.2161	0.2349	0.2534	0.2718	0.3081	0.3442	0.3800
= 1300 kPa	h_g = 2787.6	h	2809.7	2931.2	3043.1	3151.6	3259.1	3366.8	3475.2	3695.6	3920.7	4152.6
= 1.3MPa	s_g = 6.495	s	6.543	6.787	6.991	7.173	7.339	7.493'	7.638	7.906	8.151	8.377
(t_s = 191.6°C)												
p = 14.0	v_g = 0.1408	v	0.1430	0.1635	0.1823	0.2003	0.2178	0.2351	0.2521	0.2860	0.3194	0.3527
= 1400kPa	h_g = 2790.0	h	2803.3	2927.2	3040.4	3149.5	3257.5	3365.5	3474.1	3694.8	3920.1	4152.2
= 1.4 MPa	s_g = 6.469	s	6.498	6.747	6.953	7.136	7.303	7.457	7.603	7.871	8.115	8.343
p = 15.0'	v_g = 0.132	v	0.1324	0.1520	0.1697	0.1865	0.2029	0.2190	0.23501	0.2667	0.2980	0.3292

Contd. ...

Table 3: (Contd.)

(t_s)	v_g, h_g, s_g	v, h, s	50	100	150	200	250	300	350	400	450	500
= 1500 kPa = 1.5 MP $(ts = 198.3°C)$	$h_g = 272.2$ $s_g = 6.445$	h s	2796.8 6.455	2923.3 6.709	3037.6 6.918	3147.5 7.102	3255.8 7.269	3364.2 7.424	3473.1 7.507	3694.0 7.839	3919.6 8.084	4151.7 8.310
P = 16.0 = 1600 kPa = 1.6 MPa $(ts = 201.4°C)$	$v_g = 0.1238$ $h_g = 2794.0$ $s_g = 6.422$	v h s	– – –	0.1418 2919.2 6.673	0.1586 3034.8 6.884	0.1746 3145.4 7.069	0.1901 3254.2 7.237	0.2053 3362.9 7.393	0.2203 3472.0 7.539	0.250 3693.2 7.800	0.2793 3918.8 8.054	0.3085 4151.3 8.232
P = 17.0 = 1700 kPa = 1.7 MPa $(ts = 204.3°C)$	$v_g = 0.1167$ $h_g = 2795.7$ $s_g = 6.400$	250 v h s	300 0.132 2915.1 6.639	350 0.1489 3032.1 6.853	400 0.1640 3143.3 7.039	450 0.1786 3252.6 7.208	500 0.1930 3361.5 7.364	550 0.2072 3470.9 7.510	600 0.2212 3581.7 7.644	700 0.2352 3692.5 7.779	800 0.2628 3918.1 8.024	0.2904 4150.6 8.253
P = 18.0 = 1800 kPa = 1.S MPa $(t_s = 207.2°C)$	$v_g = 0.1104$ $h_g = 2792.2$ $s_g = 6.379$	v h s	0.1250 2911.0 6.607	0.1402 3029.2 6.823	0.1546 3141.2 7.010	0.1685 3250.9 7.179	0.1810 3360.2 7.336	0.1955. 3469.8 7.483	0.2087 3580.7 7.617	0.2220 3691.7 7.752	0.2481 3917.7 7.997	0.2741 4150.3 8.225
P = 19.0 = 1900 kPa = 1.9MPa $(t_s = 209.S0C)$	$v_g = 0.1048$ $h_g = 2798.4$ $s_g = 6.359$	v h s	0.1179 29.06.8 6.575	0.1324 3026.4 6.794	0.1462 3139.1 6.983	0.1594 3249.3 7.153	0.1723 3358.9 7.309	0.1851 3468.7 7.457	0.1976 3579.8 7.592	0.2102 3690.9 7.727	0.2350 3917.0 7.972	0.2596 4149.6 8.200
p = 20.0 = 20001d>a = 2.0 MPa $(t_s = 212.4°C)$	$v_g = 0.0996$ $h_g = 2799.5$ $s_g = 6.341$	v h s	0.1110 2902.5 6.545	0.1252 3023.5 6.766	0.1391 3137.0 6.956	0.1513 3247.6 7.127	0.1632 3357.5 7.285	0.1760 3467.6 7.432	0.1878 3578.8 7.567	0.1996 3690.1 7.702	0.2232 3916.3 7.948	0.2467 4149.1 8.176
p = 21.0 = 2100 kPa = 2.1 MPa $(ts = 214.9°C)$	$v_g = 0.0950$ $h_g = 2800.5$ $s_g = 6.323$	v h s	0.10561 2898.1 6.5161	0.1191 3020.6 6.740	0.1323 3135.1 6.920	0.1438 3246.0 7.103	0.1556 3356.2 7.201	0.1672 3466.5 7.408	0.1786 3577.9 7.544	0.1900 3689.4 7.680	0.2125 3916.8 7.926	0.2354 4148.6 8.154

Contd. ....

Table 3: (Contd.)

(t_s)	v_g, h_g, s_g	v, h, s	50	100	150	200	250	300	350	400	450	500
$P = 22.0$ $= 2200$ kPa $= 2.2$ MPa ($t_s = 217.3°C$)	$v_g = 0.09073$ $h_g = 2801.3$ $s_g = 6.306$	v h s	0.1004 2893.7 6.400	0.1134 3017.7 6.75	0.1255 3132.7 6.907	0.1371 3244.3 7.079	0.1484 3354.8 7.238	0.1595 3465.4 7.386	0.1704 3576.8 7.530	0.1813 3688.2 7.675	0.2028 3915.1 7.904	0.2242 4148.2 8.132
$p = 23.0$ $= 2300$ kPa $= 2.3$ MPa ($t_s = 219.6°C$)	$v_g = 0.0868$ $h_g = 2802.0$ $s_g = 6.289$	v h s	0.0955 2889.2 6.461	0.1080 3014.8 6.690	0.1198 3132.2 6.884	0.1309 3242.7 7.057	0.1417 3353.5 7.216	0.1524 3464.3 7.364	0.1628 3576.0 7.501	0.1733 3687.8 7.636	0.1939 3915.7 7.883	0.2148 4147.7 8.111
$p = 24.0$ $= 2400$ kPa $= 2.4$ MPa ($t_s = 221.8°C$)	$v_g = 0.08327$ $h_g = 2802.6$ $s_g = 6.273$	v h s	0.09109 2884.7 6.434	0.1033 3011.8 6.667'	0.1146 3128.4 6.862	0.1253 3241.0 '7.036	0.1357 3352.2 7.196	0.1459 3463.2 7.343	0.1559 3575.1 7.479	0.1660 3687.0 7.616	0.1858 3913.7 7.862	0.2054 4147.2 8.091
$P = 25.0$ $= 2500$ kPa $= 2.5$ MFa ($t_s = 224.0.$C)	$v_g = 0.080$ $h_g = 2803.8$ $s_g = 6.258$	v h s	0.0870 2880.1 6.408	0.0989 3008.8 6.644	0.1098 3126.3 6.840	0.1201 3239.3 7.015	0.1300 3350.8 7.175	0.1400 3462.1 7.323	0.1496 3574.2 7.454	0.1593 3686.3 7.596	0.1783 3914.5 7.844	0.1974 4146.7 8.071
$p = 26.0$ $= 2600$ kPa $= 2.6$ MPa ($t_s = 226.1$ °C)	$v_g = 0.07692$ $h_g = 2803.8$ $s_g = 6.243$	v h s	0.0832 2875.4 6.383	0.0948 3005.8 6.622	0.1053 3124.1 6.820	0.1153 3237.6 6.995	0.1250 3349.5 7.155	0.1345 3461.0 7.304	0.1438 3573.2 7.440	0.1531 3685.5 7.577	0.1714 3912.3 7.824	0.1895 4146.3 8.052
$P = 27.0$ $= 2700$ kPa $= 2.7$MPa ($t_s = 228.1°C$)	$v_g = 0.07092$ $h_g = 2803.8$ $s_g = 6.228$	v h s	0.0797 2870.6 5.385	0.0910 3002.8 6.500	0.1012 3122.0 6.802	0.1109 3236.0 6.976	0.1202 3348.1 7.136	0.1294 3459.9 7.286	0.1384 3572.3 7.423	0.1474 3684.7 7.560	0.1652 3911.5 7.806	0.1827 4145.7 8.034
$p = 28.0$ $= 2800$kPa $= 2.8$ MPa ($t_s = 230.1$ °C)	$v_g = 0.07145$ $h_g = 2804.0$ $s_g = 6.214$	v h s	0.0765 2865.7 6.334	0.0875 2999.7 6.579	0.0858 2998.2 6.569	0.1068 3234.3 6.957'	0.11581 3346.8 7.118	0.1247 3458.'7 7.268	0.1333 3571.3 7.405	0.1420 3683.9 7.542	0.1590 3910.7 7.788	0.1759 4145.1 8.016

Contd. ...

Table 3: (Contd.)

(t_s)	v_g, h_g, s_g	v, h, s	50	100	150	200	250	300	350	400	450	500
P = 29.0	v_g = 0.0690	v	0.0734	0.0842	0.0822	0.1029	0.1117	0.1203	0.1287	0.1371	0.1536	0.1701
= 2900 kPa	h_g = 2804.1	h	2860.8	2996.6	3070.2	3232.6	3345.4	3457.6	3570.3	3683.1	3910.4	4144.7
= 2.9 MPa	s_g = 6.200	s	6.311	6.559	6.779	6.939	7.100	7.251	7.388	7.525	7.772	8.001
(t_s = 232.0.C)												
P = 30.0	v_g = 0.0667	v	0.0706	0;0811	0.0905	0.0994	0.1079	0.1162	0.1243	0.1324	0.1483	0.1641
= 3000'kPa	h_g = 2804.2	h	2855.8	2993.5	3115.3	3230.9	3344.0	3456.5	3570.4	3684.3	3910.2	4144.3
= 3.0 MPa	s_g = 6.187	s	6.287	6.539	6.743	7.921	7.083	7.234	7.371	7.509	7.757	7.986
(t_s = 233.8.C)												
P = 32.0	$v_g = 0.0625$	v	0.0631	0.0757	0.0845	0.0929	0.1009	0.1087	0.1163	0.1240	0.1390	0.1538
= 3200 kPa	$h_g = 2804.1$	h	2844.4	2987.2	3113.2	3227.4	3341.3	3454.3	3567.5	3680.8	3908.5	4143.2
= 3.2 MPa	s_g = 6.161	s	6.178	6.501	6.712	6.888	7i051	7.202	7.339	7.477	7.725	7.954
(t_s = 237.5.2°C)												
P = 34.0	v_g = 0.0588	v	0.0688	0.0707	0.0792	0.0871	0.0948	0.1022	0.1094	0.1166	0.1307	0.1446
= 3400kPa	h_g = 2803.7	h	2833.6	2980.7	3108.7	3224.0	3338.6	3452.0	3565.6	3679.2	3907.2	4142.3
= 3.4 MPa	s_g = 6.137	s	6.197	6.464	6.679	6.856	7.020	7.172	7.311	7.448	7.698	7.928
(t_s = 240.9°C)												
P = 36.0	v_g = 0.0554	v	0.0567	0.0663	0.0745	0.0821	0.0893	0.0963	0.1031	0.1100	0.1235	0.1365
= 3600 kPa	h_g = 2803.1	h	2822.6	2974.2	3104.2	3220.5	3335.8	3449.8	3563.7	3677.6	3905.8	4141.3
= 3.6 MPa	s_g = 6.114	s	6.153	6.429	6.647	6.825	6.991	7.143	7.281	7.420	7.668	7.900
(t_s = 244.2.C)												
p = 38.0	v_g = 0.0525	v	0.0530	0.0624	0.0702	0.0775	0.0844	0.0911	0.0976	0.1041	0.1168	0.1294
= 3800 kPa	h_g = 2802.4	h	2812.0	2967.5	3099.7	3217.1	3333.0	3447.5	3561.7	3676.0	3904.9	4140.8
= 3.8 MPa	s_g = 6.091	s	6.110	6.394	6.616	6.797	6.963	7.16	7.255	7.394	7.645	7.873
(t_s = 247.4°C)												
P = 40.0	v_g = 0.0498	v	0.0588	0.0664	0.0625	0.0734	0.0801	0.1109	0.1229	0.0864	09261	0.0989
= 4000 kPa	h_g = 2801.4	h	2960'''	3092.5	3026.6	3213.6	3030.3	3445.3	3559.8	3674.4	3903.5	4139.4
= 4.0 MPa	s_g = 6.070	s	6.364	6.587	6.475	0.773	6.936	7.091	7.229	7.368	7.619	7.849
(t_s = 250AOC)												

Contd. ....

Table 3: (Contd.)

(t_s)	v_g, h_g, s_g	v, h, s	50	100	150	200	250	300	350	400	450	500
$P = 42.0$	$v_g = 0.0473$	v	0.0556	0.0630	0.0593	0.0697	0.0759	0.0821	0.0880	0.0939	0.1056	0.1170
$= 4200$ kPa	$h_g = 2800.3$	h	2955.0	3090.4	3022.7	3212.3	3327.5	3442.7	3556.9	3671.1	3902.8	4139.1
$= 4.2$ MPa	$s_g = 6.050$	s	6.332	6.558	6.445	6.747	6.908	7.066	7.205	7.344	7.595	7.826
$(t_s = 253.2°C)$												
$P = 44.0$	$v_g = 0.0451$	v	0.0527	0.0598	0.0562	0.0663	0.0722	0.0782	0.0839	0.0896	0.1007	0.1116
$= 4400$ kPa	$h_g = 2799.0$	h	2947.8	3085.7	3016.7	3208.8	3324.6	3440.5	3555.0	3669.5	3901.6	4138.2
$= 4.4$ MPa	$s_g = 6.030$	s	6.301	6.531	6.416	6.772	6.903	7.043	7.182	7.321	7.573	7.804
$(t_s = 256.0°C)$												
$P = 46.0$	$v_g = 0.0431$	v	0.0500	0.0570	0.0535	0.0632	0.0689	0.0747	0.0801	0.0856	0.0963	0.1067
$= 4600$ kPa	$h_g = 2797.6$	h	2940.5	3080.9	3010.7	3205.3	3321.7	3438.2	3553.0	3667.8	3900.3	4137.2
$= 4.6$ MPa	$s_g = 6.010$	s	6.270	6.504	6.387	6.697	6.858	7.020	7.159	7.299	7.551	7.783
$(t_s = 258.8°C)$												
$P = 48.0$	$v_g = 0.0412$	v	0.0476	0.0544	0.051	0.0604	0.0659	0.0715	0.0767	0.0820	0.0922	0.1022
$= 4800$ kPa	$h_g = 2796.0$	h	2933.1	3076.1	3004.6	3201.8	3318.8	3435.9	3551.0	3666.2	3899.1	4136.3
$= 4.8$ MPa	$s_g = 5.992$	s	6.240	6.478	6.359	6.674	6.836	6.998	7.138	7.278	7.531	7.763
$(t_s = 261.4°C)$												
$P = 50.0$	$v_g = 0.0394$	v	0.0453	0:0519	0.0486	0.0578	0.0633	0.0686	0.0736	0.0787	0.0884	0.0981
$= 5000$ kPa	$h_g = 2794.3$	h	2924.5	3068.4	2996.4	3195.7	3316.2	3433.8	3550.1	3666.5	3897.8	4135.1
$= 5.0$ MPa	$s_g = 5.973$	s	6.208	6.449	6.328	6.646	6.819	6.976	7.117	7.259	7.510	7.741
$(t_s = 264.0°C)$												
$P = 55.0$	$v_g = 0.0357$	v	0.0434	0.0467	0.0494	0.0521	0.0572	0.0620	0.0668	0.713	0.0803	0.0891
$= 5500$ kPa	$h_g = 2789.6$	h	2905.8	3055.9	3122.6	3189.3	3309.0	3427.9	3545.5	3660.4	3894.8	4133.0
$= 5.5$ MPa	$s_g = 5.930$	s	6.139	6.390	6.493	6.597	6.'167	6.928	7.073	7.210	7.464	7.697
$(t_s = 269.9°C)$												
$P = 60.0$	$v_g = 0.0324$	v	0.0362	0.0422	0.0448	0.0474	0.0521	0.0567	0.0610	0.0653	0.0735	0.0816
$= 6000$ kPa	$h_g = 2784.3$	h	2884.2	3043.0	3110.1	3177.2	3301.8	3422.2	3540.6	3658.4	3887.2	4123.8
$= 6.0$ MPa	$s_g = 5.889$	s	6.067	6.333	6.437	6.541	6.719	6.880	7.029	7.168	7.415	7.647
$(t_s = 275.8°C)$												

Contd. ...

Table 3: (Contd.)

(t_s)	v_g, h_g, s_g	v, h, s	50	100	150	200	250	300	350	400	450	500
p = 65.0	v_g = 0.0297	v	0.0326	0.0385	0.0409	0.0438	0.0479	0.0520	0.0561	0.0600	0.0677	0.0753
= 6500 kPa	h_g = 2778.6	h	2683.0	3029.7	3100.2	3170.8	3294.5	3416.4	3535.8	3652.1	3888.6	4128.8
= 6.6 MPa	s_g = 5.850	s	6.001	6.280	6.389	6.499	6.675	6.689	7.987	7.126	7.382	7.617
(t_s = 280.8°C)												
p = 70.0	v_g = 0.0274"	v	0.0295	0.0352	0.0375	0.0399	0.0442	0.0481	0.0519	0.0557	0.0628	0.0698
= 7000 kPa	h_g = 2772.1	h	2838.4	3016.0	3087.0	3158.1	3287.1	3410.3	3530.9	3650.3	3881.1	4118.9
= 7.0 MPa	s_g = 5.813	s	5.931	6.228	6.338	6.448	6.633	6.798	7.949	7.089	7.339	7.572
(t_s = 285.9°C)												
p = 75.0	v_g = 0.0253,	v	0.0267	0.0324	0.0346	0.0369	0.0410	0.0447		0.0585	0.0651	
= 7500 kPa	h_g = '765.3	h	2814.1	3001.9	3076.7	3151.6	3279.6	3404.7	3526.0	3643.7	3882.4	4123.7
= 7.5 MPa	s_g = 5.778	s	5.864	6.179	6.295	6.411	6.593	6.762	6.912	7.053	7.311	7.547
(t = 290.5°C)												
P = 80.0	v_g = 0.0235	v	0.0243	0.0299	0.0321	0.0343	0.0382	0.0417	0.0452	0.0484	0.0548	0.0609
= 8000 kPa	h_g = 2758.0	h	2785.0	2987.3	3062.8	3138.3	3272.0	3398.3	3521.0	3642.0	3874.9	4114.0
= 8.0 MPa	s_g = 5.743	s	5.791	6.130	6.246	6.363	6.555	6.724	6.878	7.021	7.273	7.507
(t_s = 295.1°C)												
P = 85.0	v_g = 0.0219	v	0.0219	0.0278	0.0299	0.0320	0.0357	0.0291	0.056	0.454	0.0515	0.0573
= 8000 kPa	h_g = 2750.3	h	2757.1	2972.2	3051.8	3131.5	3264.4	3392.8	3514.1	3635."4	3876.1	4119.0
= 8.5 MPa	s_g = 5.710	s	5.744	6.083	6.206	6.330	6.519	6.692	6.839	6.987	7.249	7.486
(t_s = 299.2°C)												
P = 90.0	v_g = 0.0205	v	-	0.0258	0.0278	0.0299	0.0335	0.0368	0.0398	0.0428	0.0486	0.0541
= 900 kPa	h_g = 2742.1	h	-	2956.6	3037.2	3117.8	3256.6	3386.1	3509.9	3633.7	3876.5	4116.7
= 9.0 MPa	s_g = 5,677	s	-	6.036	6.160	6.285	6.484	6.658	1).808	6.959	7.222	7.458
(t_s = 303.4°C)												
p = 95.0	v_g = 0.0192	v	-	0.0240	0.0260	0.0281	0.0315	0.0346	0.0376	0.0404	0.0459	0.0512
= 9500 kPa	h_g = 2733.6	h	-	2940.3	3028.6	3110.7	3248.8	3380.7	3506.0	3627.0	3869.9	4114.4
= 9.5 MPa	s_g = 5.645	s	-	5.990	6.129	6.254	6.451	6.629	6.785	6.929	7.192	7.431
(t_s = 307.2°C)												

Contd. ...

Table 3: (Contd.)

(t_s)	v_g, h_g, s_g	v, h, s	50	100	150	200	250	300	350	400	450	500
$p = 100.0$	$v_g = 0.0180$	v	0.0224	0.0245	0.0264	0.0297	0.0328	0.0356	0.0384	0.0436	0.0486	
$= 10000$ kPa	$h_g = 2724.7$	h	2923.4	3015.4	3096.5	3240.9	3373.7	3500.9	3625.3	3870.5	4112.0	
$= 10.0$ MPa	$s_g = 5.614$	s	5.944	6.089	6.212	6.419	6.597	6.756	6.903	7.169	7.396	
$(ts = 311.1°C)$												
$P = 110.0$	$v_g = 0.0160$	v	1.0196	0.0217	0.0235	0.0267	0.0295	0.0322	0.0347	0.0395	0.0441	
$= 11000$ kPa	$h_g = 2705.6$	h	2887.3	2988.2	3074.3	3224.7	3361.0	3490.7	3616.9	3864.5	4107.3	
$= 11.0$ MPa	$s_g = 5.533$	s	5.853	6.011	6.142	6.358	6.540	6.703	6.851	7.120	7.358	
$(ts = 31S.S0C)$												
$P = 120.0$	$v_g = 0.0143$	v	0.0172	0.0193	0.0211	0.0241	0.0268	0.0293	0.0316	0.0361	0.0403	
$= 12000$ kPa	$h_g = 26.84.9$	h	2487.7	2958.9	3051.3	3208.2	3348.1	3480.4	3608.3	3858.4	4102.7	
$= 12.0$ MPa	$s_g = 5.492$	s	5.760	5.935	6.075	'6.300	6.487	6.653	6.804	7.075	7.315	
$(ts = 324.88°C)$												
$P = 130.0$	$v_g = 0.0128$	$v \times 10^2$	1.511	1.725	1.900	2.194	2.450	2.684	2.905	3.322	3.716	
$= 13000$ kPa	$h_g = 2662.2$	h	2803.3	2927.9	3027.2	3191.3	3335.2	3469.9	3599.7	3852.3	4098.0	
$= 13.0$ MPa	$s_g = 5.432$	s	5.663	5.859	6.009	6.245	6.437	6.606	6.759	7.033	7.274	
$(ts = 330.9°C)$												
$P = 140.0$	$v_g = 0.0115$	$v \times 10^2$	1.322	1.546	1.722	2.007	2.252	2.474	2.683	3.075	3.444	
$= 14000$ kPa	$h_g = 2637.6$	h	2752.6	2894.5	3001.9	3174.0	3322.0	3459.3	3591.1	3846.2	4093.3	
$= 14.0$ MPa	$s_g = 5.375$	s	5.559	5.782	5.945	6.192	6.390	6.562	6.712	5.994	7.237	
$(ts = 336.8°C)$												
$p = 150.0$	$v_g = 0.0103$	$v \times 10^2$	1.145	1.388	1.565	1.845	2.080	2.293	2.491	2.861	3.209	
$= 15000$ kPa	$h_g = 2601.5$	h	2692.4	2858.4	2975.5	3156.2	3308.6	3448.6	3582.3	3840.1	4088.6	
$= 15.0$ MPa	$s_g = 5.310$	s	5.442	5.703	15.881	16.40	16.344	6.520	6.679	6.959	7.291	
$(t_s = 242.2°C)$												
$p = 160.0$	$v_g = 0.0093$	$v \times 10^2$	0.0975	1.245	1.426	1.701	1.930	2.134	2.323	2..674	3.003	
$= 16000$ kPa	$h_g = 2580.6$	h	2615.7	2818.9	2947.6	3188.0	3294.9	3437.8	3573.5	3833.9	4084.0	
$= 16.0$	$s_g = 5.245$	s	5.302	5.622	5.188	6.091	6.301	6.480	6.640	6.922	7.168	
$(t_s = 347.4°C)$												

Contd. ...

Table 3: (Contd.)

(t_s)	v_g, h_g, s_g	v, h, s	50	100	150	200	250	300	350	400	450	500
$p = 170$ $= 17000$ kPa $= 17.0$ MPa $(t_s = 352.4°C)$	$v_g = 0.0083$ $h_g = 25472$ $s_g = 5.178$	$v \times 10^2$ h s	– – –	1.117 2776.8 5.539	1.302 2918.2 5.754	1.575 3119.3 6.042	1.797 3281.1 6.259	1.993 12.174 3426.1 3564.6 6.442 6.604		2.509 3827.7 6.8890	2.821 4079.3 7.137	
$p = 180.0$ $= 18000$ kPa $= 18.0$ MPa $(t_s = 357.1°C)$	$v_g = 0.0075$ $h_g = 2509.1$ $s_g = 5.104$	$v \times 10^2$ h s	– – –	0.996 2727.9 5.448	1.190 2887.0 5.689	1.462 3100.1 5.995	1.678 3267.0 6.218	1.868 3415.9 6.405	2.042 3555.6 6.570	2.362 3821.5 6.858	2.659 4074.6 7.107	
$p = 190.0$ $= 19000$ kPa $= 19.0$ MPa $(t_s = 361.5°C)$	$v_g = 0.00666$ $h_g = 2464.5$ $s_g = 5.023$	$v \times 10^2$ h s	– – –	0.881 2671.3 5.346	1.088 2853.8 5.922	1.361 3080.4 5.948	1.572 3252.7 6.179	1.756 3404.7 6.537	1.924 3546.6 6.828	2.231 3815.3 7.078	2.515 4070.5	
$p = 200.0$ $= 20000$ kPa $= 20.0$ MPa $(t_s = 365.8C)$	$v_g = 0.00583$ $h_g = 2409.7$ $s_g = 4.927$	$v \times 10^2$ h s	– – –	0.767 2602.5 5.227	0.994 2818.1 5.554	1.269 3060.1 5902	1.477 38.2 6.140	1.655 3393.5 6.335	1.818 3537.6 6.505	2.113 3809.0 6.799	2.385 4065.3 7.051	
$p = 210.0$ $= 21000$ kPa $= 21.0$ MPa $(t_s = 369.9°C)$	$v_g = 0.00495$ $h_g = 2334.6$ $s_g = 4.801$	$v \times 10^2$ h s	– – –	0.645 2511.0 5.075	0.907 2779.6 5.483	1.186 3039.3 5.856	1.390 3223.5 6.103	1.564 3382.1 6.301	1.722 3528.4 6.474	2.006 3802.8 6.772	2.267 4060.6 7.025	
$p = 220.0$ $= 22000$ kPa $= 22.0$ MPa $(t_s = 373.8°C)$	$v_g = 0.00357$ $h_g = 2165.6$ $s_g = 4.533$	$v \times 10^2$ h s	– – –	0.482 2345.1 4.810	0.825 2737.6 5.407	1.110 3017.9 5.811	1.312 3208.6 6.066	1.481 3370.6 6.269	1.684 3519.2 6.444	1.909 3796.5 6.745	2.160 4055.9 7.001	

Table 4: Properties of super-cricital steam

p bar	v, h, s	\multicolumn{9}{c}{Temperature of steam t in °C}
		350	375	400	425	450	500	600	700	800
230	$v \times 10^2$	0.162	0.221	0.748	0.915	1.040	1.239	1.554	1.821	2.063
	h	1632.81	1912.2	2691.2	2869.2	2995.8	3193.4	3510.0	3790.2	4056.2
	s	3.137	4.137	5.327	5.587	5.765	6.030	6.415	6.719	6.980
350	$v \times 10^2$	0.160	0.197	0.600	0.788	0.916	1.112	1.414	1.665	1.891
	h	1623.5	1848.0	2580.2	2806.3	2949.7	3162.4	3491.4	3775.5	4047.1
	s	3.680	4.032	5.142	5.472	5.674	5.959	6.360	6.671	6.934
300	$v \times 10^2$	0.155	0.179	0.279	0.530	0.673	0.868	1.145	1.366	1.562
	h	1608.5	1791.5	2151.1	2614.2	2821.4	3081.1	3443.9	3745.6	4024.2
	s	3.643	3.930	4.473	5.150	5.442	5.790	6.233	6.561	6.833
400	$v \times 10^2$	0.149	0.164	0.191	0.253	0.369	0.562	0.809	0.994	1.152
	h	1588.3	1742.8	1930.9	2198.1	2512.8	2903.3	3346.4	3681.2	3978.7
	s	3.586	3.829	4.113	4.503	4.946	5.470	6.011	6.375	6.666
500	$v \times 10^2$	0.144	0.156	0.173	0.201	0.249	0.389	0.611	0.773	0.908
	h	1575.3	1716.6	1874.6	2060.0	2284.0	2720.1	3247.6	3616.8	3933.6
	s	3.542	3.764	4.003	4.273	4.588	5.173	5.818	6.219	6.529
600	$v \times 10^2$	0.140	0.150	0.163	0.182	0.209	0.296	0.483	0.627	0.746
	h	1566.4	1699.5	1843.4	2001.7	2179.01	2567.9	3151.2	3553.5	3889.1
	s	3.505	3.604	3.932	4.163	4.412	4.932	5.645	6.082	6.411
700	$v \times 10^2$	0.137	0.146	0.157	0.171	0.189	0.247	0.398	0.526	0.632
	h	1560.4	1687.7	1822.8	1967.2	2122.7	2463.2	3061.7	3492.4	3845.7
	s	3.473	3.673	3.877	4.088	4.307	4.762	5.492	5.961	6.307
800	$v \times 10^2$	0.135	0.142	0.152	0.163	0.177	0.219	0.339	0.452	0.548
	h	1556.4	1679.4	1808.3	1943.9	2086.9	2394.0	2982.7	3434.6	3803.8
	s	3.444	3.638	3.833	4.031	4.232	4.642	5.360	5.851	6.213
900	$v \times 10^2$	0.133	0.139	0.147	0.157	0.169	0.201	0.297	0.397	0.484

Contd. ...

p bar	v, h, s	Temperature of steam t in °C								
		350	375	400	425	450	500	600	700	800
1000	h	1553.9	1673.4	1797.7	1927.2	2062.0	2346.7	2915.6	3381.1	3763.8
	s	3.419	3.607	3.795	3.984	4.174	4.554	5.247	5.753	6.128
	v × 10²	0.108	a.137	0.144	0.152	0.163	0.189	0.267	0.355	0.434
	h	1552.7	1669.4	1790.0	11914.8	2043.8	2312.8	2859.8	3332.3	3726.1
	s	3.396	3.579	3.762	3.944	4.126	4.485	5.151	5.664	6.050

Index

Symbols
2-stroke SI engines
  constructional details of 361
4-strokes S.I. Engine 356

A
Abrasive 18
Absorption dynamometers 191
Air preheater
  types of 85
Air standard cycles 370
Air-conditioning
  classification of 102
Angular motion 306, 307
Antipriming pipe 89
Applied mechanics
  applications of 291
Arc welding 284
  equipment 284
  types of 285
Axial flow turbine 117

B
Babcock and Wilcox boiler 73, 74
  construction of 74
Beam 240
  types of 242
Bedded-in method 273
Belt drives 158
Belt
  length of 161
  types of 158
Belt transmission dynamometer 194

Bench and floor moulding 274
Bending process 282
Bevis Gibson flash light torsion
  dynamometer 195
Blow off cock 83
  function of 82
BMD, cantilever beam for 250
BMD supported beam 246
Boiler 64
  accessories 75, 76, 84
  classification of 65
  mountings 75
  properties of 48
  requirements of a good 64
Boiler shell material 68
Boring 12
Brakes 187
  types of 187
Brayton cycle 385
Brazing 286

C
C.I. Engines 371
  constructional details of 362
Cantilever beam 242
Carbon steel 17
  applications of 266
  mechanical properties of 266
Carnot cycle
  processes of 34
Carnot theorem 36
Carriage
  parts of 4

Cast iron 263
  varieties of 264
Casting method 275
Castings
  stages in 267
Celsius scale 27
Cemented carbides 18
Central flow condenser 342
Centrifugal casting 276
Centrifugal pump 139, 142
  parts of 139, 141
Centrifugal tension 168
Centrifuge casting 277
Ceramics 18
Chain drives 170
Chain
  types of 171
Chemical equilibrium 26
  built-up edge chip 15
  continuous chip 15
  discontinuous chip 15
  ribbon chip 15
Chip forms 16
Chip
  types of 15
Classical macroscopic
  thermodynamics 24
Clausius' statement 33
Closed cycle gas turbine unit 383
Closed system 24
Closed-loop control system 235
Clutches 182
  classification of 182
CNC 237
  classification of 234
Cochran boiler 69
Coefficient of performance 100
Column and knee type milling machine 8
Combined separating and throttling
  calorimeter 61, 62
Comfort air-conditioning 103, 104
Comfort charts 106
Compound bar under axial loading
  stress in 206
  strain in 206
Compound gear train 180

Compressive strain 205
Computer controlled system 227
Computer numerical control (CNC)
  machines 234
Condensers
  classification of 338, 342
  types of 337
Cone clutch 185
Constant pressure gas turbine cycle 384
Constant volume cycle 371
Continuous beam 244
Control system features
  classification of 235
Conveyer chain 171
Cooling pond 345
Cooling towers 345
Cooling towers
  classification of 346
Cooling water supply 344
Cope and drag pattern 271
Coplanar force system
  equilibrant of 295
Counter sinking 12
Counter boring 12
Critical temperature 50
Cross belt 163
Cross-belt drive 159
Cutting fluid 18
  purpose of 18
Cutting tool
  nomenclature of 17

D

Dead weight safety valve 76
Devices 28
Diamond 18
Die casting 275
Diesel cycle 380
Diesel cycle (constant pressure cycle)
  375
Diffuser pump 143
Direct acting lift 149, 150
Direct contact refrigeration system 97
Direct stress 203
Disc brakes 190
Double shoe brake 188

Down flow condensers 342
Down flow sleam condeñser 342
Draft tube theory 127
Drawing process 282
Drilling 10, 12
Drilling machine
    operations on 11
    parts of 11
    specification of 11
    types of 10
Dry sand mould
    advantages of 274
Dual cycle 380
Dual pressure cycle 378
Duplex feed pump 87, 88
    construction of 88
    working of 88
Dynamometer 191
Dynamometers
    types of 191

E

Economizer
    advantages of 85
    functions of 85
    working of 85
Efficiency 344
Efficiency ratio 356
Ejector condenser 340
Enthalpy-entropy chart 56
Entropy 37
    physical concept of 38
Epicyclic train dynamometer 193
Equilibrium force 295
Evaporative condenser 343
Evaporative refrigeration 98
Extension of bar (due to self weight) 210
Externally fired boiler 66
Extrusion 281

F

Fahrenheit scale 27
Feed check valve 82
Feed pump
    function of 86
Ferrous metals

    classification of 263
Fibre ropes 167
Fire tube boiler 65, 69
First law of thermodynamics 30, 40
    limitations of 32
Fixed beam 243
Fixed support 242
Flat belt 158
Flat belt drives
    types of 159
Flexible connector 157
Force, resolution of 295
Forced circulation 67
Forced draught cooling tower 347
Forging operations 281
Forging tools 280
Foundry 267
Four stroke diesel engine
    working of 366
Francis turbine 133
    construction of 125
    parts of 125
    radial curved vane of 129
    work done for 129
    working of 128
Free vibration 316
Froude transmission dynamometer 194
Fully automated machine 226
Fusible plug 83
    function of the 83

G

Gas turbine 386
    disadvantages of 387
    types of 383
    uses of 384
    working principle of 382
    welding 282
Gated pattern 271
Gear pumps 147
Gear ratio 174
Gear trains 179
Gears
    types of 175
General plane motion 310
Gravity

428 Index

acceleration due to 304
Green sand mould 272
Green's economizer 87

H
Headstock 3
Heat 29
Helical gear 176
High level jet condenser (barometric condenser) 340
High pressure boiler 66
High steam low water safety valve 79
High steam safety valve 78
Hoisting chains 171
Hooke's law
    elastic constants and 213
Hot drawing 282
Housed MCU 230
Human resources 227
Hydraulic brakes 190
Hydraulic jack 151
    function of 151
    operation of 151
Hydraulic lift 149
Hydraulic turbines
    classification of 116
Hydroelectric power plant
    layout of 115, 116
Hyperbolic cooling tower
    advantages of 348
    drawbacks of 349

I
I.C. Engine
    classification of 349
    constructional detail of 356
    performance of 354
Ice refrigeration 97
Ideal air cycle 371
Impellers
    types of 144
Impulse turbine 116, 119, 330
    compounding of 332
    parts of 330
Impulse-reaction turbine 331
Induced draught cooling tower 347

Industrial air-conditioning 102
Initial tension 170
Instantaneous axis of rotation 314
Internal energy 32
Internal expanding brakes 188
Internally fired boiler 66
Intersecting shafts 176
Inverted flow condenser 343
Investment casting process 278
Iron
    chemical compositions 264
    effect of chemical elements on 264
Irreversible process 26
Isolated system 24

J
Jet condensers 337
Junction valve
    construction of 81

K
Kaplan turbine 130, 133
    main components of 131
    working of 132

L
Lami's theorem 296
Lancashire boiler
    constructional details of 70
Lathe 1
    details of 2
    feed mechanism of 4
    types of 1
Law of moments 301
Laws of mechanics 292
Lever safety valve 77, 78
Limited pressure cycle 378
Linear motion 307
Loading
    types of 244
Locomotive boiler 73
Loose-piece pattern 271
Low level jet condenser (counter flow type) 338
Low level jet condenser (parallel flow type) 339

Low pressure boiler 66
Low water safety valve 78

M
Machine control unit 229
Manhole 83
  function of 83
Manually operated machine 226
Manufacturing system 225
  classification of 227
  components of 226
  flexibility in 228
Match plate pattern 270
Material handing system 227
Maximum work
  condition for 374
MCU
  parts of 230
  types of 230
Mean effective pressure 353
Mechanical brakes 187
Mechanical equilibrium 26
Mechanics 291
Medium alloy steels 17
Mercury boilers 67
Metal cutting process 14
Microscopic thermodynamics 24
Milling machine
  classification of 9
  parts of 9
Mixed flow turbine 117
Mobile boilers 67
Mollier chart 56
Moment
  concept of 300
Moment of centre 301
Moment of force 301
Moulding processes
  classification of 272
Moulding sand
  properties of 269
  types of 268
Multiplate disc clutch 184

N
N.C. machines 228
  advantages of 232
  constructional details of 229
  systems, advances in 234
Natural circulation 67
Natural draught cooling tower 346
Newton's first law of motion 292
Newton's laws 292
Newton's second law of motion 292
Newton's third law 293
Normal strain 204
Normal stress 203
Numerical control
  fundamentals of 228

O
Once through boiler 67
Open belt drives 159
Open cycle gas turbine 383
Open system 23
Otto cycle 371, 380
Overhanging beams 243
Oxyacetylene gas welding 283
Oxyhydrogen welding 284

P
Parallelogram law of forces 293
Particle dynamics 302
Pattern 267
  materials for 267
  types of 269
Pelton turbine 119
  parts of 119
Pelton wheel 124
Permanent mould casting 275
Pig iron 263
Pinion 178
Pit moulds 274
Pitch circle 174
Pitch point 173
Planer 7, 8
  parts of 6
Planetary gear train 182
Plasma air purifier 106
Point of contraflexure 249
Poisson ratio 205
Polygon law of equilibrium 295
Polygon law of forces 295
Portable boilers 67

Power
  calculation of 192
Power transmission 157
Pressure compound impulse turbine 333, 334
Pressure gauge 79, 80
Pressure reducing valve 89
Pressure-velocity compounded impulse turbine 334
Principle of moments 302
Production machines
  classification of 226
Prony brake dynamometrs 191
Propped beam 243
Psychrometric chart 104
Psychrometric properties 103
Psychrometry 103
Pumps 139
  classification of 140, 142

Q

Quarter-twist belt drive 159
Quasi-static process 26

R

Rack 178
Radial flow turbine 116
Radial piston pumps 148
Radial-cylinder rotary pump 149
Reaction turbine 119, 330
Reaming 12
Reciprocating pump 86, 142, 145
  classification of 145
  parts of 146
Reciprocating single acting piston pump 146
Reciprocating I.C. Engines 386
Refrigeration
  by the expansion of air 98
  by throttling of the gas 98
  definition of 97
  methods of 97
Refrigeration machine, rating of 100
Relative efficiency 356
Resistance welding methods 285
Resultant force 293

Reversed Carnot cycle 35
Reversible process 26
Reverted gear train 181
Rigid bodies
  kinematics of 303
  rotation of 306
Rigid connectors 157
Roller support 242
Rolling 281
Rope drive 166, 167
Rotary pump – external gear pump 147
Rotary pump – internal gear pump 147
Rotary pumps 145, 148
  classification of 145
Round belt 159

S

Saturation temperature
  effect of pressure on 50
S.I. Engines 371
Safety valve 76
Sand additives 269
Scalar quantities 292
Screw pump 147
Second law of thermodynamics 32, 40
Segmental chip 15
Semi-automated machine 226
Semi-centrifugal casting 276
Separating and throttling process,
  representation of 62
Separating calorimeter
  limitations of 60
Shear force diagram (SFD) 240
  cantilever beam for 250
SFD supported beam 246
Shaper 4, 8
  cutting action of 5
  parts of 5, 6
Shear strain 205
Shear stress 203
Shoe brake 187
Shuttle-block rotary pump 149
Simple band brake 188
Simple bar under axial loading
  stress in 206
  strain in 206

Simple block 187
Simple gear train 179
Simple impulse turbine 330
Simple refrigeration vapour compression cycle 100, 101
Simply supported beam 243
Single piece pattern 269
Skin-dried mould 274
Slip
   effect of 161
Slotter
   operation on 14
   parts of 13
   specification of 14
Slush casting 275
Soldering 286
Solid pattern 269
Special purpose manufacturing machines 229
Specific output 355
Specific speed 133
   significance of 135
Spiral bevel 177
   gear 177, 178
Split pattern 270
Spot facing 12
Spray pond 345
Spring loaded safety valve
   advantages of 77
Spur gears 175
Stand-alone MCU 230
Stationary boilers 67
Statistical thermodynamics 24
Steam at constant pressure 48
Steam boilers 67
Steam condenser 337
Steam
   entropy of 54
   generation of 51
   internal energy of 54
   properties of 48
Steam generator 64
Steam injector 87, 89
   thermodynamic properties of 53
Steam property charts 56

Steam separator
   function of 89
Steam stop valve 84
   construction of 81
   function of 81
Steam tables 55, 392
Steam trap 88
   function of 88
   types of 88
Steam turbines 328
Steam
   uses of 64
Steels 265
   chemical composition of 264
   processes 265
Satellites 18
Straight bevel gears 176, 177
Strain 204
   in simple bar under axial loading 206
   in compound bar under axial bar loading 206
   in uniformly tapering circular cross section bars 206
Stress
   in simple bar under axial loading 206
   in compound bar under axial bar loading 206
   in uniformly tapering circular cross section bars 206
Stress-strain curve, for brittle materials 213
Stress-strain diagram 211
Superheaters
   classification of 86
Support
   types of 242
Surface condensers 341
Suspended hydraulic lift 150
Sweep moulding 275
Swing around MCU 230

T

Tailstock 3
Tangential flow turbine 116
Tapping 12

Temperature
  measurement of 28
Temperature scale 27
Temperature-entropy chart 56
Tensile strain 204
Tensions
  ratio of 164
Thermal efficiency 355
Thermal equilibrium 26
Thermodynamic equilibrium 25
Thermodynamic properties 24
Thermodynamic system 23
Thermodynamics
  basic concepts of 23
Third law of thermodynamics 39
Three-piece split pattern 270
Throneycraft transmission
  dynamometer 194
Throttling calorimeter
  limitations of 61
Tool life
  factors affecting of 17
Tool material
  types of 17
Torque
  calculation of 192
Transmission dynamometers 193
Triangle law of equilibrium 295
Triple point 50
True centrifugal casting 276
Tubular air pre-heater 86
Turbine impulse 329
Turbines 115
  classification of 329
  terminology used for 117
Turn-over method 274
Two-stroke diesel engine
  working of 369
Two-stroke engine
  advantages of 369
  disadvantages of 369
Two-stroke petrol engine
  working of 367

U

Uniformly tapering circular
  cross-section bars
  stress in, 206
Unit discharge 137
Unit power 136
Unit quantities 135
Unit speed 136
Universe 23

V

V-belts 159, 166
Vacuum efficiency 343
Vane pump 148
Vapour compression refrigeration
  system 99
Vapour refrigeration system
  types of 98
Vapour-compression refrigerator 101
Varignon's theorem 302
Vector quantities 292
Velocity compounded impulse
  turbine 334
  advantages of 334
Velocity ratio 160
Velocity triangle 121
Vibration
  causes of 316
  effects of 317
Volute chamber pump 143
Vortex chamber pump 143

W

Water level indicator 79, 80
Water tube boiler 65, 69
Water turbines
  selection of 138
Welding 282
  equipment 284
Wire ropes 167
Work 29
Worm gear 178

Z

Zero bevel gear 178
Zeroth law of thermodynamics 29